QUANTUM ERROR CORRECTION
and FAULT TOLERANT QUANTUM COMPUTING

QUANTUM ERROR CORRECTION and FAULT TOLERANT QUANTUM COMPUTING

Frank Gaitan
Southern Illinois University, Carbondale, Illinois

CRC Press is an imprint of the
Taylor & Francis Group, an **informa** business

CRC Press
Taylor & Francis Group
6000 Broken Sound Parkway NW, Suite 300
Boca Raton, FL 33487-2742

© 2008 by Taylor & Francis Group, LLC
CRC Press is an imprint of Taylor & Francis Group, an Informa business

No claim to original U.S. Government works
Printed in the United States of America on acid-free paper
10 9 8 7 6 5 4 3 2 1

International Standard Book Number-13: 978-0-8493-7199-8 (Hardcover)

This book contains information obtained from authentic and highly regarded sources Reasonable efforts have been made to publish reliable data and information, but the author and publisher cannot assume responsibility for the validity of all materials or the consequences of their use. The Authors and Publishers have attempted to trace the copyright holders of all material reproduced in this publication and apologize to copyright holders if permission to publish in this form has not been obtained. If any copyright material has not been acknowledged please write and let us know so we may rectify in any future reprint

Except as permitted under U.S. Copyright Law, no part of this book may be reprinted, reproduced, transmitted, or utilized in any form by any electronic, mechanical, or other means, now known or hereafter invented, including photocopying, microfilming, and recording, or in any information storage or retrieval system, without written permission from the publishers.

For permission to photocopy or use material electronically from this work, please access www.copyright.com (http://www.copyright.com/) or contact the Copyright Clearance Center, Inc. (CCC) 222 Rosewood Drive, Danvers, MA 01923, 978-750-8400. CCC is a not-for-profit organization that provides licenses and registration for a variety of users. For organizations that have been granted a photocopy license by the CCC, a separate system of payment has been arranged.

Trademark Notice: Product or corporate names may be trademarks or registered trademarks, and are used only for identification and explanation without intent to infringe.

Library of Congress Cataloging-in-Publication Data

Gaitan, Frank.
    Quantum error correction and fault tolerant quantum computing / Frank Gaitan.
        p. cm.
    Includes bibliographical references and index.
    ISBN 978-0-8493-7199-8 (hardback : alk. paper)
    1. Quantum computers. 2. Error-correcting codes (Information theory) 3. Fault-tolerant computing. I. Title.

QA76.889.G35 2008
004.2--dc22   2007048869

Visit the Taylor & Francis Web site at
http://www.taylorandfrancis.com

and the CRC Press Web site at
http://www.crcpress.com

To my Mother and Father

Contents

List of Figures   xi

List of Tables   xv

Preface   xvii

1 **Introduction**   1
    1.1 Historical Background 2
    1.2 Classical Error Correcting Codes 4
       1.2.1 Linear Error Correcting Codes 4
       1.2.2 Errors, Hamming Weight and Distance 7
       1.2.3 Error Detection and Correction 10
       1.2.4 Error Probability 12
       1.2.5 Bounds on Code Parameters 14
    1.3 Using Quantum Systems to Store and Process Data 17
       1.3.1 Linear Superposition and Quantum Parallelism 17
       1.3.2 No-Cloning Theorem 20
       1.3.3 Measurement 21
       1.3.4 Distinguishing Quantum States 23
       1.3.5 Entanglement 24
       1.3.6 Noise: Errors and Decoherence 28
    1.4 Quantum Error Correcting Codes—First Pass 31
       1.4.1 Redundancy without Cloning 31
       1.4.2 Necessary and Sufficient Conditions 40
    Problems 48
    References 51

2 **Quantum Error Correcting Codes**   57
    2.1 Quantum Operations 57
       2.1.1 Operator-Sum Representation 58
       2.1.2 Depolarizing Channel 70
       2.1.3 Other Error Models 71
    2.2 Quantum Error Correcting Codes: Definitions 73
    2.3 Example: Calderbank-Shor-Steane [7, 1, 3] Code 75
    Problems 77
    References 80

3 Quantum Stabilizer Codes 83
- 3.1 General Framework 83
  - 3.1.1 Summary 83
  - 3.1.2 Errors 85
  - 3.1.3 Quantum Error Correction Reprise 90
  - 3.1.4 Encoded Operations 91
- 3.2 Examples 93
  - 3.2.1 [5,1,3] Code 93
  - 3.2.2 [4,2,2] Code 95
  - 3.2.3 [8,3,3] Code 95
- 3.3 Alternate Formulation: Finite Geometry 96
- 3.4 Concatenated Codes 100
  - 3.4.1 Single Qubit Encoding 101
  - 3.4.2 Multi-Qubit Encoding 104
- Problems 110
- References 113

4 Quantum Stabilizer Codes: Efficient Encoding and Decoding 115
- 4.1 Standard Form 115
- 4.2 Encoding 121
- 4.3 Decoding 127
- Problems 131
- References 134

5 Fault-Tolerant Quantum Computing 137
- 5.1 Fault-Tolerance 137
  - 5.1.1 To Encode or Not to Encode 139
  - 5.1.2 Fault-Tolerant Design 139
- 5.2 Error Correction 141
  - 5.2.1 Syndrome Extraction 144
  - 5.2.2 Shor State Verification 148
  - 5.2.3 Syndrome Verification 151
- 5.3 Encoded Operations in $N(\mathcal{G}_n) \cap N(\mathcal{S})$ 153
  - 5.3.1 Action of $N(\mathcal{G}_n)$ 155
  - 5.3.2 CSS Codes 158
- 5.4 Measurement 162
- 5.5 Four-Qubit Interlude 166
- 5.6 Multi-Qubit Stabilizer Codes 169
- 5.7 Operations Outside $N(\mathcal{G}_n)$—Toffoli Gate 173
- 5.8 Example: [5,1,3] Code 180
- 5.9 Example: [4,2,2] Code 186
  - 5.9.1 Fault-Tolerant Quantum Computing 186
  - 5.9.2 Fault-Tolerant Encoded Operations 187
- Problems 188
- References 192

6 Accuracy Threshold Theorem — 195
- 6.1 Preliminaries 195
  - 6.1.1 Concatenated QECCs 196
  - 6.1.2 Principal Assumptions 197
- 6.2 Threshold Analysis 199
  - 6.2.1 Recursion-Relation for $p_{op}^{(j)}$ 200
  - 6.2.2 Clifford Group Gates and Storage Registers .. 208
  - 6.2.3 Toffoli Gate 210
  - Problems 219
  - References 220

7 Bounds on Quantum Error Correcting Codes — 223
- 7.1 Quantum Hamming Bound 223
- 7.2 Quantum Gilbert-Varshamov Bound 225
- 7.3 Quantum Singleton Bound 227
- 7.4 Linear Programming Bounds for QECCs 229
  - 7.4.1 Weight Enumerators 230
  - 7.4.2 Quantum MacWilliams Identity 233
  - 7.4.3 Shadow Enumerator 236
  - 7.4.4 Bounds via Linear Programming 239
- 7.5 Entanglement Purification and QECCs 240
  - 7.5.1 Purifying Entanglement 241
  - 7.5.2 QECCs and 1-EPPs 255
  - Problems 257
  - References 258

A Group Theory — 261
- A.1 Fundamental Notions 261
  - A.1.1 Groups 261
  - A.1.2 Subgroups 262
- A.2 Group Action 265
  - A.2.1 On a Set 265
  - A.2.2 On Itself 265
- A.3 Mapping Groups 266
  - A.3.1 Homomorphisms 266
  - A.3.2 Fundamental Homomorphism Theorem 267
  - References 268

B Quantum Mechanics — 269
- B.1 States 270
- B.2 Composite Systems 271
- B.3 Observables 271
- B.4 Dynamics 272
- B.5 Measurement and State Preparation 273
- B.6 Mixed States 276

	References	280
C	**Quantum Circuits**	**283**
	C.1 Basic Circuit Elements	283
	C.2 Gottesman-Knill Theorem	285
	C.3 Universal Sets of Quantum Gates	286
	References	288
Index		**289**

List of Figures

4.1 The encoding circuit for the [5,1,3] quantum stabilizer code. Here $Y = -i\sigma_y^5$. Adapted from (i) Figure 4.2, Ref. [3]; and (ii) Ref. [4], ©Daniel Gottesman (2000), used with permission. 127

4.2 Decoding circuit for the [4,2,2] quantum stabilizer code. 131

4.3 The encoding circuit for the [4,2,2] quantum stabilizer code. Here $Y = -i\sigma_y^3$. 133

4.4 The encoding circuit for the [8,3,3] quantum stabilizer code. Here $Y = -i\sigma_y$. 134

4.5 The decoding circuit for the [5,1,3] quantum stabilizer code. . . 134

5.1 Quantum circuit for syndrome extraction for the [5,1,3] quantum stabilizer code. The upper five lines correspond to the code block and each set of lower four lines to an ancilla block. Qubit indicies increase in going from top to bottom in each block. The measured syndrome value $S(E) = l_1 l_2 l_3 l_4$ determines the recovery operation $\mathbf{R} = R_1 R_2 R_3 R_4 R_5$. 147

5.2 Quantum circuit for syndrome extraction for the [5,1,3] quantum stabilizer code following Ref. [3]. Note that each four qubit ancilla block has been compressed into a single line to make the circuit easier to examine, and the four CNOT gates applied to each ancilla block are applied transversally. The measured syndrome value $S(E) = l_1 \cdots l_4$ determines the recovery operation $\mathbf{R} = R_1 \cdots R_5$. 148

5.3 Quantum circuit to prepare the Shor state $|A\rangle$ for $w_i = 4$. The first Hadamard gate and three CNOT gates place the ancilla block in the cat-state. The remaining four Hadamard gates transform this state into $|A\rangle$. 149

5.4 Modified quantum circuit for Shor state preparation for $w_i = 4$. A new ancilla qubit is introduced to record the relative parity s of the first and fourth qubits. If measurement yields $s = 1$, the ancilla block is discarded, a new ancilla block is introduced, and the circuit is applied again. The final Hadamard gates are only applied if $s = 0$. See text for full discussion. Based on Figure 11, J. Preskill, *Proc. R. Soc. Lond. A* **454**, 385 (1998); ©Royal Society; used with permission. 151

5.5 Quantum circuit to implement the four-qubit operation in eq. (5.73). Qubit lines are labeled from 1 to 4 in going from top to bottom. Reprinted Figure 2 with permission from D. Gottesman, *Phys. Rev. A* **57**, 127 (1998). ©American Physical Society, 1998. . . 167

5.6 Quantum circuit for a CNOT gate that uses the lower qubit as control. 172

5.7 Quantum circuit for a SWAP gate. 173

5.8 Quantum circuit to apply the three-qubit operator T_3. Qubit lines are labeled from 1 to 3 in going from top to bottom. Reprinted Figure 1 with permission from D. Gottesman, *Phys. Rev. A* **57**, 127 (1998). ©American Physical Society, 1998. . . 181

5.9 Quantum circuit to measure a one-qubit Hermitian operator $\mathcal{O}$ of order 2. The control qubit is measured in the computational basis $\{|0\rangle, |1\rangle\}$. When the measurement yields $m = +1$, the control qubit is in the final state $|0\rangle$ and the target qubit is in the final state $P_+(\mathcal{O})|\psi\rangle$. The projection operator $P_+(\mathcal{O}) = (I + \mathcal{O})/2$ insures that the target qubit's final state is the eigenstate of $\mathcal{O}$ with eigenvalue $\mathcal{O}_+ = +1$. For the applications considered in this section, $|\psi\rangle$ is never an eigenstate of $\mathcal{O}$. Based on Figure 10.26, Ref. [14]; ©Cambridge University Press (2000), used with permission. 183

5.10 Quantum circuit that applies a Hadamard gate to qubit 2 conditioned on the ancilla qubit being in the state $|c_i\rangle = |1\rangle$. The lower three lines, going from top to bottom, correspond to qubits 1–3. 185

5.11 Quantum circuit that applies a controlled-phase gate to qubits 1 and 2 conditioned on the ancilla qubit being in the state $|c_i\rangle = |1\rangle$. The lower three lines, going from top to bottom, correspond to qubits 1–3. 185

5.12 Quantum circuit for syndrome extraction for the [4,2,2] quantum stabilizer code. 190

6.1 Quantum circuit for Shor state preparation for $w_i = 4$. A fifth ancilla qubit is introduced to record the relative parity s of the first and fourth qubits. If measurement yields $s = 1$, the ancilla block is discarded, a new ancilla block is introduced, and the circuit is applied again. The final Hadamard gates are only applied if $s = 0$. See Section 5.2.2 for a full discussion. 202

List of Figures xiii

6.2 Error correction circuit for the [7,1,3] CSS code in standard form. The seven top-most lines correspond to the data block and each of the six lower lines corresponds to a block of four ancilla qubits. The four qubit lines in each ancilla block have been compressed into a single line to make the circuit easier to examine, and the four CNOT gates applied to each ancilla block are applied transversally. The measured syndrome value $S(E) = l_1 \cdots l_6$ determines the recovery operation $\mathbf{R} = R_1 \cdots R_7$ that is applied to the data block. If no errors occur during ancilla block preparation, each ancilla block enters the circuit in the Shor state $|A\rangle$. 204

6.3 Quantum circuit to implement a Toffoli gate U_T on level-j qubits: $|\psi'\rangle = U_T|\psi\rangle$. The top (bottom) three lines correspond to qubits 1–3 (4–6). Three ancilla blocks (denoted A_1–A_3 from top to bottom) are prepared in cat-states $|\psi_{cat}\rangle$, which are used to insure that (to $\mathcal{O}(p)$) qubits 1–3 are in the state $|A\rangle$ after the boxed NOT gate. Majority voting on the measurement outcomes $\{m_1, m_2, m_3\}$ determines whether the boxed NOT gate is applied. See text for further discussion. Based on Figure 12, J. Preskill, *Proc. R. Soc. Lond. A* **454**, 385 (1998). ©Royal Society, used with permission. 211

6.4 Quantum circuit to place qubits 1–3 (top three lines) in the state $|A\rangle$. The ancilla block contains seven level-$(j-1)$ qubits prepared in the cat-state $|\psi_{cat}\rangle$. The seven ancilla qubit lines have been collapsed into a single line to make the circuit easier to examine. The operator $\mathcal{O}^{(j-1)} = X_1^{(j-1)} \cdots X_7^{(j-1)}$ is measured on the ancilla block. The measurement outcome m determines whether a NOT gate is applied to qubit 3. See text for a discussion of how this circuit must be modified when phase errors are possible during cat-state preparation. 212

6.5 Alternative quantum circuit for preparation of $|A\rangle$. 212

7.1 Plot of the change in fidelity $\Delta F = F' - F$ per iteration of the recurrence method versus the initial fidelity F. 246

7.2 Plot of the pair yield $D_H(W_F)$ for the hashing method when the pair state is the Werner state W_F. The yield is positive for $F > 0.8107$. 255

List of Tables

1.1 Set of errors $E_0, \ldots, E_7$ produced by the phase-flip channel on three qubits. Listed with each error is its associated error probability and error syndrome. Note that E_0 is the identity operator I that leaves all three qubits alone. 38

1.2 List of error syndrome values for the errors produced by the phase-flip channel acting on three qubits. For each syndrome value we list the expected error E_S and its associated recovery operator $R_S = E_S^\dagger$. Note that I is the identity operator that leaves all three qubits alone. 38

7.1 Unilateral and bilateral operations used by Alice and Bob to map Bell states to Bell states. For brevity, the Bell basis states $|ij\rangle$ are denoted ij with $i,j = 0,1$. Each entry of the BCNOT table has two lines, the first showing the output state for the source-pair, the second showing the output state of the target-pair. Reprinted Table 1 with permission from C. H. Bennett et. al., *Phys. Rev. A* **54**, 3824 (1996). ©American Physical Society, 1996. 254

Preface

It is difficult to overstate the significance of the years 1994–1997 for the field of quantum computing. During those years the best-known quantum algorithms were invented, the theory of quantum error correcting codes was worked out, and protocols for doing a quantum computation fault-tolerantly were established. The culmination of these developments was the demonstration of the accuracy threshold theorem. It established that, under appropriate conditions, an arbitrarily long quantum computation could be done with arbitrarily small error probability in the presence of noise and imperfect quantum gates. The theorem made it clear that no physical law stands in the way of building a quantum computer. The various proofs of the theorem made it clear that the task would be technically difficult, but not impossible. This came as quite a surprise to many and set in motion the enormous research effort currently underway whose aim is the construction of a working, scalable quantum computer. This book is an introduction to the theory of quantum error correcting codes and fault-tolerant quantum computing. It is written for students at the graduate and advanced undergraduate level, and its central application is a proof of the accuracy threshold theorem for a particular error model. The book will also be useful to researchers in other fields who would like to understand how quantum error correction and fault-tolerant quantum computing are possible, and how the accuracy threshold theorem is proved. A careful reading of the book will provide the reader with the background needed to meaningfully enter the literature on quantum error correcting codes and fault-tolerant quantum computing, and to begin working in this important new research area.

The structure of this book is as follows. Chapter 1 introduces key ideas from the theory of classical linear error correcting codes, and important properties of quantum systems that impact their use as sites for storage and processing of computational data. The chapter closes with a first look at quantum error correcting codes. A detailed presentation of quantum error correcting codes then follows in Chapters 2 through 4. Chapter 2 discusses these codes in general, while Chapter 3 zooms in on quantum stabilizer codes, which are the most important class of quantum codes discovered so far. Chapter 4 explains how a quantum stabilizer code can be efficiently encoded and decoded.

Chapter 5 discusses the question of fault-tolerant quantum computing in detail. It begins by explaining why fault-tolerant protocols are needed in quantum computing, and then goes on to show how error correction and a universal set of quantum gates can be implemented fault-tolerantly on data encoded using any quantum stabilizer code.

Having laid out the theory of quantum error correcting codes and fault-tolerant quantum computing, Chapter 6 pulls these two themes together to give a proof of the accuracy threshold theorem for a particular error model. Specifically, this theorem establishes that quantum computing is possible using imperfect quantum gates and in the presence of noise if one: (a) encodes the data using a sufficiently layered concatenated quantum error correcting code; (b) uses fault-tolerant protocols to carry out quantum computation and error correction; and (c) only uses quantum gates in the computation that have error probabilities that fall below a value known as the accuracy threshold. The accuracy threshold is calculated for the assumed error model.

Chapter 7 closes the book with derivations of a number of well-known bounds on the parameters of a quantum error correcting code. It also establishes the connection between quantum error correcting codes and one-way entanglement purification protocols. Finally, to make the book more self-contained, three Appendices are included that summarize essential results from group theory (Appendix A), quantum mechanics (Appendix B), and quantum circuit theory (Appendix C).

Interspersed throughout the 7 chapters are 35 Exercises and 40 Problems for which a solution manual is available. The Exercises tend to be straightforward calculations meant to highlight a point in the discussion. The Problems often introduce tools, ideas, results, and even topics that will be presented later in the text. The book also contains 17 worked Examples, and 8 further Examples that are sufficiently important as to warrant presentation as a section or subsection of a chapter. All together, the Exercises and Problems provide the reader with ample opportunities to test his or her understanding, while the Examples provide detailed illustrations of how the formalism is applied.

We have endeavored to give a careful presentation of the theory of quantum error correcting codes and fault-tolerant quantum computing. When working out the presentation of a topic, we have sought to meet the needs of the graduate student, rather than those of the expert in the field, no doubt to the occasional annoyance of the latter. It is hoped this book will help train some of the next generation of experts.

Frank Gaitan

August 2007

1

Introduction

In its short history, the field of quantum computing has produced a number of truly startling revelations about the computational power inherent in quantum systems. These discoveries generated a great deal of excitement, though this excitement was tempered by the realization that the quantum correlations responsible for this power are extremely fragile. It became clear that to properly assess the performance of a quantum computer, the effects of noise and imperfect quantum gates would have to be included in the analysis. It was thought by many that the virtues of a quantum computer would not survive the presence of these effects. Remarkably, in three years (1995–1997), theoretical tools were developed that stood this expectation on its head. Beginning with the groundbreaking work of P. W. Shor and A. M. Steane, a very general framework for constructing quantum error correcting codes was established, and protocols were developed that allowed quantum computations to be done in a fault-tolerant manner. This line of development culminated in a demonstration of the accuracy threshold theorem. This theorem showed that fault-tolerant protocols for quantum computation, in conjunction with concatenated quantum error correcting codes, would allow a quantum computation of arbitrary duration to be done with arbitrarily small error probability, provided all quantum gates used in the computation had error probabilities that fell below a value known as the accuracy threshold P_a. This threshold was calculated for a number of simple error models yielding values in the range $10^{-6} \leq P_a \leq 10^{-3}$. The good news is that $P_a \neq 0$ so that reliable quantum computing in the presence of noise and imperfect quantum gates is possible. The bad news is that P_a is small so that building quantum gates that satisfy the tolerances of the threshold theorem will be technically challenging.

The aim of this book is to explain how the accuracy threshold theorem is proved. To do this we will need to examine: (1) how computational data can be encoded into the states of a quantum system; (2) how quantum error correcting codes can be used to detect and correct errors introduced in the encoded data by noise and imperfect quantum gates; and (3) how quantum gates can be implemented fault-tolerantly on encoded data. Chapters 2–5 provide an explanation of how all this is done. With these tools in hand, we present a proof of the accuracy threshold theorem for a particular error model in Chapter 6. Chapter 7 then presents upper and lower bounds that must be satisfied by the parameters of a quantum error correcting code. Summaries of the essentials of finite group theory, quantum mechanics, and the quantum circuit model appear in Appendices A, B, and C, respectively.

In the rest of this chapter we set the stage for the detailed arguments presented in the remainder of this book. We begin with a brief historical introduction in Section 1.1. Essential ideas from the theory of classical error correcting codes are presented in Section 1.2, while Section 1.3 discusses the pros and cons of using quantum systems to store and process computational data. Finally, Section 1.4 describes quantum error correcting codes in broad strokes. The aim here is to bring out the broader issues before diving into the details in Chapters 2 and 3.

1.1 Historical Background

The idea of encoding will be central to this book. It is worth noting that this idea is a relatively recent one. Although the human family is thought to have split off from a common ancestor with the apes some 7 million years ago, it is no more than 35,000 years ago that evidence first appears of human beings using images to represent the contents of their mental world. It is as little as 6000 years ago that humans are first found to possess a written language [1]. It has taken 99.9% of the time since we parted with the apes for evolution to have produced human beings who are capable of encoding their experiences and discoveries in written strings of abstract symbols.

Tacit in the existence of a written language is the existence of a sufficiently advanced technology that can produce writing instruments and a medium for storing texts. As technology advances, so do the writing instruments and storage media. In our current age of electronic digital computers, the nature of this technology dictates that the symbols used in writing and mathematics be encoded into strings of binary digits (bits). In the ASCII character code, letters, punctuation, decimal digits, and other useful symbols and control characters are encoded into binary strings of length 7. A computer usually stores such strings in a register containing 8 storage elements. The extra bit is often set to 0, though it can also be used as a parity check on the other 7 bits. Each storage element in the register can only exist in one of two states, with one state representing 0 and the other representing 1. Computational data is thus encoded into the states of a set of registers, and a computation is effected by applying an appropriate sequence of digital logic gates to the registers that hold the computational data [2]. The essential lesson here is that machine computations are carried out by manipulating the states of physical systems in which the computational data is stored.

Eventually, the question arose as to whether there might be advantages to doing machine computations with physical systems whose dynamics is quantum mechanical. Benioff took the first step in this direction [3]. He considered

Introduction 3

a classical Turing machine made out of quantum components. Feynman [4] examined the connection between the laws of physics and what is computable. He pointed to the gains in computational efficiency that could be expected if a quantum system was used to simulate the dynamics of another quantum system. Deutsch [5] further explored the connection between the laws of physics and computability, defining for the first time a quantum Turing machine. In another important paper, Deutsch [6] introduced the quantum network/circuit model of quantum computation, which has become the paradigm for subsequent developments. Universal sets of quantum logic gates were found by a number of authors [5–9], allowing an arbitrary quantum computation to be implemented using a quantum circuit whose gates all belong to the universal set.

The next milestone for quantum computing occurred in 1994 when Shor published computationally efficient quantum algorithms for factoring integers and for evaluating discrete logarithms [10, 11]. With these algorithms, the owner of a quantum computer could crack popular/highly utilized public-key cryptosystems. Grover [12] discovered a quantum algorithm for the important problem of searching an unstructured database that yields a substantial speed-up over classical search algorithms. The performance of these quantum algorithms drew a great deal of attention, though it was thought that the effects of noise and imperfectly applied quantum gates would quash their performance advantages. Shor [13] and Steane [14] were the first to realize how data could be redundantly encoded in the states of a quantum system, and how the redundancy could be used to protect the data. These were the first examples of a quantum error correcting code (QECC). This work was extended in a remarkably short period of time [15–21], culminating in the general framework of quantum stabilizer codes [22–25].

The discovery of QECCs was an enormous step forward. Still, it was recognized that the operations of detection and correction of errors had to be done using imperfect quantum gates and in the presence of noise. The question arose as to whether the process of error detection and correction might produce more errors than were removed. If so, more errors would eventually appear in the data than a QECC could handle. In seminal work, Shor showed how this problem could be solved [26]. He introduced fault-tolerant protocols for error correction, and for applying a universal set of quantum gates on data encoded using a particular class of QECCs. This work was further developed to show how fault-tolerant quantum computation could be implemented on data encoded using a quantum stabilizer code [27–29]. Important contributions were also made by Steane [30, 31]. These efforts to determine whether a quantum computer could operate reliably using imperfect quantum gates, and in the presence of noise, culminated in the proof of the accuracy threshold theorem by a number of authors [32–41]. In the following chapters we give a systematic presentation of the framework of quantum stabilizer codes and fault-tolerant quantum computing. As with the historical development, the climax of our story will be a demonstration of how the accuracy threshold theorem is proved.

1.2 Classical Error Correcting Codes

In any real machine computation, noise and imperfectly applied logic gates can lead to errors in the final computational results. The unavoidable coupling of a computer to its environment, and our inability to apply gates with infinite precision, means that a way must be found to protect the integrity of the computational data. Similar issues arise in the problem of communication through a noisy channel. Here reliable communication is possible when the transmitted data is protected by an error correcting code whose rate is less than the channel capacity [42–47]. These codes add redundancy into the encoding that allows a set of errors to be detected and removed from the received data. In this Section we review the essential theory of linear error correcting codes. These codes prove to be the most important for our later discussion of quantum error correcting codes. Nonlinear error correcting codes have also been developed [44,45], though we will not discuss them in this book.

1.2.1 Linear Error Correcting Codes

In the problem of noisy communication, one is interested in sending a stream of encoded symbols through a noisy channel as reliably as possible. As a first step, message symbols are encoded into a block of k bits: $\mathbf{m} = (m_1, \ldots, m_k)$ ($m_i = 0, 1$). The 2^k message symbols $\mathbf{m}$ are the elements of the k-dimensional vector space $F_2^k = F_2 \otimes \cdots \otimes F_2$ over the finite field $F_2 = 0, 1$. Addition $+$ and multiplication $\cdot$ in F_2 are modulo 2 (binary arithmetic):

+	0	1
0	0	1
1	1	0

$\cdot$	0	1
0	0	0
1	0	1

[Note that for all $a \in F_2$, $a + a = 0$ and so $a = -a$. Thus $a - b = a + b$ for all $a, b \in F_2$.] The message symbols $\mathbf{m}$ are then further encoded into 2^k codewords $\mathbf{c} = (c_1, \ldots, c_n)$, which are elements of the n-dimensional vector space F_2^n. In the simplest encoding, the first k bits in a codeword $\mathbf{c}$ are the corresponding message bits m_j in $\mathbf{m}$:

$$c_j = m_j \quad (j = 1, \ldots, k) \ .$$

The remaining $n - k$ bits c_j ($j = k+1, \ldots, n$) are parity check bits whose values are set by requiring that all codewords $\mathbf{c}$ obey $n-k$ linearly independent constraints known as parity checks. Each of the $n - k$ parity checks can be

Introduction

thought of as a vector in F_2^n: $\mathbf{H}(i) = (H_1(i), \ldots, H_n(i))$, and the constraint is that all codewords **c** must have vanishing scalar product (modulo 2) with $\mathbf{H}(i)$:

$$\mathbf{H}(i) \cdot \mathbf{c} \equiv \sum_{j=1}^n H_j(i) c_j = 0 \quad (i = 1, \ldots, n-k) \ . \tag{1.1}$$

[*Note:* In the remainder of this section vectors will be written as column vectors. $\mathbf{v}^T$ is then the row-vector that is the transpose of $\mathbf{v}$.] All parity check vectors $\mathbf{H}(i)$ can be collected into an $(n-k) \times n$ matrix H known as the parity check matrix whose i^{th} row is $\mathbf{H}^T(i)$:

$$H = \begin{pmatrix} \text{---} & \mathbf{H}^T(1) & \text{---} \\ & \vdots & \\ \text{---} & \mathbf{H}^T(n-k) & \text{---} \end{pmatrix} \ . \tag{1.2}$$

The set of parity checks (eq. (1.1)) can then be written as a single matrix equation:

$$H \mathbf{c} = 0 \ . \tag{1.3}$$

As the parity checks $\mathbf{H}(i)$ are assumed to be linearly independent, the rank of H is $n-k$, and eq. (1.3) uniquely fixes the $n-k$ parity check bits c_j ($j = k+1, \ldots, n$).

DEFINITION 1.1 *The linear code $\mathcal{C}$ with $(n-k) \times n$ parity check matrix H consists of the 2^k vectors $\mathbf{c} \in F_2^n$ that satisfy eq. (1.3). The vectors $\mathbf{c}$ are referred to as codewords.*

Note that if $\mathbf{c}_1$ and $\mathbf{c}_2$ are codewords, so is $\mathbf{c}_1 + \mathbf{c}_2$ since $H(\mathbf{c}_1 + \mathbf{c}_2) = H\mathbf{c}_1 + H\mathbf{c}_2 = 0 + 0 = 0$. Also if $s \in F_2$ is a scalar and **c** is a codeword, then $s\mathbf{c}$ is also a codeword since $H(s\mathbf{c}) = s(H\mathbf{c}) = s \cdot 0 = 0$. Thus a linear code $\mathcal{C}$ with 2^k codewords is closed under vector addition and scalar multiplication and thus forms a k-dimensional subspace of F_2^n. $\mathcal{C}$ is said to have dimension k, the codewords to have length n, and $\mathcal{C}$ is referred to as an $[n,k]$ code.

Since a k-dimensional linear code $\mathcal{C}$ is a k-dimensional subspace of F_2^n, any set of k linearly independent codewords $\mathbf{b}_1, \ldots, \mathbf{b}_k$ forms a basis for the code:

$$\mathbf{c} = \sum_{j=1}^k m_j \mathbf{b}_j \ , \tag{1.4}$$

where $m_j \in F_2$. It proves useful to introduce an $n \times k$ matrix G known as the generator matrix whose columns are the k basis vectors:

$$G = \begin{pmatrix} | & & | \\ \mathbf{b}_1 & \cdots & \mathbf{b}_k \\ | & & | \end{pmatrix} \ . \tag{1.5}$$

The code $\mathcal{C}$ is then the column space of G. Equation (1.4) can then be viewed as a linear transformation $G : F_2^k \to F_2^n$ that encodes message symbols $\mathbf{m}$ into codewords $\mathbf{c}$:

$$\mathbf{c} = G\mathbf{m} , \qquad (1.6)$$

where

$$\mathbf{m} = \begin{pmatrix} m_1 \\ \vdots \\ m_k \end{pmatrix} . \qquad (1.7)$$

The generator matrix G and eq. (1.6) provide an alternative means for defining a linear error correcting code $\mathcal{C}$. Using eq. (1.6) in eq. (1.3) gives

$$0 = HG\mathbf{m} .$$

Since this equation must be true for all $\mathbf{m} \in F_2^k$, it follows that

$$HG = 0 , \qquad (1.8)$$

where the RHS is the $(n - k) \times k$ zero-matrix.

Note that the columns of H^T are the $(n - k)$ linearly independent parity check vectors $\mathbf{H}(i)$. Thus the column space of H^T is an $n - k$ dimensional vector space $\mathcal{C}^\perp \subset F_2^n$, and H^T is its generator matrix $G_\perp$. Taking the transpose of eq. (1.8), we see that G^T acts as the parity check matrix $H_\perp$ for $\mathcal{C}_\perp$:

$$(HG)^T = G^T H^T = G^T G_\perp = H_\perp G_\perp = 0 . \qquad (1.9)$$

Since the rows of G^T are a basis set for $\mathcal{C}$, eq. (1.9) says that any basis for $\mathcal{C}$ is orthogonal to any basis for $\mathcal{C}^\perp$. Thus the codewords of $\mathcal{C}$ are orthogonal to the codewords of $\mathcal{C}^\perp$. $\mathcal{C}^\perp$ is known as the dual code of $\mathbf{C}$.

DEFINITION 1.2 Let $\mathcal{C}$ be an $[n, k]$ code. Its dual code $\mathcal{C}^\perp$ is the set of vectors that are orthogonal to all codewords in $\mathcal{C}$:

$$\mathcal{C}^\perp = \{ \mathbf{d} \in F_2^n \,|\, \mathbf{d} \cdot \mathbf{c} = 0 \text{ for all } \mathbf{c} \in \mathcal{C} \} .$$

Note that a vector in F_2^n will be orthogonal to itself, $\mathbf{c} \cdot \mathbf{c} = 0$, if it has an even number of non-zero components. Thus a codeword $\mathbf{c}$ can belong to both $\mathcal{C}$ and $\mathcal{C}^\perp$! If $\mathcal{C} = \mathcal{C}^\perp$, we say that $\mathcal{C}$ is a self-dual code. If $\mathcal{C} \subset \mathcal{C}^\perp$, then $\mathcal{C}$ is said to be weakly self-dual.

Exercise 1.1

(a) Let H be an arbitrary parity check matrix. Argue that by appropriate use of elementary row operations, H can always be put into the form

$$H = (\, A \,|\, I_{n-k} \,) ,$$

where A is an $(n - k) \times k$ matrix and I_{n-k} is the $(n - k) \times (n - k)$ identity matrix.

Introduction

(b) Use eq. (1.8) to show that if H has the form appearing in part (a), then the generator matrix G has the form

$$G = \begin{pmatrix} I_k \\ -A \end{pmatrix} .$$

Exercise 1.2

Let C be a $[6,3]$ code with the following parity check matrix H:

$$H = \left(\begin{array}{ccc|ccc} 1 & 1 & 0 & 1 & 0 & 0 \\ 1 & 0 & 1 & 0 & 1 & 0 \\ 0 & 1 & 1 & 0 & 0 & 1 \end{array} \right) .$$

(a) Find the generator matrix G for C.

(b) Find the codewords for C.

1.2.2 Errors, Hamming Weight and Distance

Our discussion so far has described how to encode 2^k message symbols $\mathbf{m} \in F_2^k$ into 2^k codewords $\mathbf{c} \in F_2^n$. In the problem of communication through a noisy channel, a codeword $\mathbf{c}$ would be sent through the channel and a vector $\mathbf{y} \in F_2^n$ would be received. Because of the noise, $\mathbf{y}$ may differ from the transmitted codeword $\mathbf{c}$. We define the error vector $\mathbf{e}$ as

$$\mathbf{e} = \mathbf{y} - \mathbf{c} = \begin{pmatrix} e_1 \\ \vdots \\ e_n \end{pmatrix} . \tag{1.10}$$

If $e_i = 0$, the ith bit of the received vector $\mathbf{y}$ is correct (incorrect), which is assumed to occur with probability $1 - p_i$ (p_i). It is further assumed that errors occurring on separate bits are uncorrelated; the error probability p_i is the same for all bits, $p_i = p$; and $0 \leq p < 1/2$. A channel with these properties is referred to as a binary symmetric channel (BSC). The decoder at the receiving end must decide from $\mathbf{y}$ what error $\mathbf{e}_*$ is most likely to have occurred. It then returns its best guess $\bar{\mathbf{c}}$ for the transmitted codeword:

$$\bar{\mathbf{c}} = \mathbf{y} - \mathbf{e}_* . \tag{1.11}$$

This decoding strategy is known as maximum likelihood decoding, and it minimizes the probability of a decoding error ($\bar{\mathbf{c}} \neq \mathbf{c}$, or equivalently, $\mathbf{e}_* \neq \mathbf{e}$) being made in the case where the codewords are equally likely [44].

DEFINITION 1.3 *The Hamming weight wt($\mathbf{x}$) of a vector $\mathbf{x}$ is equal to the number of its non-zero components x_i.*

It is possible to define a distance measure (metric) on F_2^n known as the Hamming distance.

DEFINITION 1.4 *The Hamming distance $d(\mathbf{x}, \mathbf{y})$ between two vectors $\mathbf{x}$ and $\mathbf{y}$ is the number of places in which they differ (viz. the number of positions i for which $x_i \neq y_i$).*

Example 1.1
If $\mathbf{x}^T = (011001)$ and $\mathbf{y}^T = (110011)$, then $d(\mathbf{x}, \mathbf{y}) = 3$. ☐

Exercise 1.3 *Show that $d(\mathbf{x}, \mathbf{y}) = wt(\mathbf{x} - \mathbf{y})$.*

Exercise 1.4 *A metric function on a set S is a single-valued, non-negative, real-valued function $d(x, y)$ defined for all $x, y \in S$ that has the following three properties:*

1. $d(x, y) = 0$ if and only if $\mathbf{x} = \mathbf{y}$; (Reflexivity)

2. $d(x, y) = d(y, x)$; (Symmetry)

3. $d(x, y) \leq d(x, z) + d(z, y)$. (Triangle Inequality)

Show that the Hamming distance is a metric function.

If ν bit errors occur, the error vector $\mathbf{e}$ has weight $wt(\mathbf{e}) = \nu$. For a BSC, the probability that $\mathbf{e}$ occurs is $\text{Prob}(\mathbf{e}) = p^\nu (1-p)^{n-\nu}$. Since $0 \leq p < 1/2$, an error of weight 1 is more likely than one of weight 2, etc. In nearest neighbor decoding, the received vector $\mathbf{y}$ is decoded as the codeword $\bar{\mathbf{c}}$ that is the least Hamming distance from $\mathbf{y}$.

An important property of a code $\mathcal{C}$ is the minimum distance between codewords d:

$$d = \min_{\mathbf{c}, \mathbf{c}' \in \mathcal{C}} d(\mathbf{c}, \mathbf{c}').$$

Since $d(\mathbf{c}, \mathbf{c}') = wt(\mathbf{c} - \mathbf{c}')$ (Exercise 1.3), and $\mathbf{c} - \mathbf{c}' \in \mathcal{C}$ when $\mathcal{C}$ is a linear code, it follows that

$$d = \min_{\substack{\mathbf{c} \in \mathcal{C} \\ \mathbf{c} \neq 0}} wt(\mathbf{c}) \ . \tag{1.12}$$

This result simplifies the task of finding the minimum distance d for a linear code $\mathcal{C}$.

A linear code $\mathcal{C}$ with length n, dimension k, and minimum distance d will be referred to as an $[n, k, d]$ code.

The following theorem will be needed in Section 1.2.5.

Introduction

THEOREM 1.1
If H is the parity check matrix for a linear code C, then C has minimum distance d if and only if every $d-1$ columns of H are linearly independent and some d columns are linearly dependent.

PROOF Let $\mathbf{x}$ be a codeword of weight w. Then

$$H\mathbf{x} = \sum_{i=1}^{n} x_i H_i = 0 \ ,$$

where x_i is the ith component of $\mathbf{x}$, H_i is the ith column of H, and n is the codeword length. Thus, the w columns of H associated with the w non-zero components of $\mathbf{x}$ are linearly dependent. If $\mathbf{x}_*$ is a non-zero codeword of minimum weight, then there are d columns of H that are linearly dependent. If some $d-1$ columns of H are linearly dependent, we could define a vector $\mathbf{y}$ of weight $d-1$ that has $y_i = 1$ for i corresponding to the column indices of these $d-1$ columns. Then $H\mathbf{y} = 0$ and $\mathbf{y}$ would be a codeword of weight $d-1$. This is a contradiction since the minimum distance for the code is d. Thus all sets of $d-1$ columns are linearly independent. The converse can be proved by reversing the steps of this argument. ∎

The following theorem establishes an important connection between the minimum distance d of a linear code C and the maximum number of bit errors that C can correct. Let $\lfloor x \rfloor$ denote the largest integer less than x.

THEOREM 1.2
A linear code C with minimum distance d can correct $t = \lfloor (d-1)/2 \rfloor$ bit errors.

PROOF Let $\mathbf{s} \in F_2^n$. The sphere centered on $\mathbf{s}$ of radius r is the set of vectors $\mathbf{v}$ whose distance from $\mathbf{s}$ satisfies $d(\mathbf{s}, \mathbf{v}) \leq r$. Let $t = \lfloor (d-1)/2 \rfloor$. If a sphere of radius t is placed around each codeword, the spheres will not overlap since the minimum distance between codewords is d and $2t \leq d - 1$.

Let $\mathbf{c}$ be the transmitted codeword, and assume that no more than t errors occurred during transmission. The received vector $\mathbf{y}$ then lies inside the sphere centered on the codeword $\mathbf{c}$, and will be closer to $\mathbf{c}$ than to any other codeword. Nearest neighbor decoding will then correctly identify $\mathbf{c}$ as the transmitted codeword when no more than t errors occur.

If spheres of radius $r \geq t + 1$ are placed around the codewords, some of the spheres will overlap since the codewords have minimum distance d and $d/2 \leq t + 1$. Let $\mathbf{c}$ and $\mathbf{c}'$ correspond to codewords for which $d(\mathbf{c}, \mathbf{c}') = d$ so that their associated spheres of radius r overlap. Imagine that $\mathbf{c}$ is the transmitted codeword, that $t + 1$ errors occur, and that the received vector $\mathbf{y}$

lies in the overlap region. If $d(\mathbf{y}, \mathbf{c}') < t+1$, nearest neighbor decoding will incorrectly decode $\mathbf{y}$ as $\mathbf{c}'$, so that a decoding error occurs. Thus errors of weight $t+1$ can occur that cannot be corrected by $\mathcal{C}$. ∎

1.2.3 Error Detection and Correction

The parity check matrix H for a code $\mathcal{C}$ produces a linear transformation from F_2^n to F_2^{n-k}:

$$\mathbf{s} = H\mathbf{y} , \qquad (1.13)$$

where $\mathbf{y} \in F_2^n$ and $\mathbf{s} \in F_2^{n-k}$. The image vector $\mathbf{s}$ is called the error syndrome. If $\mathbf{y} \in \mathcal{C}$, then by eq. (1.3), the parity check matrix H annihilates it, $H\mathbf{y} = 0$, and its error syndrome vanishes: $\mathbf{s} = H\mathbf{y} = 0$. Since $\mathcal{C}$ contains all vectors $\mathbf{y}$ that are annihilated by H, $\mathcal{C}$ is the kernel of the transformation produced by H (see Appendix A for a summary of group theory):

$$\ker(H) = \mathcal{C} .$$

The significance of this remark will become clear below. First we must introduce the cosets of $\mathcal{C}$.

DEFINITION 1.5 *Let $\mathcal{C}$ be an $[n, k, d]$ code. Let $\mathbf{x}$ be an arbitrary vector in F_2^n. The coset of $\mathcal{C}$ containing $\mathbf{x}$ is the set*

$$\mathbf{x} + \mathcal{C} = \{\mathbf{x} + \mathbf{c} \,|\, \mathbf{c} \in \mathcal{C}\} .$$

It is clear from this definition that every vector in F_2^n belongs to a coset. Cosets have the following useful property.

THEOREM 1.3
Two cosets are either disjoint or equal.

PROOF Let $\mathbf{u}$ and $\mathbf{w}$ be distinct elements of F_2^n. They are elements of the cosets $\mathbf{v} + \mathcal{C}$ and $\mathbf{w} + \mathcal{C}$, respectively. If these cosets are disjoint, we are done. Thus, assume that the cosets overlap and let $\mathbf{g}$ be an element in their intersection. Then $\mathbf{g} = \mathbf{v} + \mathbf{x} = \mathbf{w} + \mathbf{x}'$, for some $\mathbf{x}$ and $\mathbf{x}'$ in $\mathcal{C}$. Thus $\mathbf{w} = \mathbf{v} + (\mathbf{x} - \mathbf{x}')$ so that $\mathbf{w} \in \mathbf{v} + \mathcal{C}$ since $\mathbf{x} - \mathbf{x}' \in \mathcal{C}$. It follows that $\mathbf{w} + \mathcal{C} \subset \mathbf{v} + \mathcal{C}$. Similarly, $\mathbf{v} = \mathbf{w} + (\mathbf{x}' - \mathbf{x})$, which implies that $\mathbf{v} + \mathcal{C} \subset \mathbf{w} + \mathcal{C}$. Thus $\mathbf{v} + \mathcal{C} = \mathbf{w} + \mathcal{C}$. ∎

Theorem 1.3 indicates that the cosets of $\mathcal{C}$ partition F_2^n:

$$F_2^n = \mathcal{C} \cup (\mathbf{l}_1 + \mathcal{C}) \cup \cdots \cup (\mathbf{l}_c + \mathcal{C}) , \qquad (1.14)$$

where $c = 2^{n-k} - 1$. The partitioning of F_2^n into the cosets of the code $\mathcal{C}$ is referred to as the standard array for $\mathcal{C}$ [44, 45].

Introduction

DEFINITION 1.6 *Let* $g + C$ *be a coset of* C. *The vector* l *of minimum weight in this coset is called the coset leader. If more than one vector in* $g + C$ *has minimum weight, then randomly pick one to be the coset leader.*

The cosets of C allow us to define the quotient group F_2^n/C since F_2^n forms an abelian group under vector addition and C is a normal subgroup of F_2^n. Addition of cosets is defined as $(x + C) + (y + C) = (x + y) + C$: the sum of the cosets containing x and y is the coset containing their sum $x + y$. By the Fundamental Homomorphism Theorem (see Appendix A), the image F_2^{n-k} of F_2^n under H is isomorphic to the quotient group $F_2^n/\ker(H) = F_2^n/C$. There is thus a one-to-one correspondence between the cosets of C and the error syndromes $s \in F_2^{n-k}$. This allows the following maximum likelihood decoding scheme to be implemented. Let y be the received vector, and let $g + C$ be the coset to which it belongs. Thus $y = g + x$ for some $x \in C$. Let c be the transmitted codeword so that the error $e = y - c = g + (x - c)$. Thus $e \in g + C$ since $x - c \in C$. The error e thus belongs to the same coset as the received vector y. The most probable error e_p is the vector in $g + C$ that has minimum weight. If this vector is unique, it will be the coset leader l. If it is not unique, the decoding protocol sets e_p equal to the coset leader l, which will always be one of the minimum weight vectors in $g + C$. In both cases, the decoder returns $\bar{c} = y - l$ as the most probable transmitted codeword. Thus if the actual error e differs from the coset leader l, a decoding error occurs. Clearly, one wants this situation to be a low probability occurrence.

[Note that one can establish the one-to-one correspondence between cosets and error syndromes without having to resort to group theory. Let v and w both belong to the coset $l + C$. Thus $v = l + x$ and $w = l + x'$, where $x, x' \in C$. Thus $s_v \equiv Hv = Hl + Hx = Hl$, where we have used that H annihilates codewords. Similarly, $s_w \equiv Hw = Hl + Hx' = Hl = s_v$. Thus vectors belonging to the same coset have the same error syndrome.]

The following example shows how standard array decoding works for a given code, and also shows how some errors cause this type of decoding to fail.

Example 1.2
Exercise 1.2 examined the code C defined by the parity check matrix

$$H = \begin{pmatrix} 1 & 1 & 0 & 1 & 0 & 0 \\ 1 & 0 & 1 & 0 & 1 & 0 \\ 0 & 1 & 1 & 0 & 0 & 1 \end{pmatrix}.$$

The eight codewords (written as bit strings of increasing weight) are:

000000; 100110; 111000; 010101; 001011; 110011; 101101; 011110 .

Since the minimum distance of the code d is equal to the minimum weight of the non-zero codewords, it follows that $d = 3$ and this code is a $[6, 3, 3]$ code.

It follows from Theorem 1.2 that this code can correct all weight 1 errors: $t = 1$.

The standard array for this code is

```
000000  100110  111000  010101  001011  110011  101101  011110
000001  100111  111001  010100  001010  110010  101100  011111
000010  100100  111010  010111  001001  110001  101111  011100
000100  100010  111100  010001  001111  110111  101001  011010
001000  101110  110000  011101  000011  111011  100101  010110
010000  110110  101000  000101  011011  100011  111101  001110
100000  000110  011000  110101  101011  010011  001101  111110
100001  000111  011001  110100  101010  010010  001100  111111
```

The first row are the elements of $\mathcal{C}$, and the remaining rows are the cosets of $\mathcal{C}$. The elements of the first column are the coset leaders. Notice that all errors of weight 1 are coset leaders and one error of weight 2 is a coset leader.

Imagine that the transmitted codeword is $\mathbf{c} = 110011$ and that the received vector is $\mathbf{y} = 110001$. $\mathbf{y}$ lies in the coset with coset leader $\mathbf{l} = 000010$, which is the only weight 1 vector in this coset. Thus the most probable error is unique for this coset and standard array decoding returns $\bar{\mathbf{c}} = \mathbf{y} - \mathbf{l} = 110011$, which is in fact the transmitted codeword $\mathbf{c}$. (Recall that we are using binary arithmetic so that $\mathbf{y} - \mathbf{l} = \mathbf{y} + \mathbf{l}$.) This is consistent with $t = 1$ since 000010 is a weight 1 error, and by Theorem 1.2, it is expected to be a correctable error.

Again, let the transmitted codeword be $\mathbf{c} = 110011$ and let the received vector be $\mathbf{y} = 111111$. $\mathbf{y}$ lies in the coset with coset leader $\mathbf{l} = 100001$. Standard array decoding returns $\bar{\mathbf{c}} = 111111 - 100001 = 011110 \neq \mathbf{c}$! Thus a decoding error occurs. The actual error is $\mathbf{e} = 111111 - 110011 = 001100$, which, as expected, belongs to the same coset as $\mathbf{y}$, though it is not the coset leader. As expected, this situation causes a decoding error to occur. □

1.2.4 Error Probability

We now examine how an error correcting code enhances the reliability of data transmission through a noisy channel. If a receiver uses standard array decoding, the decoding will be correct whenever the error is a coset leader. Otherwise a decoding error occurs and the wrong codeword is returned.

DEFINITION 1.7 *The error probability P_e for a particular decoding scheme is the probability that the decoder outputs the wrong codeword.*

If the error correcting code contains M codewords $\mathbf{c}_1, \ldots, \mathbf{c}_M$ of length n, and $\bar{\mathbf{c}}$ is the decoder output, then

$$P_e = \sum_{i=1}^{M} \text{Prob}(\bar{\mathbf{c}} \neq \mathbf{c}_i | \mathbf{c}_i \text{ was sent}) \cdot \text{Prob}(\mathbf{c}_i \text{ was sent}) \ . \tag{1.15}$$

Introduction

If all codewords occur with equal probability, Prob(c_i was sent) $= 1/M$ and

$$P_e = \frac{1}{M} \sum_{i=1}^{M} \text{Prob}(\bar{c} \neq c_i \mid c_i \text{ was sent}) \ . \tag{1.16}$$

With standard array decoding, an error occurs if and only if the error e is not a coset leader:

$$P_e = \text{Prob}(e \neq \text{coset leader}) \ . \tag{1.17}$$

Let α_i be the number of coset leaders with weight i. If the channel is a BSC with error probability p, then the probability that the error e is a coset leader l_i of weight i is

$$\text{Prob}(e = \text{coset leader of weight } i) = \alpha_i p^i (1-p)^{n-i} \ . \tag{1.18}$$

The probability that decoding is done correctly is then

$$P_{ne} = \sum_{i=0}^{n} \text{Prob}(e = \text{coset leader of weight } i).$$

Since $P_e + P_{ne} = 1$, we have finally

$$P_e = 1 - \sum_{i=0}^{n} \alpha_i p^i (1-p)^{n-i} \ . \tag{1.19}$$

Example 1.3

For the code in Example 1.2, $\alpha_0 = 1$, $\alpha_1 = 6$, and $\alpha_2 = 1$. Thus, eq. (1.19) gives

$$P_e = 1 - (1-p)^6 - 6p(1-p)^5 - p^2(1-p)^4 \ .$$

If $p = 0.05$,

$$P_e = 0.031 < p = 0.05 \ .$$

Thus encoding the data using this code only produces a modest improvement in reliability when the bit error probability for the BSC is $p = 0.05$. Note, however, that if p is reduced to $p = 0.01$, then $P_e = 0.0013$, which is an order of magnitude improvement over the unencoded error probabiltity $p = 0.01$. This is an important lesson: the degree to which reliability is improved by using an error correcting code depends on both the code used (via α_i) and on the specifics of the communication channel (here the error probability p for the BSC). □

Imagine that our error correcting code has minimum distance d so that it can correct $t = \lfloor (d-1)/2 \rfloor$ errors. Then every error e of wt(e) $\leq t$ will be a coset leader, and α_i is equal to the number of ways in which i objects can be chosen from n objects independent of order:

$$\alpha_i = \binom{n}{i} = \frac{n!}{i!(n-i)!} \quad , \quad 0 \leq i \leq t \ . \tag{1.20}$$

DEFINITION 1.8 An $[n, k, d]$ error correcting code is called a perfect error correcting code if $\alpha_i = 0$ for $i > t = \lfloor (d-1)/2 \rfloor$. For a perfect code,

$$P_e = 1 - \sum_{i=0}^{t} \binom{n}{i} p^i (1-p)^{n-i} \ .$$

DEFINITION 1.9 An $[n, k, d]$ error correcting code is called a quasi-perfect error correcting code if $\alpha_i = 0$ for $i > t+1$, and $t = \lfloor (d-1)/2 \rfloor$. For a quasi-perfect error correcting code,

$$P_e = 1 - \sum_{i=0}^{t} \binom{n}{i} p^i (1-p)^{n-i} - \alpha_{t+1} p^{t+1} (1-p)^{n-t-1} \ .$$

Note that the code used in Examples 1.2 and 1.3 is a quasi-perfect code.

1.2.5 Bounds on Code Parameters

Error correcting codes must serve two masters. On the one hand, an $[n, k, d]$ code needs to maximize the rate k/n so that as many message symbols are sent per codeword as possible. On the other hand, one wants the minimum distance d to be large so that as many errors as possible can be corrected. For codes of finite length n, these are conflicting demands that do not allow the code parameters n, k, and d to be assigned arbitrarily. In this subsection we derive three well known bounds that the code parameters must satisfy.

THEOREM 1.4 (Singleton Bound)
If C is an $[n, k, d]$ code, then $n - k \geq d - 1$.

PROOF Any codeword $\mathbf{c} \in C$ will have at most $n - k$ non-zero parity check bits. Thus a codeword $\mathbf{x}$ with only one information bit will have a weight $\text{wt}(\mathbf{x}) \leq n - k + 1$. Since the code has minimum distance d, $\text{wt}(\mathbf{x}) \geq d$, and so $n - k + 1 \geq d$. ∎

THEOREM 1.5 (Hamming Bound)
Let C be an $[n, k, d]$ binary code that corrects $t = \lfloor (d-1)/2 \rfloor$ errors. The code

Introduction

parameters must satisfy

$$\sum_{i=0}^{t} \binom{n}{i} \leq 2^{n-k} .$$

PROOF Since the code corrects t errors, we can place a sphere of radius t around each codeword and be certain that the spheres will not overlap. The total number of vectors inside one of these spheres is equal to the number of vectors whose Hamming distance from the center is less than or equal to t. From eq. (1.20), this number is

$$\sum_{i=0}^{t} \binom{n}{i} .$$

Since there are 2^k codewords, and one sphere per codeword, the total number of vectors within these spheres is

$$2^k \sum_{i=0}^{t} \binom{n}{i} .$$

Since the total number of vectors in F_2^n is 2^n, it follows that

$$2^k \sum_{i=0}^{t} \binom{n}{i} \leq 2^n ,$$

which proves the theorem. ∎

Exercise 1.5 *Show that a perfect error correcting code saturates the Hamming bound:*

$$\sum_{i=0}^{t} \binom{n}{i} = 2^{n-k} .$$

THEOREM 1.6 (Gilbert-Varshamov Bound)
There exists a binary $[n, k, d]$ code provided the code parameters satisfy

$$\sum_{i=0}^{d-2} \binom{n-1}{i} < 2^{n-k} .$$

PROOF To prove this theorem we construct an $(n-k) \times n$ parity check matrix H for which no choice of $d-1$ columns yields a linearly dependent set. By Theorem 1.1, H yields a linear code $\mathcal{C}$ with minimum distance d.

To begin, let column 1 be any non-zero vector $c_1 \in F_2^{n-k}$. Let column 2 be any other non-zero vector $c_2 \in F_2^{n-k}$ such that c_1 and c_2 are linearly independent. Now assign column 3 to be any non-zero vector $c_3 \in F_2^{n-k}$ that is not a linear combination of c_1 and c_2 so that the three vectors form a linearly independent set. Continuing in this fashion, let column i be any non-zero vector $c_i \in F_2^{n-k}$ that is not a linear combination of any $d-2$ or fewer of the previous $i-1$ columns. This procedure insures that no choice of $d-1$ or fewer of the i columns forms a linearly dependent set.

As long as the set of all linear combinations of $d-2$ or fewer columns does not exhaust all non-zero vectors in F_2^{n-k}, another column can be added to H. If there are $i-1$ columns, the total number of distinct linear combinations that can be formed from j columns (with coefficients in F_2) is given by the binomial coefficient

$$\binom{i-1}{j}.$$

The largest number of vectors is obtained when each linear combination gives a distinct element of F_2^{n-k}. If j takes all values from 1 to $d-2$, then there will be at most

$$\sum_{j=1}^{d-2} \binom{i-1}{j}$$

distinct linear combinations of $d-2$ or fewer columns drawn from the set of $i-1$ columns. If this number is less than $2^{n-k}-1$, the number of non-zero vectors in F_2^{n-k}, we can add another column to H. Thus another column can be added to H if

$$\sum_{j=1}^{d-2} \binom{i-1}{j} < 2^{n-k} - 1 . \tag{1.21}$$

This yields an $(n-k) \times i$ parity check matrix H for which every $d-1$ columns are linearly independent. Let n be the largest value of i for which eq. (1.21) is true. Then,

$$\sum_{j=0}^{d-2} \binom{n-1}{j} < 2^{n-k} , \tag{1.22}$$

which proves the theorem. ∎

Notice that by the definition of n, letting $i = n+1$ in eq. (1.21) causes this inequality to be violated. Thus one often sees the Gilbert-Varshamov bound written as

$$\sum_{j=0}^{d-2} \binom{n}{j} \geq 2^{n-k} . \tag{1.23}$$

1.3 Using Quantum Systems to Store and Process Data

This section describes some of the advantages and disadvantages of using quantum systems to store and process computational data. For readers not familiar with quantum mechanics, a summary of the needed background is given in Appendix B.

1.3.1 Linear Superposition and Quantum Parallelism

The states of a quantum system satisfy the principle of linear superposition, which says that if $|\chi\rangle$ and $|\phi\rangle$ are allowed states, then so is the linear superposition $a|\chi\rangle + b|\phi\rangle$, where a and b are complex numbers. As is well known [48], this principle restricts the quantum system's state space to be a vector space (in fact, a Hilbert space), and its dynamics to be linear. Superselection rules, when present, restrict the validity of the superposition principle. See Appendix B for an explanation of why they can be safely ignored in discussions of quantum computing.

The simplest non-trivial quantum system is a qubit. Its state space H_2 is a two-dimensional Hilbert space. As a result, any basis that spans H_2 must contain two linearly independent states. A general way to construct a basis is to use the eigenvectors of a physical observable $\mathcal{O}$:

$$\mathcal{O}|\lambda_i\rangle = \lambda_i|\lambda_i\rangle \ . \tag{1.24}$$

Here λ_i are the eigenvalues of $\mathcal{O}$; and it can be shown that the eigenvectors $|\lambda_i\rangle$ form a basis that can always be made into a mutually orthogonal set [48]. For a qubit whose states are elements of H_2, we can choose the two index values to be $i = 0, 1$. We can identify the computational basis states $|0\rangle$ and $|1\rangle$ with the orthogonal eigenvectors of $\mathcal{O}$, say $|0\rangle = |\lambda_0\rangle$ and $|1\rangle = |\lambda_1\rangle$. The states $|0\rangle$ and $|1\rangle$ are the quantum analogs of the classical bit states 0 and 1. Unlike a bit that at any moment must be in either the 0 or 1 state, a qubit can exist in a linear superposition of the computational basis states $|0\rangle$ and $|1\rangle$:

$$|\psi\rangle = a|0\rangle + b|1\rangle \ . \tag{1.25}$$

The coefficients a and b are known as probability amplitudes since, should a measurement of $\mathcal{O}$ be done, $|a|^2$ ($|b|^2$) gives the probability that: (i) the measurement outcome will be λ_0 (λ_1), and (ii) the qubit will be left in the state $|0\rangle$ ($|1\rangle$) immediately after the measurement. As the measurement outcome must yield one of the eigenvalues of $\mathcal{O}$ with certainty $|a|^2 + |b|^2 = 1$, and $|\psi\rangle$ is said to be normalized.

In quantum computing, one uses n qubits to form a quantum register. The n-qubit state space H_2^n is the direct product of the n single-qubit Hilbert spaces H_2:

$$H_2^n = H_2 \otimes \cdots \otimes H_2 \ .$$

The n-qubit computational basis states are constructed by forming all possible direct products of the single-qubit computational basis states:

$$|i_1 \cdots i_n\rangle = |i_1\rangle \otimes \cdots \otimes |i_n\rangle , \qquad (1.26)$$

where $i_1, \ldots, i_n = 0, 1$. There are thus a total of 2^n such states, and this number is then the dimension of H_2^n. Note that this dimension grows exponentially with the number of qubits n. By comparison, for n bits, the state space is F_2^n. This space is n-dimensional since the n vectors $\mathbf{e}_1 = (1, 0, \cdots, 0)$; $\mathbf{e}_2 = (0, 1, 0, \cdots, 0)$; $\ldots$; $\mathbf{e}_n = (0, \cdots, 0, 1)$ form a basis. Thus the dimension of F_2^n only grows linearly with the number of bits n.

Notice that the n-qubit computational basis states $|i_1 \cdots i_n\rangle$ are labeled by the bits strings $i_1 \cdots i_n$. Each bit string can be thought of as the binary decomposition of an integer i in the range $[0, 2^n - 1]$: $i = \sum_{j=1}^{n} 2^{j-1} i_j$. The states of an n-bit classical register are exactly this set of 2^n bit strings $i_1 \cdots i_n$. A computation $\mathcal{C}$ implemented on this register maps the input string $i_1 \cdots i_n$ into the output string $O_1(i) \cdots O_n(i)$:

$$\begin{pmatrix} O_1(i) \\ \vdots \\ O_n(i) \end{pmatrix} = \mathcal{C} \begin{pmatrix} i_1 \\ \vdots \\ i_n \end{pmatrix} , \qquad (1.27)$$

where i on the LHS is the integer whose binary decomposition is $i_1 \cdots i_n$. In general, to determine all possible output strings $O_1(i) \cdots O_n(i)$, the n-bit register would have to apply $\mathcal{C}$ serially to all 2^n input strings $i_1 \cdots i_n$. Imagine now that we can program a $2n$-qubit quantum register to implement $\mathcal{C}$ through a unitary transformation $U(\mathcal{C})$:

$$U(\mathcal{C}) |i_1 \cdots i_n\rangle \otimes |0 \cdots 0\rangle = |i_1 \cdots i_n\rangle \otimes |O_1(i) \cdots O_n(i)\rangle . \qquad (1.28)$$

An appropriate measurement of the last n qubits returns the output string $O_1(i) \cdots O_n(i)$. The need for $2n$ qubits is demanded by the unitary character of $U(\mathcal{C})$, which preserves inner products. A problem arises when only n qubits are used when the computation $\mathcal{C}$ is many-to-one. Then there exist $i \neq j$ for which $O_1(i) \cdots O_n(i) = O_1(j) \cdots O_n(j)$. If we try to use n qubits to implement $|x\rangle \to |O_1(x) \cdots O_n(x)\rangle$ in this case, the image states $|O_1(i) \cdots O_n(i)\rangle$ and $|O_1(j) \cdots O_n(j)\rangle$ will be equal and so

$$\langle O_1(i) \cdots O_n(i) | O_1(j) \cdots O_n(j) \rangle = 1 .$$

But this violates unitarity since for $i \neq j$,

$$\langle i_1 \cdots i_n | j_1 \cdots j_n \rangle = 0 \implies \langle O_1(i) \cdots O_n(i) | O_1(j) \cdots O_n(j) \rangle = 0 ,$$

and so we have a contradiction. The extra n qubits in eq. (1.28) insure that unitarity is maintained even when $\mathcal{C}$ is a many-to-one map. For the case just

Introduction 19

considered, $i \neq j$ and $O_1(i) \cdots O_n(i) = O_1(j) \cdots O_n(j)$,

$$\begin{aligned}
\langle i_1 \cdots i_n \, &O_1(i) \cdots O_n(i) \, | \, j_1 \cdots j_n \, O_1(j) \cdots O_n(j) \rangle \\
&= \langle i_1 \cdots i_n | j_1 \cdots j_n \rangle \langle O_1(i) \cdots O_n(i) | O_1(j) \cdots O_n(j) \rangle \\
&= 0 \times 1 \\
&= 0 \ .
\end{aligned} \quad (1.29)$$

This is the same as the inner product between the initial orthogonal states $|i_1 \cdots i_n 0 \cdots 0\rangle$ and $|j_1 \cdots j_n 0 \cdots 0\rangle$. Eq. (1.28) is thus consistent with unitarity even when $\mathcal{C}$ is a many-to-one map.

Linear superposition allows us to form the $2n$-qubit state:

$$|\psi_{in}\rangle = \left[\frac{1}{\sqrt{2^n}} \sum_i |i_1 \cdots i_n\rangle \right] \otimes |0 \cdots 0\rangle \ . \quad (1.30)$$

Notice that the first n qubits are in a uniform superposition of states labeled by all 2^n classical input strings $i_1 \cdots i_n$. Because the dynamics of a quantum system is linear,

$$\begin{aligned}
|\psi_{out}\rangle &= U(\mathcal{C}) |\psi_{in}\rangle \\
&= \frac{1}{\sqrt{2^n}} \sum_i U(\mathcal{C}) |i_1 \cdots i_n\rangle \otimes |0 \cdots 0\rangle \\
&= \frac{1}{\sqrt{2^n}} \sum_i |i_1 \cdots i_n\rangle \otimes |O_1(i) \cdots O_n(i)\rangle \ .
\end{aligned} \quad (1.31)$$

We see that $|\psi_{out}\rangle$ is a uniform superposition of direct product states in which the state of the first n qubits is labeled by the string $i_1 \cdots i_n$ and that of the last n qubits by $O_1(i) \cdots O_n(i)$, the image of $i_1 \cdots i_n$ under $\mathcal{C}$. The quantum register (a. k. a. quantum computer) has managed to encode all the output strings produced by $\mathcal{C}$ into $|\psi_{out}\rangle$. Naively, it appears as though our quantum computer has simultaneously pursued 2^n classical computational paths. We say 'naively' because the 2^n output strings are embedded in a ghostly quantum mechanical superposition. How one goes about retrieving all/some of the 2^n output strings in a single measurement is an extremely non-trivial question that must be addressed when designing a quantum algorithm [49]. This ability of a quantum computer to encode multiple computational results into a quantum state in a single quantum computational step is known as quantum parallelism. It is an advantage that quantum computers have over classical computers, though as has been noted, it is not sufficient to make a quantum computer useful. It is essential to find a way to extract the multiple computational results generated by this parallelism from the quantum computer.

1.3.2 No-Cloning Theorem

The No-Cloning Theorem [50, 51] states that a device cannot be constructed that produces an exact copy of an arbitrary quantum state $|\psi\rangle$: $|\psi\rangle \otimes |C_0\rangle \to |\psi\rangle \otimes |\psi\rangle$. This theorem has a number of important consequences. (1) It rules out the possibility of using EPR correlations to transmit signals faster than light. (2) It prevents an eavesdropper from obtaining the key to a quantum cryptosystem by copying the quantum state used to distribute this key over a public channel. (3) It does not allow the state of a quantum computer to be copied at intermediate times during a computation as a way of backing up the data should errors occur. For our purposes, this latter point is the most important. It indicates that quantum error correction will not have recourse to copies of the computational data and something more subtle will have to be done to protect the data from errors. We give a simple proof of the No-Cloning Theorem in the case where a pure state is to be copied and the quantum dynamics is unitary. For references to papers that extend the range of applications of this theorem, see [52].

THEOREM 1.7 (No-Cloning Theorem)
A device D cannot be constructed whose action is to make an exact copy of an arbitrary quantum state.

PROOF We will assume the theorem is false and show that this leads to a contradiction. Let Q and C be the quantum systems upon which D acts. Initially Q carries the quantum state to be copied, and upon completion of the copying action, C will carry an exact copy of this state. Initially C is prepared in the (normalized) fiducial state $|C_0\rangle$. Let $|\psi\rangle$ and $|\xi\rangle$ be a pair of quantum states. By assumption, D can copy each of these states:

$$|\psi\rangle \otimes |C_0\rangle \to |\psi\rangle \otimes |\psi\rangle \tag{1.32}$$

and

$$|\xi\rangle \otimes |C_0\rangle \to |\xi\rangle \otimes |\xi\rangle \ . \tag{1.33}$$

Unitarity requires that the overlap of the initial states in eqs. (1.32) and (1.33) equal the overlap of the final states. Thus $\langle\xi|\psi\rangle = \langle\xi|\psi\rangle^2$, which requires that $\langle\xi|\psi\rangle = 1$ or 0. The first case says that $|\psi\rangle = |\xi\rangle$, which is not very interesting. The second case says that $|\psi\rangle$ and $|\xi\rangle$ are orthogonal. Now consider a superposition of these states: $|\rho\rangle = a|\psi\rangle + b|\xi\rangle$, where $a, b \neq 0$ and $|\rho\rangle$ is normalized. To satisfy unitarity, we have just shown that $|\rho\rangle$ must either be equal to $|\psi\rangle$ or $|\xi\rangle$ (which it is not), or it must be orthogonal to these two states (which it is not). Thus $|\rho\rangle$ cannot be copied as this would violate the unitary dynamics imposed by D. This contradicts our initial assumption that an arbitrary quantum state can be copied, and the theorem is thus proved.

Introduction

As we have just seen, the theorem does not rule out the possibility of building a device that can copy a particular set of orthonormal quantum states. What the theorem does rule out, however, is the hope of ever building a device that can take an arbitrary quantum state as input and produce an exact copy of it.

1.3.3 Measurement

Measurement of quantum systems will play an important role in detecting and correcting errors in a quantum computation. In the standard formulation of quantum mechanics, the postulates of the theory spell out how a quantum system responds when measured. The question of whether the measurement postulates can be derived from the other postulates has long been a source of controversy. Fortunately, it will not be necessary to enter into that discussion. For our purposes, the standard approach to measurement will do just fine. Here we summarize the essential points. Interested readers will find useful introductions to the quantum measurement problem in References [53–56].

Suppose we want to measure the observable O of a quantum system Q. O is represented by an Hermitian operator $\mathcal{O}$ that has eigenvalues O_λ and eigenvectors $|O_\lambda; r\rangle$, where r is a degeneracy index. Since $\mathcal{O}$ is Hermitian, its eigenvalues are real, and its eigenvectors form a basis that can always be made orthonormal. Prior to measurement, let Q be in the state

$$|\psi\rangle = \sum_{\lambda,r} c_{\lambda,r} |O_\lambda; r\rangle \ . \tag{1.34}$$

In the standard formulation of quantum mechanics, measurement of O is postulated to produce the following effects. (1) The measurement outcome will always be one of the eigenvalues of $\mathcal{O}$. (2) The probability $p(\lambda)$ that the eigenvalue O_λ is found is

$$p(\lambda) = \sum_{r} |c_{\lambda,r}|^2 \ , \tag{1.35}$$

and immediately after the measurement, Q will be in the state

$$|\psi_f\rangle = \frac{1}{\sqrt{p(\lambda)}} \sum_{r} c_{\lambda,r} |O_\lambda; r\rangle \ . \tag{1.36}$$

A concise formal description of measurement is possible through the introduction of projection operators [57]. Since $\mathcal{O}$ is Hermitian, it can be diagonalized:

$$\mathcal{O} = \sum_{\lambda,r} O_\lambda |O_\lambda; r\rangle\langle O_\lambda; r|$$

$$= \sum_{\lambda} O_\lambda P_\lambda \ , \tag{1.37}$$

where
$$P_\lambda = \sum_r |O_\lambda; r\rangle\langle O_\lambda; r| \ . \tag{1.38}$$

The operators $\{P_\lambda\}$ are projection operators since they satisfy $P_\lambda P_\rho = \delta_{\lambda\rho} P_\lambda$ and $P_\lambda^\dagger = P_\lambda$. Note that since the eigenvectors $\{|O_\lambda; r\rangle\}$ form a complete set (basis), it follows that
$$\sum_\lambda P_\lambda = I \ . \tag{1.39}$$

One can write the probability $p(\lambda)$ (see eq. (1.35)) as
$$p(\lambda) = \langle\psi|P_\lambda|\psi\rangle \ , \tag{1.40}$$

and the state immediately after measurement (see eq. (1.36)) as
$$|\psi_f\rangle = \frac{1}{\sqrt{p(\lambda)}} P_\lambda |\psi\rangle \ . \tag{1.41}$$

From eqs. (1.38) and (1.41) we see that if measurement of O yields the r-fold degenerate eigenvalue O_λ, the pre-measurement state $|\psi\rangle$ will be projected onto the subspace spanned by the r associated eigenvectors $\{|O_\lambda; k\rangle : k = 1, \cdots, r\}$. This measurement-induced projection is often referred to as the collapse/reduction of the wavefunction.

Example 1.4 *Measurement in the Computational Basis*
As an application, consider an n qubit system and let $\{|i_1 \cdots i_n\rangle : i_1, \cdots, i_n = 0, 1\}$ be the computational basis. As in the remarks preceding eq. (1.27), let i be the integer whose binary decomposition is $(i_1, \cdots, i_n)$. Measurement of the n qubit system in the computational basis will be an operation we will make use of often in the remainder of this book. It corresponds to choosing the projection operators associated with the measurement to be
$$P_i = |i_1 \cdots i_n\rangle\langle i_1 \cdots i_n| \ . \tag{1.42}$$

The probability that the measurement outcome will be the bit-string $i_1 \cdots i_n$ when the pre-measurement state was $|\psi\rangle$ is then
$$P(i_1 \cdots i_n) = \langle\psi|P_i|\psi\rangle \ , \tag{1.43}$$

and the state immediately after the measurement will be
$$|\psi_f\rangle = |i_1 \cdots i_n\rangle \ . \tag{1.44}$$

☐

Introduction 23

1.3.4 Distinguishing Quantum States

When constructing a quantum error correcting code that can detect and correct a set of errors $\{E_n\}$, we must be able to distinguish the error E_a acting on codeword $|\psi_i\rangle$ from the error E_b acting on a different codeword $|\psi_j\rangle$. Here we show that quantum theory does not allow us to unambiguously distinguish non-orthogonal quantum states. Thus the erroneous images $E_a|\psi_i\rangle$ and $E_b|\psi_j\rangle$ must be orthogonal if the code is to correctly distinguish these errors. The proof below applies to pure states and is inspired by the one given in Ref. [49]. Peres [58] gives a proof using mixed states. He also shows that the ability to distinguish non-orthogonal quantum states could be used to construct a cyclic process that would violate the second law of thermodynamics.

THEOREM 1.8
It is impossible to unambiguously distinguish non-orthogonal quantum states.

PROOF To prove the theorem we assume that non-orthogonal states can be distinguished and show that this leads to a contradiction. Let M be an observable whose measurement allows us to unambiguously distinguish two non-orthogonal states $|\psi_a\rangle$ and $|\psi_b\rangle$. Let $\mathcal{M}$ be the Hermitian operator that represents this observable with eigenvalues m_λ and associated projection operators P_λ. By assumption, eigenvalues m_α and m_β exist such that observation of m_α (m_β) unambiguously identifies $|\psi_a\rangle$ ($|\psi_b\rangle$) as the pre-measurement state. Formally, this means that the probability to observe m_α (m_β) when the pre-measurement state is $|\psi_a\rangle$ ($|\psi_b\rangle$) is one:

$$\langle\psi_a|P_\alpha|\psi_a\rangle = 1 \quad \text{and} \quad \langle\psi_b|P_\beta|\psi_b\rangle = 1 \ , \qquad (1.45)$$

and thus the probability to observe m_β (m_α) when the pre-measurement state is $|\psi_a\rangle$ ($|\psi_b\rangle$) is zero:

$$\langle\psi_a|P_\beta|\psi_a\rangle = 0 \quad \text{and} \quad \langle\psi_b|P_\alpha|\psi_b\rangle = 0 \ . \qquad (1.46)$$

Since $|\psi_b\rangle$ and $|\psi_a\rangle$ are assumed to be non-orthogonal, we can write

$$|\psi_b\rangle = c|\psi_a\rangle + d|\xi\rangle \ , \qquad (1.47)$$

where $|c|^2 + |d|^2 = 1$, and $|\xi\rangle$ is orthogonal to $|\psi_a\rangle$.

Note that eq. (1.46) requires $P_\beta|\psi_a\rangle = 0$ since

$$\begin{aligned} 0 &= \langle\psi_a|P_\beta|\psi_a\rangle \\ &= \langle\psi_a|P_\beta P_\beta|\psi_a\rangle \\ &= \|P_\beta|\psi_a\rangle\|^2 \ , \end{aligned} \qquad (1.48)$$

and the only state with zero norm is the null state 0. Combining this result with eq. (1.47) allows us to explicitly evaluate $\langle\psi_b|P_\beta|\psi_b\rangle$:

$$\begin{aligned} \langle\psi_b|P_\beta|\psi_b\rangle &= [\, c^*\langle\psi_a| + d^*\langle\xi|\,]\,[\,cP_\beta|\psi_a\rangle + dP_\beta|\xi\rangle\,] \\ &= |d|^2\,\langle\xi|P_\beta|\xi\rangle \ . \end{aligned} \qquad (1.49)$$

Note that $\langle\xi|P_\lambda|\xi\rangle \geq 0$ for all λ since this gives the probability that m_λ is the measurement outcome when the pre-measurement state is $|\xi\rangle$. As we saw in Section 1.3.3, $\sum_\lambda P_\lambda = I$ so that

$$\begin{aligned} 1 &= \langle\xi|\xi\rangle \\ &= \langle\xi| \sum_\lambda P_\lambda |\xi\rangle \\ &= \sum_\lambda \langle\xi|P_\lambda|\xi\rangle \geq \langle\xi|P_\beta|\xi\rangle \; , \end{aligned} \quad (1.50)$$

where the inequality appears since all terms in the sum are non-negative. Combining eqs. (1.49) and (1.50) gives

$$\langle\psi_b|P_\beta|\psi_b\rangle \leq |d|^2 \; . \quad (1.51)$$

The only way eq. (1.51) can be consistent with $\langle\psi_b|P_\beta|\psi_b\rangle = 1$ is if $|d|^2 = 1$. From eq. (1.47), this requires that $|\psi_b\rangle = |\xi\rangle$. Thus for $|\psi_b\rangle$ to be unambiguously distinguishable from $|\psi_a\rangle$, the two states must be orthogonal. But we assumed that these states were non-orthogonal so that we have arrived at a contradiction that proves the theorem. The lesson here is that only orthogonal quantum states can be unambiguously distinguished. ∎

1.3.5 Entanglement

Entanglement is currently a large and active area of research. We will only focus on those aspects of it that are relevant to the purposes of this book. Entanglement is a physical property of composite quantum systems. It manifests as non-local correlations that do not arise in composite classical systems. These correlations lie at the heart of the Einstein-Podolsky-Rosen (EPR) paradox [59]. A measure of their non-classical character is provided by the degree to which they violate Bell-type inequalities [53,60]. Experimental tests of Bell inequality violations have been carried out since the 1970s [53]. Although efforts to close loopholes continue [61], the prevailing consensus is that the results of these tests are consistent with quantum mechanics, and not with theories that incorporate local hidden variables. In quantum computing and quantum information, entanglement is treated as a physical resource that is at our disposal. Quantum teleportation makes use of it to transfer an unknown quantum state between distant observers [62]. It is used to construct unbreakable codes [63], and to identify and correct errors arising in a quantum computation. This latter application will be discussed in detail in the following chapters.

Consider a bipartite quantum system S composed of two subsystems E and F. The subsystems may themselves be composite systems. Let H_E and H_F be the N-dimensional and M-dimensional Hilbert spaces for E and F, respectively. If $\{|e_i\rangle : i = 1, \ldots, N\}$ and $\{|f_j\rangle : j = 1, \ldots, M\}$ are

Introduction

orthonormal bases for H_E and H_F, respectively, then an orthonormal basis for S is $\{\,|e_i f_j\rangle \equiv |e_i\rangle \otimes |f_j\rangle : i = 1,\ldots,N;\ j = 1,\ldots,M\,\}$.

DEFINITION 1.10 *Let $|\psi\rangle$ be a normalized pure state of S:*

$$|\psi\rangle = \sum_{i,j} c_{ij}\,|e_i f_j\rangle \ . \tag{1.52}$$

We say that $|\psi\rangle$ represents an entangled state if it cannot be written as a direct product of states $|\phi\rangle \in H_E$ and $|\chi\rangle \in H_F$:

$$|\psi\rangle \neq |\phi\rangle \otimes |\chi\rangle \ .$$

Not surprisingly, a pure state $|\psi\rangle$ is said to be non-entangled if it can be written as a direct product $|\psi\rangle = |\phi\rangle \otimes |\chi\rangle$, with $|\phi\rangle \in H_E$ and $|\chi\rangle \in H_F$.

THEOREM 1.9
The state $|\psi\rangle$ defined in eq. (1.52) is non-entangled if and only if $c_{ij} = \alpha_i \beta_j$.

PROOF
($\Longrightarrow$) Assume that $|\psi\rangle$ is non-entangled. Then there exist $|\nu\rangle \in H_E$ and $|\omega\rangle \in H_F$ such that

$$\begin{aligned}|\psi\rangle &= |\nu\rangle \otimes |\omega\rangle \tag{1.53}\\ &= \left[\sum_i \alpha_i |e_i\rangle\right] \otimes \left[\sum_j \beta_j |f_j\rangle\right] \\ &= \sum_{ij} \alpha_i \beta_j\, |e_i f_j\rangle \ . \tag{1.54}\end{aligned}$$

Comparing eqs. (1.52) and (1.54) identifies $c_{ij} = \alpha_i \beta_j$.

($\Longleftarrow$) Assume that $c_{ij} = \alpha_i \beta_j$. Inserting this into eq. (1.52) gives

$$\begin{aligned}|\psi\rangle &= \sum_{ij} \alpha_i \beta_j |e_i f_j\rangle \\ &= \left[\sum_i \alpha_i |e_i\rangle\right] \otimes \left[\sum_j \beta_j |f_j\rangle\right] \\ &= |\nu\rangle \otimes |\omega\rangle \ ,\end{aligned}$$

where $|\nu\rangle \in H_E$ and $|\omega\rangle \in H_F$. Thus $|\psi\rangle$ is a non-entangled state. Note that if $|\nu\rangle$ and $|\omega\rangle$ are normalized states, then $\sum_i |\alpha_i|^2 = \sum_j |\beta_j|^2 = 1$. ∎

We now derive two useful results. (1) If the composite system S is in a pure non-entangled state, then the subsystems E and F will each be in pure states. (2) If the composite system S is in a pure entangled state, then the subsystems E and F will each be in mixed states.

Let S be in the state $|\psi\rangle$ given in eq. (1.52). The associated density operator $\rho = |\psi\rangle\langle\psi|$ is:

$$\rho = \sum_{ij,lm} c_{ij}c_{lm}^* |e_i f_j\rangle\langle e_l f_m| \ . \tag{1.55}$$

(1) If $|\psi\rangle$ represents a normalized non-entangled state, then we have just seen that $c_{ij} = \alpha_i \beta_j$, and $\sum_i |\alpha_i|^2 = \sum_j |\beta_j|^2 = 1$. The reduced density operator ρ_E for E is (see Appendix B)

$$\rho_E = \sum_n \langle f_n | \rho | f_n \rangle$$

$$= \sum_{ij} \alpha_i \alpha_j^* |e_i\rangle\langle e_j|$$

$$= \left[\sum_i \alpha_i |e_i\rangle \right] \left[\sum_j \alpha_j^* \langle e_j| \right]$$

$$= |\nu\rangle\langle\nu| \ .$$

Thus if S is in a non-entangled pure state, the subsystem E is in a pure state. Repeating this argument for F leads to the same conclusion about F.

(2) Now let $|\psi\rangle$ represent an entangled state. The reduced density operator ρ_E for E is now:

$$\rho_E = \sum_{ijl} c_{il} c_{jl}^* |e_i\rangle\langle e_j|$$

$$= \sum_{ij} (\rho_E)_{ij} |e_i\rangle\langle e_j| \ , \tag{1.56}$$

where $(\rho_E)_{ij} \equiv \sum_l c_{il} c_{jl}^*$. Since ρ_E is a positive Hermitian operator, it can be diagonalized, and its eigenvalues are non-negative. Carrying out the diagonalization reduces eq. (1.56) to:

$$\rho_E = \sum_i p_i |p_i\rangle\langle p_i| \ , \tag{1.57}$$

where the $\{p_i\}$ are the eigenvalues and the $\{|p_i\rangle\}$ are the corresponding eigenvectors. Since $\mathrm{Tr}\rho_E = \mathrm{Tr}\rho = 1$, we have that $\sum_i p_i = 1$. It follows that $0 \leq p_i \leq 1$ for all i, since the $\{p_i\}$ are non-negative, and they sum to 1. Because $|\psi\rangle$ was assumed to be an entangled state, at least two of the $\{p_i\}$ are non-zero and so eq. (1.57) represents a mixed state. Thus, if S is in an entangled pure state, E will be in a mixed state. A similar analysis for F leads to the same conclusion about F.

Introduction

Since $\text{Tr}\rho^2 < 1$ for a quantum system in a mixed state, the preceding analysis shows that a simple test for whether a composite system is in an entangled pure state is to check whether $\text{Tr}\rho_E^2 < 1$ for subsystem E. If yes, then the state of the composite system is entangled.

Testing whether $\text{Tr}\rho_E^2 < 1$ gives a simple yes-no test for entanglement. It does not however allow us to compare whether one entangled pure state has more or less entanglement than another. A quantitative measure of the entanglement of a pure state $|\psi\rangle$ is known. It is the entropy of entanglement [64]:

$$E(|\psi\rangle) = S(\rho_E) = S(\rho_F) \ . \tag{1.58}$$

Here $S(\rho) = -\text{Tr}\rho\log\rho$ is the von-Neumann entropy. $E(|\psi\rangle)$ vanishes for a non-entangled pure state, and can be as large as $\log N$ for a maximally entangled state of two N-state systems.

Exercise 1.6

(a) Let $|\psi\rangle = |\phi\rangle \otimes |\chi\rangle$, where $|\phi\rangle$ and $|\chi\rangle$ are normalized states in H_E and H_F, respectively. Show that $E(|\psi\rangle) = 0$.

(b) Let $\{|ij\rangle\}$ be the computational basis for $H_E \otimes H_F$ and let

$$|\psi\rangle = \frac{1}{\sqrt{N}} \sum_i |ii\rangle \ .$$

Show that $E(|\psi\rangle) = \log N$.

Entanglement is used to detect and correct errors that may arise in a quantum computation. We describe the basic protocol here. A discussion of how this protocol is implemented fault-tolerantly will be given in Chapter 5. Assume that initially our quantum computer Q and its environment E are not entangled, and that Q's initial state is $|\chi\rangle$ and that of E is $|e\rangle$. Due to the unavoidable coupling between Q and E, the two systems become entangled:

$$|e\rangle|\chi\rangle \longrightarrow \sum_s |e_s\rangle \{ E_s|\chi\rangle \} \ . \tag{1.59}$$

The final states of the environment $\{|e_s\rangle\}$ need not form an orthonormal set. The operators $\{E_s\}$ formally describe the errors introduced by the environment into a quantum computation. At this point in the protocol, ancilla qubits are introduced that are coupled to the qubits in Q. The ancilla are initially prepared in the fiducial state $|a_0\rangle$. The unitary interaction U that couples them to Q is designed to produce the following effect:

$$U\,[\,|a_0\rangle\,\{\,E_s|\chi\rangle\,\}\,] = |s\rangle\{E_s|\chi\rangle\} \ . \tag{1.60}$$

The final ancilla state $|s\rangle$ is assumed to depend on E_s, but not on $|\chi\rangle$. The ancilla states $\{|s\rangle\}$ are assumed to form an orthonormal set. Because the quantum dynamics is linear, U produces the following transformation:

$$U\left[\sum_s |e_s\rangle|a_0\rangle\{E_s|\chi\rangle\}\right] = \sum_s |e_s\rangle|s\rangle\{E_s|\chi\rangle\} \ . \tag{1.61}$$

Since the $\{|s\rangle\}$ are orthonormal, they can be distinguished. If we measure the ancilla in the s-basis, and find the result S, then the post-measurement state is the *non-entangled* pure state $|\psi_{pm}\rangle$:

$$|\psi_{pm}\rangle = |e_S\rangle|S\rangle\{E_S|\chi\rangle\} \ . \tag{1.62}$$

S is referred to as the error syndrome, and the process of determining it through measurement of the ancilla is known as syndrome extraction. Since S is known from the ancilla measurement, and it is assumed that E_S can be identified from S (more on this below), we can apply E_S^{-1} to Q, leaving the composite system in the final state:

$$|e_S\rangle|S\rangle|\chi\rangle \ . \tag{1.63}$$

This sequence of operations and measurements has (1) unentangled the quantum computer Q from its environment E and from the ancilla, and (2) put Q back into its initial state $|\chi\rangle$ (viz. removed the errors introduced by the environment). Thus the unwanted entanglement between the quantum computer and its environment has been removed by further entangling the quantum computer with a set of ancilla qubits. Appropriate measurement of the ancilla then removes the unwanted entanglement and indicates what operation must be applied to correct errors (see below).

Two further comments are in order before moving on. (1) It is not necessary that the error syndrome S identify E_S. It is enough if S allows us to identify an operation that corrects the quantum computer's state and disentangles it from its environment. (2) In general, it will not be possible to correct all the E_s. Error correcting codes, classical or quantum, are only able to fix a definite set of errors. For useful codes, the correctable errors are also the most probable errors.

1.3.6 Noise: Errors and Decoherence

Just as no man is an island, no physical system is ever truly isolated from its surroundings. This is certainly true for a quantum computer, which will have both desired and undesired interactions with the other physical systems that make up its environment. The interaction that couples a quantum computer to the classical control fields that drive the quantum gates used in a quantum computation is certainly a desired interaction. If these gates operate perfectly, the state of the quantum computer evolves exactly as specified by the quantum algorithm that underlies the computation. Any other interactions between the quantum computer and its environment cause its state to deviate from

Introduction

the evolution spelled out by the quantum algorithm, and so cause errors to appear in the quantum computation. These undesired interactions with the environment manifest as noise in the operation of the quantum computer.

In Section 1.3.5 we saw that the unwanted interactions with the environment can be described by a set of error operators $\{E_s\}$. The effect of the unwanted interactions is to entangle the quantum computer with its environment. Thus if our quantum computer Q and its environment E are initially unentangled, with Q in the state $|\chi\rangle$ and E in the state $|e_0\rangle$, the unwanted interactions map the initial state $|\psi_0\rangle = |e_0\rangle \otimes |\chi\rangle$ into the entangled state (eq. (1.59)):

$$|\psi_f\rangle = \sum_s |e_s\rangle \otimes E_s|\chi\rangle \ . \tag{1.64}$$

If we denote the composite state for Q and E by $|\psi(t)\rangle$, then the associated density operator is $\rho_c(t) = |\psi(t)\rangle\langle\psi(t)|$. As the state of the environment is not observed, Q's state will be described by the reduced density operator $\rho(t)$, which is obtained by tracing out the environment $\rho(t) = \text{tr}_E \rho_c$. The unwanted interactions thus produce the following map on the reduced density operator $\rho(t)$: $\rho_0 = |\chi\rangle\langle\chi| \to \rho_f = \text{tr}_E|\psi_f\rangle\langle\psi_f|$, where

$$\rho_f = \text{tr}_E \sum_{s,s'} |e_s\rangle\langle e_{s'}| \otimes \left[E_s|\chi\rangle\langle\chi|E_{s'}^\dagger \right]$$

$$= \sum_n E_n|\chi\rangle\langle\chi|E_n^\dagger$$

$$= \sum_n E_n \rho_0 E_n^\dagger \ . \tag{1.65}$$

The final expression for ρ_f in eq. (1.65) is an example of an operator-sum representation for a quantum operation. This is an extremely useful way to describe the effects of environmental noise on the dynamics of a quantum computer. We will discuss quantum operations in Chapter 2. As will be seen there, a quantum operation will be trace-preserving, $\text{tr}\rho_f = \text{tr}\rho_0$, if the error operators satisfy

$$\sum_n E_n^\dagger E_n = I \ . \tag{1.66}$$

Not surprisingly, a non-trace-preserving quantum operation has error operators $\{E_s\}$ that do not satisfy eq. (1.66).

As an example, we consider an environment acting on a single qubit that with probability $1-p$ leaves the qubit alone, while with probability p rotates the qubit by 180° about the z-axis. The error operators for this quantum operation are

$$E_0 = \sqrt{1-p}\, I \quad ; \quad E_1 = -i\sqrt{p}\, \sigma_z \ . \tag{1.67}$$

It is easy to check that $\sum_n E_n^\dagger E_n = I$ so that this quantum operation is trace-preserving. As is well known [65, 66], the density operator for a qubit

can be written as
$$\rho = \frac{1}{2}(I + \mathbf{R} \cdot \boldsymbol{\sigma}) \ , \tag{1.68}$$

where $\mathbf{R} = \text{tr}(\boldsymbol{\sigma}\rho)$ is the Bloch vector for the state (also known as the polarization). Let $\mathbf{R}_0 = [R_x, R_y, R_z]$ correspond to the Bloch vector for the initial qubit state ρ_0. The environment then maps $\rho_0 \to \rho_f$, where

$$\begin{aligned}\rho_f &= \sum_{n=0,1} E_n \rho_0 E_n^\dagger \\ &= \frac{1}{2}(I + \mathbf{R}_f \cdot \boldsymbol{\sigma}) \ , \end{aligned} \tag{1.69}$$

and a straightforward calculations gives

$$\mathbf{R}_f = [(1-2p)R_x, \ (1-2p)R_y, \ R_z] \ . \tag{1.70}$$

Notice that the environment has altered the x,y components of $\mathbf{R}_f$, but left the z-component intact. When $p = 0$, we see that $\mathbf{R}_f = \mathbf{R}_0$. This is to be expected since in this case the environment leaves the qubit alone, and so the Bloch vector should not change. When $p = 1$, the environment carries out (with certainty) a $180°$ rotation about $\hat{z}$ so that $R_x \to -R_x$, $R_y \to -R_y$, in agreement with eq. (1.70). For $0 < p < 1$, R_{fx} is simply the average of these two cases: $R_{fx} = (1-p)R_x + p(-R_x) = (1-2p)R_x$, and similarly for R_{fy}. In the case where $p = 1/2$, the two error operators are equally likely to occur, and so the average yields $R_{fx} = R_{fy} = 0$ and $\mathbf{R}_f = R_z \hat{z}$. Using eq. (1.69), it follows that

$$\begin{aligned}\text{tr}\,\rho_f^2 &= \frac{1}{2}(1 + R_f^2) \\ &= 1 + 2p(p-1)\sin^2\theta \ , \end{aligned} \tag{1.71}$$

where $\mathbf{R}_f = R_f[\sin\theta\cos\phi, \ \sin\theta\sin\phi, \ \cos\theta]$. For $p = 0, 1$, either no interaction with the environment occurs, or a $180°$ rotation is done, both of which yield a pure state. This agrees with eq. (1.71), which gives $\text{tr}\,\rho_f^2 = 1$ for these values of p. In all other cases, ρ_f corresponds to a mixed state. The mixed state is due to the entangling of the qubit with the environment. Notice that the eigenstates of σ_z, $|\uparrow_z\rangle$ and $|\downarrow_z\rangle$ are eigenstates of the error operators E_0 and E_1. This is why R_{fz} was left intact by the qubit/environment interaction. On the other hand, linear superpositions of these two states are modified by the environment. Suppose the initial (normalized) state is $|\psi_0\rangle = a|\uparrow_z\rangle + b|\downarrow_z\rangle$ so that $\rho_0 = |\psi_0\rangle\langle\psi_0|$ is

$$\rho_0 = \begin{pmatrix} |a|^2 & ab^* \\ a^*b & |b|^2 \end{pmatrix} \ . \tag{1.72}$$

The corresponding Bloch vector has components $R_{0x} = 2\,\text{Re}(a^*b)$, $R_{0y} = 2\,\text{Im}(a^*b)$, and $R_{0z} = |a|^2 - |b|^2$. The environment sends $\rho_0 \to \rho_f$, where ρ_f

Introduction

follows from eqs. (1.69) and (1.70):

$$\rho_f = \begin{pmatrix} |a|^2 & (1-2p)ab\exp[-i\phi] \\ (1-2p)ab\exp[i\phi] & |b|^2 \end{pmatrix} . \qquad (1.73)$$

We see that for $0 < p < 1$, the off-diagonal elements of ρ_f (in the σ_z-basis) are suppressed by a factor of $1 - 2p$, and for $p = 1/2$, these elements vanish. The diagonal elements of ρ_f are not altered by the interaction with the environment. This environment-induced selection of a special basis, here $|\uparrow_z\rangle$ and $|\downarrow_z\rangle$, and of suppression of the off-diagonal matrix elements of the density operator when referred to this special basis, is known as decoherence, or sometimes, phase decoherence [56]. An environment that has error operators given by eq. (1.67) is referred to as a phase-flip channel. Decoherence is the mortal enemy of quantum computing. It causes a quantum computer to lose its quantum properties, and thus destroys its performance advantages over a classical computer. It is for quantum computers what kryptonite is to Superman. Removing the unwanted entanglement with the environment that is responsible for decoherence is one of the principal tasks of fault-tolerant quantum error correction.

1.4 Quantum Error Correcting Codes—First Pass

1.4.1 Redundancy without Cloning

We have seen that the No-Cloning Theorem (Section 1.3.2) forbids making an exact copy of an arbitrary quantum state. Thus one cannot construct a quantum error correcting code (QECC) by encoding data into a quantum state and then cloning that state. Prior to the discovery of the first QECCs, many thought that the No-Cloning Theorem was a no-go theorem for quantum error correction. The essential breakthrough made by the first QECCs was the discovery of how data could be redundantly encoded into a quantum state without violating this theorem.

Even at the classical level, we have seen that linear error correcting codes (Section 1.2.1) do not add redundancy by simply making copies of the data. Such codes map k bits into n bits in such a way that each codeword $\mathbf{c}$ satisfies a set of parity checks $\mathbf{H}(i) : \mathbf{H}^T(i) \cdot \mathbf{c} = 0$ $(i = 1, \ldots, n-k)$. The realization that parity checks could be carried out on data stored in entangled quantum states proved to be a significant theoretical development. These generalized parity checks established a fruitful connection to the theory of classical error correcting codes, with the classical theory providing useful guidance in the construction of the new quantum codes. Just as in the classical theory, the generalized parity checks allow an error syndrome to be defined that identifies the error expected to have occurred and the recovery operation that corrects it.

Here we sketch out how parity checks are implemented in quantum stabilizer codes and use them to construct a simple quantum code that corrects the one-qubit errors produced by the phase-flip channel (Section 1.3.6). A full discussion of quantum stabilizer codes will be given in Chapter 3.

Preliminaries

A QECC that encodes k qubits into n qubits is defined through an encoding map ξ from the k-qubit Hilbert space H_2^k onto a 2^k-dimensional subspace $\mathcal{C}_q$ of the n-qubit Hilbert space H_2^n. Just as with classical linear codes, a QECC is identified with the image space $\mathcal{C}_q$.

To set notation, let the computational basis (CB) states for the jth qubit be denoted by $|\delta_j\rangle$ ($\delta_j = 0, 1$). Usually we will choose the single-qubit CB states to be the eigenstates of σ_z^j:

$$\sigma_z^j |\delta_j\rangle = (-1)^{\delta_j} |\delta_j\rangle , \tag{1.74}$$

where the index $j = 1, \ldots, k$ labels the qubits. The CB states for H_2^k are formed by taking all possible direct products of the single-qubit CB states (Section 1.3.1):

$$|\delta\rangle \equiv |\delta_1 \ldots \delta_k\rangle$$
$$= |\delta_1\rangle \otimes \cdots \otimes |\delta_k\rangle . \tag{1.75}$$

The encoding operation $\xi : H_2^k \to \mathcal{C}_q$ is required to be unitary. This establishes a 1-1 correspondence between the unencoded and encoded CB states $|\delta\rangle = |\delta_1 \cdots \delta_k\rangle$ and $|\overline{\delta}\rangle = |\overline{\delta_1 \cdots \delta_k}\rangle$, respectively:

$$|\overline{\delta_1 \cdots \delta_k}\rangle = \xi |\delta_1 \cdots \delta_k\rangle . \tag{1.76}$$

For example, $|\overline{0}\rangle$ is the encoded image of $|0\rangle$, and $|\overline{011}\rangle$ is the encoded image of $|011\rangle$. ξ also encodes $\sigma_z^j \to Z_j = \xi \sigma_z^j \xi^\dagger$. Not surprisingly, the encoded CB states $\{|\overline{\delta}\rangle\}$ are the simultaneous eigenstates of the $\{Z_j : j = 1, \ldots, k\}$:

$$Z_j |\overline{\delta}\rangle = \left[\xi \sigma_z^j \xi^\dagger\right] \left[\xi |\delta\rangle\right]$$
$$= \xi \sigma_z^j |\delta\rangle$$
$$= (-1)^{\delta_j} \xi |\delta\rangle$$
$$= (-1)^{\delta_j} |\overline{\delta}\rangle , \tag{1.77}$$

and the encoding preserves the eigenvalue $(-1)^{\delta_j}$. Exactly which states in H_2^n the $\{|\overline{\delta}\rangle\}$ are (viz. which subspace $\mathcal{C}_q$ is in H_2^n) depends on the details of ξ. In quantum stabilizer codes, $\mathcal{C}_q$ is identified with the unique subspace of H_2^n that is fixed by the elements of an Abelian group $\mathcal{S}$ known as the stabilizer (of $\mathcal{C}_q$). Specifically, for all $s \in \mathcal{S}$ and $|c\rangle \in \mathcal{C}_q$,

$$s|c\rangle = |c\rangle. \tag{1.78}$$

We now introduce the essential properties of the stabilizer group $\mathcal{S}$. Chapter 3 will give a careful presentation.

Introduction   33

Stabilizer Group $\mathcal{S}$

The stabilizer group $\mathcal{S}$ is constructed from a set of $n-k$ operators $g_1, \ldots, g_{n-k}$ known as the generators of $\mathcal{S}$. Each element $s \in \mathcal{S}$ can be written as a unique product of powers of the generators

$$s = g_1^{p_1} \cdots g_{n-k}^{p_{n-k}} \; . \tag{1.79}$$

Because $\mathcal{S}$ is Abelian, the generators are a mutually commuting set of operators. As we shall see (Chapter 3), they are also Hermitian, unitary, and of order 2 ($g_i^2 = I$). Two important consequences follow from the property that the generators are of order 2.

(1) The eigenvalues $\{\lambda_i\}$ of g_i are $\lambda_i = (-1)^{l_i}$, ($l_i = 0, 1$). To see this, let $|l_i\rangle$ be an eigenstate of g_i with eigenvalue λ_i: $g_i |l_i\rangle = \lambda_i |l_i\rangle$. Then,

$$\begin{aligned} |l_i\rangle &= I \, |l_i\rangle \\ &= g_i^2 \, |l_i\rangle \\ &= \lambda_i^2 \, |l_i\rangle \; . \end{aligned}$$

Thus $\lambda_i^2 = 1$ and so $\lambda_i = (-1)^{l_i}$, ($l_i = 0, 1$).

(2) The elements of $\mathcal{S}$ can be labeled by bit strings of length $n - k$: $p = p_1 \cdots p_{n-k}$. As noted above, each element $s \in \mathcal{S}$ can be written as a product of powers of the generators $\{g_i\}$ (eq. (1.79)). Since the generators are of order 2, each p_i can only take the values 0 and 1. Thus each $s \in \mathcal{S}$ has a unique bit string $p = p_1 \cdots p_{n-k} \in F_2^{n-k}$ associated with it that specifies how it is constructed from the generators of $\mathcal{S}$. In fact, it is possible to show that $\mathcal{S}$ is isomorphic to F_2^{n-k}.

Exercise 1.7 *Show that $\mathcal{S}$ is isomorphic to F_2^{n-k}.*

Since $\mathcal{S}$ is an Abelian group, its elements can be simultaneously diagonalized. As the parent space H_2^n for the quantum code $\mathcal{C}_q$ is 2^n-dimensional, we need n commuting operators of order 2 to specify a unique state $|\psi\rangle \in H_2^n$. We shall see (Chapter 3) that the set of $\{Z_j\}$ can be chosen to commute with all of the generators $\{g_i\}$, as well as amongst themselves. Thus the 2^n simultaneous eigenstates of $\{g_1, \ldots, g_{n-k}; Z_1, \ldots, Z_k\}$ provide a basis for H_2^n. These eigenstates can be labeled by the bit strings $l = l_1 \cdots l_{n-k}$ and $\overline{\delta} = \overline{\delta}_1 \cdots \overline{\delta}_k$, where

$$\begin{aligned} g_i \, |l; \overline{\delta}\rangle &= (-1)^{l_i} \, |l; \overline{\delta}\rangle \\ Z_j \, |l; \overline{\delta}\rangle &= (-1)^{\overline{\delta}_j} \, |l; \overline{\delta}\rangle \; , \end{aligned} \tag{1.80}$$

with $i = 1, \ldots, n - k$; $j = 1, \ldots, k$; and $l_i, \overline{\delta}_j = 0, 1$. Using eqs. (1.79) and (1.80), we see that for $s(p) \in \mathcal{S}$,

$$s(p) \, |l; \overline{\delta}\rangle = (-1)^{l \cdot p} \, |l; \overline{\delta}\rangle \; , \tag{1.81}$$

where
$$l \cdot p = l_1 p_1 + \cdots + l_{n-k} p_{n-k} \pmod{2} . \tag{1.82}$$

Note that for a given bit string $l = l_1 \cdots l_{n-k} \in F_2^{n-k}$, the set of 2^k eigenstates $\{|l; \bar{\delta}\rangle : \bar{\delta} \in F_2^k\}$ span a subspace of H_2^n, which we will denote $\mathcal{C}(l)$. Said another way, $\mathcal{C}(l)$ is the subspace of H_2^n whose elements are simultaneous eigenvectors of the generators $g_1, \ldots, g_{n-k}$ with respective eigenvalues $(-1)^{l_1}, \ldots, (-1)^{l_{n-k}}$. The exponents of these eigenvalues specify the label l for $\mathcal{C}(l)$: $l = l_1 \cdots l_{n-k}$. Furthermore, the set of 2^{n-k} subspaces $\{\mathcal{C}(l) : l \in F_2^{n-k}\}$ provide a partition of H_2^n. Since the quantum stabilizer code $\mathcal{C}_q$ is fixed by the stabilizer $\mathcal{S}$, the states $|c\rangle \in \mathcal{C}_q$ are eigenstates of the generators $g_1, \ldots, g_{n-k}$ with associated eigenvalues $\lambda_1 = \cdots = \lambda_{n-k} = +1$. Thus $\mathcal{C}_q \subset \mathcal{C}(l = 0 \cdots 0)$. Since both $\mathcal{C}_q$ and $\mathcal{C}(l = 0 \cdots 0)$ are 2^k-dimensional, it follows that $\mathcal{C}_q = \mathcal{C}(l = 0 \cdots 0)$. We see that the stabilizer $\mathcal{S}$ establishes a partition of H_2^n into subspaces $\{\mathcal{C}(l) : l \in F_2^{n-k}\}$, and the quantum stabilizer code $\mathcal{C}_q$ associated with $\mathcal{S}$ corresponds to the subspace $\mathcal{C}(l = 0 \cdots 0)$. Note that the encoded CB states defined in eqs. (1.76) and (1.77) are the eigenstates $\{|l = 0 \cdots 0; \bar{\delta}\rangle\}$.

Generalized Parity Checks

Throughout this book we will consider an error model in which (1) errors on different qubits occur independently; (2) the single-qubit errors σ_x^j, σ_y^j, and σ_z^j are equally likely; and (3) the single-qubit error probability ϵ_j is the same for all qubits: $\epsilon_j = \epsilon$. For this model (see Chapter 2), it is sufficient to focus on error operators E that are unitary and can be represented by direct products of the identity operator I and the single-qubit Pauli operators σ_x^j, σ_y^j, σ_x^j, where the index $j = 1, \ldots, n$ labels the qubits.

Exercise 1.8 *Let E be an error operator that maps a quantum stabilizer code $\mathcal{C}_q$ onto the set of states $E(\mathcal{C}_q) = \{E|c\rangle : |c\rangle \in \mathcal{C}_q\}$. Show that $E(\mathcal{C}_q)$ is a subspace of H_2^n by showing that it is closed under vector addition and scalar multiplication.*

The following theorem states under what conditions E maps the code-space $\mathcal{C}_q$ onto an orthogonal subspace $E(\mathcal{C}_q) \subset H_2^n$. Note that since E is unitary, $E(\mathcal{C}_q)$ and $\mathcal{C}_q$ both have dimension 2^k.

THEOREM 1.10
Let E be an error and $\mathcal{S}$ the stabilizer group for a quantum stabilizer code. If $\mathcal{S}$ contains an element s that anticommutes with E, then for all $|c\rangle, |c'\rangle \in \mathcal{C}_q$, $E|c\rangle$ is orthogonal to $|c'\rangle$:

$$\langle c'|E|c\rangle = 0 .$$

Introduction

PROOF Let $|c\rangle \in \mathcal{C}_q$. Since $\{E, s\} = 0$ by assumption, we have that

$$\begin{aligned} E|c\rangle &= Es|c\rangle \\ &= -sE|c\rangle \end{aligned} \quad , \qquad (1.83)$$

where we have used that s fixes states $|c\rangle \in \mathcal{C}_q$. Taking the inner product of eq. (1.83) with $|c'\rangle \in \mathcal{C}_q$ gives

$$\begin{aligned} \langle c'|E|c\rangle &= -\langle c'|sE|c\rangle \\ &= -\langle c'|E|c\rangle \end{aligned} \quad . \qquad (1.84)$$

Thus $\langle c'|E|c\rangle = 0$ for all $|c\rangle, |c'\rangle \in \mathcal{C}_q$, and so E maps the code $\mathcal{C}_q$ onto the orthogonal subspace $E(\mathcal{C}_q) = \{E|c\rangle : |c\rangle \in \mathcal{C}_q\}$. ∎

We will see in Chapter 3 that, for our noise model, an error E will either commute or anticommute with each element $s \in \mathcal{S}$. This property allows us to determine the corrupted image $E(\mathcal{C}_q)$ of the stabilizer code $\mathcal{C}_q$.

THEOREM 1.11
Let E be an error operator and $\mathcal{C}_q$ a quantum stabilizer code with generators $g_1, \ldots, g_{n-k}$. The image $E(\mathcal{C}_q)$ of $\mathcal{C}_q$ under E is $\mathcal{C}(l)$,

$$E(\mathcal{C}_q) = \mathcal{C}(l) \quad ,$$

where $l = l_1 \cdots l_{n-k}$ and

$$l_i = \begin{cases} 0 & \text{if } [E, g_i] = 0 \\ 1 & \text{if } \{E, g_i\} = 0 \end{cases} \quad (i = 1, \ldots, n-k) \quad .$$

PROOF Recall that $\mathcal{C}(l)$ is the 2^k-dimensional subspace of H_2^n spanned by the eigenstates $\{|l; \overline{\delta}\rangle\}$ defined in eq. (1.80) with $l = l_1 \cdots l_{n-k}$. As noted above, E either commutes or anticommutes with each of the elements of $\mathcal{S}$, and hence with each generator g_i. Let $|c\rangle \in \mathcal{C}_q$. If g_i is a generator of $\mathcal{S}$, then

$$\begin{aligned} g_i E|c\rangle &= (-1)^{l_i} E g_i |c\rangle \\ &= (-1)^{l_i} E|c\rangle \end{aligned} \quad , \qquad (1.85)$$

where

$$l_i = \begin{cases} 0 & \text{if } [E, g_i] = 0 \\ 1 & \text{if } \{E, g_i\} = 0 \end{cases} \quad (i = 1, \ldots, n-k) \quad . \qquad (1.86)$$

Thus $E|c\rangle$ is a simultaneous eigenvector of each of the g_i with corresponding eigenvalue $(-1)^{l_i}$, for $i = 1, \ldots, n-k$. Since the 2^n eigenvectors $\{|l'; \overline{\delta}\rangle : l' \in F_2^{n-k}; \overline{\delta} \in F_2^k\}$ span H_2^n, $E|c\rangle$ can be written as a linear combination of these vectors

$$E|c\rangle = \sum_{l'} \sum_{\overline{\delta}} a(l'; \overline{\delta}) |l'; \overline{\delta}\rangle \quad . \qquad (1.87)$$

From eq. (1.80), we have

$$g_i E|c\rangle = \sum_{l'} \sum_{\bar{\delta}} (-1)^{l'_i} a(l'; \bar{\delta}) |l'; \bar{\delta}\rangle \ . \tag{1.88}$$

Since $E|c\rangle$ is an eigenvector of the generators $(g_1, \ldots, g_{n-k})$ with corresponding eigenvalues $((-1)^{l_1}, \ldots, (-1)^{l_{n-k}})$, consistency of eq. (1.88) with eq. (1.85) requires that the RHS of eqs. (1.87) and (1.88) only contain the eigenvectors $\{|l; \bar{\delta}\rangle\}$, where $l = l_1 \cdots l_{n-k}$, and the l_i are specified by eq. (1.86). Eq. (1.87) then becomes

$$E|c\rangle = \sum_{\bar{\delta}} a(l; \bar{\delta}) |l; \bar{\delta}\rangle \ . \tag{1.89}$$

Thus, $E|c\rangle \in \mathcal{C}(l)$ for all $|c\rangle \in \mathcal{C}_q$, and consequently, $E(\mathcal{C}_q) \subset \mathcal{C}(l)$. Since both $E(\mathcal{C}_q)$ and $\mathcal{C}(l)$ are 2^k-dimensional, it follows that $E(\mathcal{C}_q) = \mathcal{C}(l)$. ∎

Theorem 1.11 shows that an error E can be associated with the subspace $\mathcal{C}(l)$ to which it maps the code $\mathcal{C}_q$. Thus we can define the error syndrome $S(E)$ of E to be the bit string $l = l_1 \cdots l_{n-k}$ that labels $\mathcal{C}(l)$:

$$S(E) = l_1 \cdots l_{n-k} \ . \tag{1.90}$$

The generators of $\mathcal{S}$ thus enable an error sydrome to be defined through eq. (1.86), and thus play the role of generalized parity checks. The similarity to classical parity checks will become more apparent in Chapter 3 where we shall see that both generators and errors can be represented by bit strings of length $2n$:

$$\begin{aligned} g_i^T &= (-a^i-|-b^i-) \\ E^T &= (-e-|-f-) \ , \end{aligned} \tag{1.91}$$

where a^i, b^i, e, and f are bit strings of length n. (g_i and E are column vectors and so their transposes g_i^T and E^T are row vectors.) The bit string g_i^T can be used to define a parity check $\mathbf{H}^T(i)$:

$$\mathbf{H}^T(i) = (-b^i-|-a^i-) \ . \tag{1.92}$$

(We will explain in Chapter 3 why $\mathbf{H}^T(i)$ contains the bit strings a^i and b^i in reverse order.) The set of $\{\mathbf{H}^T(i) : i = 1, \ldots, n-k\}$ can be collected into a parity check matrix H:

$$H = \begin{pmatrix} - & \mathbf{H}^T(1) & - \\ & \vdots & \\ - & \mathbf{H}^T(n-k) & - \end{pmatrix} \tag{1.93}$$

such that the error syndrome $S(E)$ is given by the product of H with the column vector $E = (-e-|-f-)^T$,

$$S(E) = HE \ , \tag{1.94}$$

just as is done with classical linear error correcting codes.

Introduction

Quantum Error Correction

As discussed in Section 1.3.5, quantum error correction uses ancilla qubits to extract the error syndrome. It was assumed there that measurement of the syndrome allowed the expected error to be identified. We now describe how this identification is made. The protocol begins by measuring the value of the error syndrome. The measured syndrome value S determines a set of errors $\Sigma = (E_1, E_2, \cdots)$ whose syndromes are equal to the measured value: $S(E_i) = S$ for $E_i \in \Sigma$. We define the expected error for the measured syndrome value S to be the most probable of the errors in Σ. If more than one error qualifies as most probable, one arbitrarily chooses one of these most probable errors to be the expected error E_S. We then use $E_S^\dagger$ as the recovery operation. If the actual error $E \in \Sigma$ differs from E_S, the error recovery operation $E_S^\dagger$ will not remove the error. Since the actual error E and the expected error E_S both belong to Σ, $S(E) = S(E_S) = S$. We will see (Sections 3.1.2–3.1.3) that because of this, the resulting state $E_S^\dagger E |c\rangle$ is an element of $\mathcal{C}_q$, but (usually) will differ from the uncorrupted state $|c\rangle$. To illustrate these ideas, we show how to construct a quantum stabilizer code that corrects the 1-qubit errors produced by the phase-flip channel introduced in Section 1.3.6.

Quantum Stabilizer Code for Phase-Flip Channel

As we saw in Section 1.3.6, the action of the phase-flip channel on a single qubit is to either: (1) leave the qubit alone with probability $1 - p$, or (2) rotate it by 180° about the z-axis with probability p. A quantum stabilizer code that maps 1 qubit into 3 qubits proves sufficient to correct all one-qubit errors produced by this channel.

We want our encoding map ξ to send $H_2^1 \to \mathcal{C}_q \subset H_2^3$. For this encoding, $k = 1$ and $n = 3$, and so we will need $n - k = 2$ generators (g_1, g_2) to define our 2-dimensional subspace $\mathcal{C}_q$ in H_2^3. For our error model (page 34), the phase-flip channel produces eight possible errors on three qubits. These errors $\{E_i : i = 0, \ldots, 7\}$ are listed in Table 1.1, along with their associated error probabilities. We want to protect against the set of single-qubit errors (E_1, E_2, E_3). Thus at least one of the two generators g_1, g_2 must anticommute with each of these errors. One possible choice for the generators is:

$$g_1 = \sigma_x^1 \sigma_x^2$$
$$g_2 = \sigma_x^1 \sigma_x^3 \ . \tag{1.95}$$

Having the generators, the error syndrome $S(E)$ is easily evaluated using eqs. (1.90) and (1.86). The result for each error E is also included in Table 1.1. Notice that two errors correspond to each of the four error syndrome values. For example, the errors E_3 and E_4 both have syndrome 01. Table 1.1 indicates that E_3 is the most probable of the two errors for $p < 1/2$. Thus the expected error for the syndrome value 01 is $E_{01} = E_3$, and the associated recovery

TABLE 1.1
Set of errors $E_0, \ldots, E_7$ produced by the phase-flip channel on three qubits. Listed with each error is its associated error probability and error syndrome. Note that E_0 is the identity operator I that leaves all three qubits alone.

Error E	Error Probability $P(E)$	Error Syndrome $S(E)$
$E_0 = I$	$(1-p)^3$	00
$E_1 = \sigma_z^1$	$p(1-p)^2$	11
$E_2 = \sigma_z^2$	$p(1-p)^2$	10
$E_3 = \sigma_z^3$	$p(1-p)^2$	01
$E_4 = \sigma_z^1 \sigma_z^2$	$p^2(1-p)$	01
$E_5 = \sigma_z^1 \sigma_z^3$	$p^2(1-p)$	10
$E_6 = \sigma_z^2 \sigma_z^3$	$p^2(1-p)$	11
$E_7 = \sigma_z^1 \sigma_z^2 \sigma_z^3$	p^3	00

operator is $R_{01} = E_{01}^\dagger$. Table 1.2 lists the error syndrome values, and the corresponding expected error and recovery operators for $p < 1/2$.

TABLE 1.2
List of error syndrome values for the errors produced by the phase-flip channel acting on three qubits. For each syndrome value we list the expected error E_S and its associated recovery operator $R_S = E_S^\dagger$. Note that I is the identity operator that leaves all three qubits alone.

Error Syndrome S	Expected Error E_S	Recovery Operator R_S
00	$E_{00} = I$	$R_{00} = I$
01	$E_{01} = \sigma_z^3$	$R_{01} = \sigma_z^3$
10	$E_{10} = \sigma_z^2$	$R_{10} = \sigma_z^2$
11	$E_{11} = \sigma_z^1$	$R_{11} = \sigma_z^1$

The stabilizer group $\mathcal{S}$ is the collection of elements $s(p) = g_1^{p_1} g_2^{p_2}$, with $p_1, p_2 = 0, 1$. Thus $\mathcal{S} = \{I, \sigma_x^1 \sigma_x^2, \sigma_x^1 \sigma_x^3, \sigma_x^2 \sigma_x^3\}$. For our choice of generators (eq. (1.95)), it proves convenient to define the single-qubit CB states to be the eigenstates of σ_x:

$$|0\rangle \equiv |\sigma_x = +1\rangle \quad ; \quad |1\rangle \equiv |\sigma_x = -1\rangle \ . \tag{1.96}$$

The stabilizer $\mathcal{S}$ must fix the codespace $\mathcal{C}_q$, and consequently, $\mathcal{S}$ must fix the encoded CB states $|\bar{0}\rangle$ and $|\bar{1}\rangle$ that span $\mathcal{C}_q$. For our choice of g_1 and g_2, the

Introduction

following encoding of the single-qubit CB states insures this:

$$|0\rangle \to |\bar{0}\rangle = |000\rangle \quad ; \quad |1\rangle \to |\bar{1}\rangle = |111\rangle \ . \tag{1.97}$$

Note that this choice of encoding is consistent with the No-Cloning Theorem since this theorem allows orthogonal states to be cloned (Section 1.3.2).

We see from Table 1.2 that the expected errors are the zero- and one-qubit errors produced by the phase-flip channel. Thus whenever the actual error is one of the expected errors, the associated recovery operation will succesfully remove it, and our encoding corrects zero- and one-qubit errors as desired. As noted earlier (page 37), the error correction protocol fails whenever the actual error is not the expected error. Thus the failure probability for error correction P_f is the probability that one of the non-expected errors (E_4, E_5, E_6, E_7) occurs. From Table 1.1 we have

$$P_f = 3p^2(1-p) + p^3 \ . \tag{1.98}$$

In the absence of encoding, the probability that the channel will cause an error in a single qubit is p. Our encoding improves the reliability of the encoded qubit if $P_f < p$. Using eq. (1.98), we see that this occurs when $p < 1/2$. For example, if $p = 0.1$, then $P_f = 0.028$, and encoding reduces the failure probability by almost an order of magnitude.

Finally, imagine that the measured error syndrome is 10, and that the actual error is $E_5 = \sigma_z^1 \sigma_z^3$. For this syndrome value, Table 1.2 lists the expected error as $E_{10} = \sigma_z^2$ and the recovery operator as $R_{10} = E_{10}^\dagger$. Since R_{10} is a one-qubit operator, it cannot correct the two-qubit error E_5. We noted earlier (page 37) that, in this situation, the state after R_{10} is applied will be an element of $\mathcal{C}_q$, but (usually) will not be the state initially sent into the channel. To see this in this specific example, let the original uncorrupted state be $|c\rangle = a|\bar{0}\rangle + b|\bar{1}\rangle$. The error E_5 sends $|c\rangle \to E_5|c\rangle$. Applying the recovery operator $R_{10} = E_{10}^\dagger$ yields the final state $|\psi\rangle = E_{10}^\dagger E_5 |c\rangle = \sigma_z^1 \sigma_z^2 \sigma_z^3 |c\rangle$. Using eqs. (1.96) and (1.97), together with

$$\sigma_z \begin{cases} |\sigma_x = +1\rangle \\ |\sigma_x = -1\rangle \end{cases} = \begin{cases} |\sigma_x = -1\rangle \\ |\sigma_x = +1\rangle \end{cases} ,$$

gives

$$E_{10}^\dagger E_5 \begin{cases} |\bar{0}\rangle \\ |\bar{1}\rangle \end{cases} = \begin{cases} |\bar{1}\rangle \\ |\bar{0}\rangle \end{cases} \ .$$

Thus the final state $|\psi\rangle$ is

$$\begin{aligned} |\psi\rangle &= E_{10}^\dagger E_5 |c\rangle \\ &= a|\bar{1}\rangle + b|\bar{0}\rangle \ . \end{aligned} \tag{1.99}$$

This is clearly a linear combination of the encoded CB states, and so $|\psi\rangle \in \mathcal{C}_q$ as promised. Furthermore, if $a \neq b$, we see that $|\psi\rangle \neq |c\rangle$ so that the error correction protocol has failed, though it has managed to return the state to the codespace.

1.4.2 Necessary and Sufficient Conditions

Here we present necessary and sufficient conditions that a quantum code $\mathcal{C}_q$ must satisfy if it is to correct a given set of errors [18,21]. The central result is Theorem 1.13. We begin with a heuristic discussion of the physical significance of these conditions. This discussion is followed by a proof of the central theorem.

Heuristic Discussion

As discussed by Feynman [67], the terms in a linear superposition of quantum states represent interfering alternatives according to which a quantum system can evolve. Should any physical system (measuring apparatus, environment) determine which of the alternatives is followed, the superposition collapses, and all quantum interference effects disappear. We have just seen that QECCs use code states that are linear superpositions of encoded CB states. In such states, the CB states are thus interfering alternatives for the quantum register. Should the environment's interaction with the quantum register enable it to distinguish between these alternatives, the code state superposition would collapse, with consequent loss of the entanglement that protects the encoded data. When designing a quantum code to protect against a given environment, the encoded CB states must be chosen so that the environment is unable to differentiate between them. This design criteria will lead to one of the two conditions that are the focus of this subsection. The second condition will result from requiring that the choice of encoded CB states always allow us to distinguish the erroneous image of one CB state from the erroneous image of a different CB state. We begin with a discussion of this latter condition.

We model the effect of the environment on the quantum register by a set of linear operators $\{E_a\}$, which are the errors that the environment can produce in the register's state. This model is an example of a quantum operation (see Chapter 2). We restrict ourselves here to trace-preserving environments for which

$$\sum_a E_a^\dagger E_a = I \ . \tag{1.100}$$

If initially the register and the environment are not entangled, and the register's initial state is given by the density operator ρ_0, then its final state ρ will be

$$\rho = \sum_a E_a \rho_0 E_a^\dagger \ . \tag{1.101}$$

Introduction

Let $|\bar{i}\rangle$ and $|\bar{j}\rangle$ be distinct encoded CB states, and E_a and E_b two possible errors produced by the environment E. If these errors are to be correctable, one might expect that the corrupted states $E_a|\bar{i}\rangle$ and $E_b|\bar{j}\rangle$ should be distinguishable so that the recovery operation will not confuse the corrupted image of one codeword with that of another. As seen in Section 1.3.4, the corrupted images will be distinguishable if they are orthogonal. Thus we might expect the encoded CB states for $\mathcal{C}_q$ to satisfy the condition

$$\langle \bar{i}|E_a^\dagger E_b|\bar{j}\rangle = 0 \ , \qquad (\bar{i} \neq \bar{j}) \tag{1.102}$$

for any pair of correctable errors E_a, E_b.

Now consider the encoded CB state $|\bar{i}\rangle$ and two errors E_a and E_b. These errors produce the corrupted states $E_a|\bar{i}\rangle$ and $E_b|\bar{i}\rangle$. Surprisingly, in this case, the corrupted images of $|\bar{i}\rangle$ do not have to be distinguishable. As pointed out above, what is essential is that the environment be unable to distinguish between the encoded CB states. Note that the squared modulus of the inner product of the two corrupted states $|\langle\bar{i}|E_a^\dagger E_b|\bar{i}\rangle|^2$ is observable. It gives the probability that the state $E_b|\bar{i}\rangle$ will pass a measurement test for the state $E_a|\bar{i}\rangle$ [58]. Thus the environment will be unable to distinguish the encoded CB states if this squared modulus is the same for all such states. Thus we might expect the code $\mathcal{C}_q$ to satisfy

$$|\langle\bar{i}|E_a^\dagger E_b|\bar{i}\rangle|^2 = |\langle\bar{j}|E_a^\dagger E_b|\bar{j}\rangle|^2 \ .$$

In fact, we will see below that the more stringent condition of equality of the inner products is needed:

$$\langle\bar{i}|E_a^\dagger E_b|\bar{i}\rangle = \langle\bar{j}|E_a^\dagger E_b|\bar{j}\rangle \ , \tag{1.103}$$

for all pairs of correctable errors E_a, E_b and all encoded CB states. This condition insures that the inner product of $E_a|c\rangle$ and $E_b|c\rangle$ is independent of $|c\rangle$ for all $|c\rangle \in \mathcal{C}_q$. To see this, we write $\lambda_{ab} = \langle\bar{i}|E_a^\dagger E_b|\bar{i}\rangle$ and $|c\rangle = \sum_{\bar{i}} a_{\bar{i}}|\bar{i}\rangle$, and evaluate the inner product:

$$\begin{aligned}\langle c|E_a^\dagger E_b|c\rangle &= \sum_{\bar{i},\bar{j}} a_{\bar{i}}^* a_{\bar{j}} \langle\bar{i}|E_a^\dagger E_b|\bar{j}\rangle \\ &= \sum_{\bar{i}} |a_{\bar{i}}|^2 \langle\bar{i}|E_a^\dagger E_b|\bar{i}\rangle \\ &= \lambda_{ab} \sum_{\bar{i}} |a_{\bar{i}}|^2 \\ &= \lambda_{ab} \ . \end{aligned} \tag{1.104}$$

We used eq. (1.102), eq. (1.103), and the normalization of $|c\rangle$ to go from the first to the second line, the second to the third line, and the third to the fourth line, respectively. The final result is seen to be independent of the $\{a_{\bar{i}}\}$ and hence of $|c\rangle$.

Having gotten some insight into the physical issues underlying eqs. (1.102) and (1.103), we now show that they are necessary and sufficient conditions for C_q to correct the set of errors $\{E_a\}$ produced by the environment E.

Proper Discussion

Our discussion follows Knill and Laflamme [21]. We have seen that a quantum code that maps k qubits into n qubits is a 2^k-dimensional subspace C_q of the n-qubit Hilbert space H_2^n. The environment E introduces unwanted interactions into the dynamics of the n-qubit register, which is modeled by a trace-preserving quantum operation with error operators $\{E_a\}$. Error correction is implemented using a recovery operation R that is also a trace-preserving quantum operation with recovery operators $\{R_r\}$. The task of R is to diagnose the most probable error and to correct it. Ideally, R is designed to correct all environmental errors $\{E_a\}$, though resource constraints may limit R to correcting only a subset of these errors. Here we assume that no constraints are present and that R corrects the full set $\{E_a\}$. A quantum error correcting code is defined to be a pair (C_q, R) composed of a 2^k-dimensional subspace $C_q \subset H_2^n$ and a trace-preserving recovery operation R.

Suppose that initially the quantum register and the environment are not entangled, and that the register is initially in the code state $|c\rangle \in C_q$. The environment then interacts with the register and leaves it in the state ρ_1:

$$\rho_1 = \sum_a E_a |c\rangle\langle c| E_a^\dagger . \tag{1.105}$$

To correct for the effect of the environment we apply the recovery operator R, which produces the state

$$\rho_f = \sum_{r,a} R_r E_a |c\rangle\langle c| E_a^\dagger R_r^\dagger . \tag{1.106}$$

The success probability for the recovery operation is the probability that R leaves the register in the state $|c\rangle\langle c|$. Since R and E are both trace-preserving, this probability $\mathcal{F}(|c\rangle, R, E)$ is

$$\mathcal{F}(|c\rangle, R, E) = \text{Tr}\,(|c\rangle\langle c|\rho_f) = \sum_{r,a} |\langle c|R_r E_a|c\rangle|^2 . \tag{1.107}$$

The error $\mathcal{E}(|c\rangle, R, E)$ associated with R is then the failure probability $1 - \mathcal{F}(|c\rangle, R, E)$. Using eq. (1.107) one finds

$$\mathcal{E}(|c\rangle, R, E) = \sum_{r,a} |(R_r E_a - \langle c|R_r E_a|c\rangle)\,|c\rangle|^2 . \tag{1.108}$$

Exercise 1.9 *Show that for trace-preserving R and E,*

$$1 - \mathcal{F}(|c\rangle, R, E) = \sum_{r,a} |(R_r E_a - \langle c|R_r E_a|c\rangle)\,|c\rangle|^2 .$$

Introduction

So far we have restricted ourselves to a specific state $|c\rangle \in \mathcal{C}_q$. If we let $|c\rangle$ range over all states in $\mathcal{C}_q$, then the error for the code $\mathcal{E}(\mathcal{C}_q, R, E)$ is defined to be the largest error value found:

$$\mathcal{E}(\mathcal{C}_q, R, E) = \max_{|c\rangle \in \mathcal{C}_q} \sum_{r,a} |(R_r E_a - \langle c|R_r E_a|c\rangle)\,|c\rangle|^2 \ . \tag{1.109}$$

Notice that since the error $\mathcal{E}(\mathcal{C}_q, R, E)$ is a sum of non-negative terms, the condition that this error vanish requires that every term on the RHS of eq. (1.109) vanish for all states $|c\rangle \in \mathcal{C}_q$:

$$|(R_r E_a - \langle c|R_r E_a|c\rangle)\,|c\rangle| = 0 \ .$$

The worst-case (i.e., minimum) success probability defines the code fidelity $\mathcal{F}(\mathcal{C}_q, R, E)$:

$$\mathcal{F}(\mathcal{C}_q, R, E) = \min_{|c\rangle \in \mathcal{C}_q} \sum_{r,a} |\langle c|R_r E_a|c\rangle|^2 \ . \tag{1.110}$$

A quantum error correcting code $(\mathcal{C}_q, R)$ is said to be E-correcting if its error for E vanishes:

$$\mathcal{E}(\mathcal{C}_q, R, E) = 0 \ . \tag{1.111}$$

Let $E_a \in E$. It proves convenient to define the error $\mathcal{E}(\mathcal{C}_q, R, E_a)$ for the code $(\mathcal{C}_q, R)$ to correct E_a:

$$\mathcal{E}(\mathcal{C}_q, R, E_a) = \max_{|c\rangle \in \mathcal{C}_q} \sum_{r} |(R_r E_a - \langle c|R_r E_a|c\rangle)\,|c\rangle|^2 \ . \tag{1.112}$$

Then,

$$\mathcal{E}(\mathcal{C}_q, R, E) = \sum_{a} \mathcal{E}(\mathcal{C}_q, R, E_a) \ .$$

Thus if the code error $\mathcal{E}(\mathcal{C}_q, R, E)$ vanishes, so must the error $\mathcal{E}(\mathcal{C}_q, R, E_a)$ for all $E_a \in E$. Let $E(\mathcal{C}_q, R)$ denote the maximal set of operators e for which $(\mathcal{C}_q, R)$ is $E(\mathcal{C}_q, R)$-correcting. Thus all operators e belonging to $E(\mathcal{C}_q, R)$ have zero error $\mathcal{E}(\mathcal{C}_q, R, e) = 0$, and since $E(\mathcal{C}_q, R)$ is maximal, it contains all operators e for which $\mathcal{E}(\mathcal{C}_q, R, e) = 0$. The following theorem gives a necessary and sufficient algebraic condition that an operator E_a must satisfy if it is to belong to $E(\mathcal{C}_q, R)$.

THEOREM 1.12
The operator E_a is in $E(\mathcal{C}_q, R)$ if and only if, when restricted to $\mathcal{C}_q$, $R_r E_a = \gamma_{ra} I$ for each $R_r \in R$. The family $E(\mathcal{C}_q, R)$ is linearly closed.

PROOF ($\Longleftarrow$) If $R_r E_a = \gamma_{ra} I$ for all $R_r \in R$, then for all $|c\rangle \in \mathcal{C}_q$,

$$|(R_r E_a - \langle c|R_r E_a|c\rangle)\,|c\rangle| = 0 \ . \tag{1.113}$$

Squaring, summing over all $R_r \in R$, and noting that this sum vanishes for all $|c\rangle \in \mathcal{C}_q$ shows that $\mathcal{E}(\mathcal{C}_q, R, E_a) = 0$, and so $E_a \in E(\mathcal{C}_q, R)$.

($\Longrightarrow$) By assumption $(\mathcal{C}_q, R)$ is $E(\mathcal{C}_q, R)$-correcting and $E_a \in E(\mathcal{C}_q, R)$. Thus $\mathcal{E}(\mathcal{C}_q, R, E_a) = 0$, which implies that eq. (1.113) is true for E_a. Since the norm of a state only vanishes if the state is the null state, it follows that

$$R_r E_a |c\rangle = \langle c|R_r E_a|c\rangle |c\rangle$$
$$= \gamma_{ra}(|c\rangle) |c\rangle \quad , \quad (1.114)$$

where $\gamma_{ra}(|c\rangle) = \langle c|R_r E_a|c\rangle$. To see that $\gamma_{ra}(|c\rangle)$ is independent of $|c\rangle$, consider two linearly independent states $|c_1\rangle, |c_2\rangle \in \mathcal{C}_q$. Applying eq. (1.114) to the state $\alpha_1|c_1\rangle + \alpha_2|c_2\rangle$ gives

$$R_r E_a [|\alpha_1|c_1\rangle + \alpha_2|c_2\rangle)] = \gamma_{ra}(\alpha_1|c_1\rangle + \alpha_2|c_2\rangle) [\alpha_1|c_1\rangle + \alpha_2|c_2\rangle] \quad . \quad (1.115)$$

Since $R_r E_a$ is a linear operator, we also have that

$$R_r E_a [\alpha_1|c_1\rangle + \alpha_2|c_2\rangle] = \alpha_1 \gamma_{ra}(|c_1\rangle) |c_1\rangle + \alpha_2 \gamma_{ra}(|c_2\rangle) |c_2\rangle \quad . \quad (1.116)$$

Equating eqs. (1.115) and (1.116), and collecting terms gives

$$\alpha_1 [\gamma_{ra}(\alpha_1|c_1\rangle + \alpha_2|c_2\rangle) - \gamma_{ra}(|c_1\rangle)] |c_1\rangle +$$
$$\alpha_2 [\gamma_{ra}(\alpha_1|c_1\rangle + \alpha_2|c_2\rangle) - \gamma_{ra}(|c_2\rangle)] |c_2\rangle = 0 \quad . \quad (1.117)$$

Since $|c_1\rangle$ and $|c_2\rangle$ are linearly independent, this equation can only be true if the coefficients vanish. Thus,

$$\gamma_{ra}(\alpha_1|c_1\rangle + \alpha_2|c_2\rangle) = \gamma_{ra}(|c_1\rangle)$$
$$\gamma_{ra}(\alpha_1|c_1\rangle + \alpha_2|c_2\rangle) = \gamma_{ra}(|c_2\rangle) \quad .$$
$$(1.118)$$

Since both LHSs depend on α_1 and α_2, which are arbitrary, while the RHSs do not, it follows that both sides must be equal to a constant γ_{ra}. Eq. (1.114) then gives $R_r E_a = \gamma_{ra} I$.

Having established the theorem in both directions, we can use it to prove that $E(\mathcal{C}_q, R)$ is linearly closed. To that end, let E_a and E_b belong to $E(\mathcal{C}_q, R)$ so that $R_r E_a = \gamma_{ra} I$ and $R_r E_b = \gamma_{rb} I$ for all $R_r \in R$. Now define $F = \alpha E_a + \beta E_b$. Then

$$R_r F = \alpha R_r E_a + \beta R_r E_b$$
$$= \gamma_{rf} I \quad ,$$

where $\gamma_{rf} = \alpha \gamma_{ra} + \beta \gamma_{rb}$. Thus $F \in E(\mathcal{C}_q, R)$, and $E(\mathcal{C}_q, R)$ is linearly closed. ∎

We are now in a position to state and prove our central theorem.

Introduction

THEOREM 1.13
The code $\mathcal{C}_q$ can be extended to an E-correcting code if and only if for all encoded CB states $|\bar{i}\rangle$, $|\bar{j}\rangle$ ($\bar{i} \neq \bar{j}$) and operators $E_a, E_b \in E$:

$$\langle \bar{i}|E_a^\dagger E_b|\bar{i}\rangle = \langle \bar{j}|E_a^\dagger E_b|\bar{j}\rangle \qquad (1.119)$$

and

$$\langle \bar{i}|E_a^\dagger E_b|\bar{j}\rangle = 0 \ . \qquad (1.120)$$

PROOF ($\Longrightarrow$) Since $(\mathcal{C}_q, R)$ is an E-correcting code, Theorem 1.12 tells us that for all $R_r \in R$ and $E_a \in E$, $R_r E_a = \gamma_{ra} I$. Recall that R is trace-preserving so the recovery operators $\{R_r\}$ satisfy

$$\sum_r R_r^\dagger R_r = I \ .$$

We calculate $\langle \bar{i}|E_a^\dagger E_b|\bar{j}\rangle$:

$$\begin{aligned}
\langle \bar{i}|E_a^\dagger E_b|\bar{j}\rangle &= \langle \bar{i}|E_a^\dagger I E_b|\bar{j}\rangle \\
&= \sum_r \langle \bar{i}|E_a^\dagger R_r^\dagger R_r E_b|\bar{j}\rangle \\
&= \sum_r \langle \bar{i}|\gamma_{ra}^* \gamma_{rb}|\bar{j}\rangle \\
&= \left(\sum_r \gamma_{ra}^* \gamma_{rb}\right) \delta_{\bar{i},\bar{j}} \\
&= \lambda_{ab}\, \delta_{\bar{i},\bar{j}} \ . \qquad (1.121)
\end{aligned}$$

Note that λ_{ab} is independent of the CB states. The Kronecker delta insures that eq. (1.119) is satisfied and eq. (1.120) follows since $\langle \bar{i}|E_a^\dagger E_b|\bar{i}\rangle$ and $\langle \bar{j}|E_a^\dagger E_b|\bar{j}\rangle$ both equal λ_{ab}.

($\Longleftarrow$) We are given the code $\mathcal{C}_q$ and the validity of eqs. (1.119) and (1.120). Let $\mathcal{V}^{\bar{i}}$ be the subspace of H_2^n spanned by the corrupted images $\{E_a|\bar{i}\rangle\}$ of the CB state $|\bar{i}\rangle$, and let $\{|v_r^{\bar{i}}\rangle\}$ be an orthonormal basis for $\mathcal{V}^{\bar{i}}$. We define such a subspace for each of the CB states. Eq. (1.120) indicates that $\mathcal{V}^{\bar{i}}$ and $\mathcal{V}^{\bar{j}}$ ($\bar{i} \neq \bar{j}$) are orthogonal subspaces. In the event that the direct sum $\oplus_{\bar{i}} \mathcal{V}^{\bar{i}}$ is a proper subset of H_2^n, we denote its orthogonal complement by $\mathcal{O}$. Let $\{|\mathcal{O}_k\rangle\}$ be an orthonormal basis for $\mathcal{O}$. Then the set of states $\{|v_r^{\bar{i}}\rangle, |\mathcal{O}_k\rangle\}$ constitutes an orthonormal basis for H_2^n. We introduce the quantum operation R with operation elements

$$R = \{\hat{\mathcal{O}}, R_1, \ldots, R_r, \ldots\} \ .$$

Here $\hat{\mathcal{O}}$ projects onto $\mathcal{O}$,

$$\hat{\mathcal{O}} = \sum_k |\mathcal{O}_k\rangle\langle \mathcal{O}_k|$$

and
$$R_r = \sum_{\bar{i}} |\bar{i}\rangle\langle v_r^{\bar{i}}| \ . \qquad (1.122)$$

It follows immediately that R is trace-preserving:

$$\hat{\mathcal{O}}^\dagger \hat{\mathcal{O}} + \sum_r R_r^\dagger R_r = \sum_k |\mathcal{O}_k\rangle\langle\mathcal{O}_k| + \sum_r \sum_{\bar{i}} |v_r^{\bar{i}}\rangle\langle v_r^{\bar{i}}|$$
$$= I \ ,$$

since the set of states $\{|v_r^{\bar{i}}\rangle, |\mathcal{O}_k\rangle\}$ is an orthonormal basis for H_2^n.

To show that R is the recovery operator that transforms $\mathcal{C}_q$ into an E-correcting code, we introduce an operator U_i that maps $\mathcal{V}^{\bar{0}}$ onto $\mathcal{V}^{\bar{i}}$. Let $|v_\alpha\rangle \in \mathcal{V}^{\bar{0}}$:

$$|v_\alpha\rangle = \sum_a \alpha_a E_a |\bar{0}\rangle \ ,$$

where the $\{\alpha_a\}$ are complex numbers. We define U_i to be the map that sends $|v_\alpha\rangle \to |\mathsf{V}_\alpha\rangle \in \mathcal{V}^{\bar{i}}$, where

$$|\mathsf{V}_\alpha\rangle = \sum_a \alpha_a E_a |\bar{i}\rangle \ .$$

We now show that U_i (1) is a 1-1 correspondence, (2) maps $E_a|\bar{0}\rangle \to E_a|\bar{i}\rangle$, (3) is unitary, and (4) maps $|v_r^{\bar{0}}\rangle \to |v_r^{\bar{i}}\rangle$.

(1) To see that U_i is a 1-1 map, suppose that U_i maps $|v_\alpha\rangle = \sum_a \alpha_a E_a|\bar{0}\rangle$ and $|v_\beta\rangle = \sum_a \beta_a E_a|\bar{0}\rangle$ onto the same state $|\mathsf{V}\rangle$. By definition of U_i,

$$|\mathsf{V}\rangle = \sum_a \alpha_a E_a|\bar{i}\rangle$$

and

$$|\mathsf{V}\rangle = \sum_a \beta_a E_a|\bar{i}\rangle \ .$$

Equality of these two expressions indicates that $\alpha_a = \beta_a$ for all a and so $|v_\alpha\rangle = |v_\beta\rangle$. The proof that U_i is onto is immediate. Let $|\mathsf{V}\rangle = \sum_a \alpha_a E_a|\bar{i}\rangle \in \mathcal{V}^{\bar{i}}$. By definition of U_i, $|\mathsf{V}\rangle$ is the image of $|v_\alpha\rangle = \sum_a \alpha_a E_a|\bar{0}\rangle$. Thus U_i is a 1-1 correspondence.

(2) Let $|v_\alpha\rangle = E_c|\bar{0}\rangle$ so that $\alpha_a = 0$ for $a \neq c$ and $\alpha_c = 1$. U_i then maps $|v_\alpha\rangle$ onto the state $E_c|\bar{i}\rangle$. This argument applies for all c so that (letting $c \to a$)

$$U_i E_a |\bar{0}\rangle = E_a |\bar{i}\rangle \ .$$

Introduction 47

(3) Let $|V\rangle, |W\rangle \in \mathcal{V}^{\bar{i}}$, with

$$|V\rangle = \sum_a v_a E_a |\bar{i}\rangle$$

$$|W\rangle = \sum_a w_a E_a |\bar{i}\rangle .$$

By definition of U_i, these states are the images of $|v\rangle, |w\rangle \in \mathcal{V}^{\bar{0}}$:

$$|v\rangle = \sum_a v_a E_a |\bar{0}\rangle$$

$$|w\rangle = \sum_a w_a E_a |\bar{0}\rangle .$$

We evaluate $\langle w|U_i^\dagger U_i|v\rangle$:

$$\langle w|U_i^\dagger U_i|v\rangle = \langle W|V\rangle$$
$$= \sum_{a,b} w_b^* v_a \langle \bar{i}|E_b^\dagger E_a|\bar{i}\rangle$$
$$= \sum_{a,b} w_b^* v_a \langle \bar{0}|E_b^\dagger E_a|\bar{0}\rangle$$
$$= \langle w|v\rangle .$$

Eq. (1.119) was used to go from the second to the third line. Thus $\langle w|U_i^\dagger U_i|v\rangle = \langle w|v\rangle$ for all $|v\rangle, |w\rangle \in \mathcal{V}^{\bar{0}}$ so that U_i is unitary [68].

(4) Recalling that $\mathcal{V}^{\bar{0}}$ is the subspace of H_2^n spanned by the states $\{E_a|\bar{0}\rangle\}$, it follows that the orthonormal basis elements $\{|v_r^{\bar{0}}\rangle\}$ for $\mathcal{V}^{\bar{0}}$ can be written as a linear combination of the $\{E_a|\bar{0}\rangle\}$:

$$|v_r^{\bar{0}}\rangle = \sum_a \alpha_{a,r} E_a |\bar{0}\rangle .$$

U_i maps $|v_r^{\bar{0}}\rangle$ onto

$$U_i |v_r^{\bar{0}}\rangle = \sum_a \alpha_{a,r} E_a |\bar{i}\rangle .$$

Since U_i is unitary and a 1-1 correspondence, the image states $\{U_i|v_r^{\bar{0}}\rangle\}$ that we denote by $\{|v_r^{\bar{i}}\rangle\}$ form an orthonormal basis for $\mathcal{V}^{\bar{i}}$, and so we have shown that U_i maps $|v_r^{\bar{0}}\rangle \to |v_r^{\bar{i}}\rangle$.

We now have all the ingredients needed to show that $(\mathcal{C}_q, R)$ is an E-correcting code. If $(\mathcal{C}_q, R)$ corrects E, it must recover a code state $|c\rangle = \sum_{\bar{i}} c_{\bar{i}} |\bar{i}\rangle$ with vanishing error for each $E_a \in E$: $\mathcal{E}(\mathcal{C}_q, R, E_a) = 0$. From Theorem 1.12, this occurs only if $R_r E_a = \gamma_{ra} I$ for all $R_r \in R$. To show this is the case, we first evaluate $E_a|c\rangle$:

$$E_a|c\rangle = \sum_{\bar{i}} c_{\bar{i}} E_a |\bar{i}\rangle$$

$$= \sum_{\bar{i}} c_{\bar{i}} U_i E_a |\bar{0}\rangle$$

$$= \sum_{\bar{i}} c_{\bar{i}} U_i \left[\sum_{r'} \beta^{\bar{0}}_{ar'} |v^{\bar{0}}_{r'}\rangle \right]$$

$$= \sum_{\bar{i},r'} c_{\bar{i}} \beta^{\bar{0}}_{ar'} |v^{\bar{i}}_{r'}\rangle .$$

Applying R_r gives

$$R_r E_a|c\rangle = \sum_{\bar{i},r'} c_{\bar{i}} \beta^{\bar{0}}_{ar'} R_r |v^{\bar{i}}_{r'}\rangle$$

$$= \sum_{\bar{i},r'} c_{\bar{i}} \beta^{\bar{0}}_{ar'} \sum_{\bar{j}} |\bar{j}\rangle\langle v^{\bar{j}}_r | v^{\bar{i}}_{r'}\rangle$$

$$= \sum_{\bar{i}} c_{\bar{i}} \beta^{\bar{0}}_{ar} |\bar{i}\rangle$$

$$= \beta^{\bar{0}}_{ar} |c\rangle .$$

We used eq. (1.122) to go from the first to the second line. We see that $R_r E_a = \beta^{\bar{0}}_{ar} I$ on $\mathcal{C}_q$. Finally, consider the operation element $\hat{O} \in R$ that projects onto the orthogonal complement of $\oplus_{\bar{i}} \mathcal{V}^{\bar{i}}$. Since $E_a|\bar{i}\rangle \in \mathcal{V}^{\bar{i}}$, it follows that

$$\hat{O} E_a |\bar{i}\rangle = 0$$

$$= 0 \cdot I .$$

We see that all recovery operators in $R = \{\hat{O}, R_1, \ldots, R_r, \ldots\}$ satisfy the only-if premise of Theorem 1.12 and so $(\mathcal{C}_q, R)$ is E-correcting. ∎

Problems

1.1 *The intersection of two binary vectors* $\mathbf{x} = (x_1, \ldots, x_n)$ *and* $\mathbf{y} = (y_1, \ldots, y_n)$ *is the binary vector*

$$\mathbf{x} * \mathbf{y} = (x_1 y_1, \ldots, x_n y_n) .$$

Show that

$$wt(\mathbf{x} + \mathbf{y}) = wt(\mathbf{x}) + wt(\mathbf{y}) - 2wt(\mathbf{x} * \mathbf{y}) .$$

Introduction

1.2 Let C be a classical linear binary code. Show that either all codewords have even weight, or else half have even weight and half have odd weight.

1.3 Let C be a classical linear binary code and let $|C|$ be the number of codewords in C. Show the following.

(a) If $x \in C_\perp$,
$$\sum_{y \in C} (-1)^{x \cdot y} = |C| .$$

(b) If $x \notin C_\perp$,
$$\sum_{y \in C} (-1)^{x \cdot y} = 0 .$$

1.4 The binary Hamming code C_H^r has parity check matrix H_H^r whose columns consist of all non-zero binary vectors of length r, each used only once.

(a) Show that C_H^r is an $[n, k, d]$ code with $n = 2^r - 1$, $k = 2^r - 1 - r$, and $d = 3$.

(b) Show that C_H^r is a perfect single-error correcting code.

(c) Write out the parity check matrix for the $[7, 4, 3]$ binary Hamming code and determine its codewords.

1.5 Let C be an $[n, k, d]$ binary code in which some codewords have odd weight. Define a new code C_e by adding a 0 to every codeword of C of even weight and a 1 to every codeword of C of odd weight.

(a) Show that the distance between every pair of codewords in C_e is even, and that if the minimum distance of C is odd, then the minimum distance of C_e is $d + 1$.

(b) Explain why the codewords of C_e satisfy the new parity check
$$x_1 + \cdots + x_{n+1} = 0 .$$

(c) Argue that C_e is an $[n+1, k, d+1]$ code when d is odd. This construction is known as extending the code C.

1.6 The binary simplex code C_s^r is the dual of the $[2^r - 1, 2^r - 1 - r, 3]$ Hamming code C_H^r (see Problem 1.4). Thus the generator matrix G_s^r for C_s^r is the parity check matrix H_H^r of C_H^r. Let $r = 3$.

(a) Write out the codewords for C_s^3.

(b) Show that C_s^3 is a $[7, 3, 4]$ binary code.

1.7 *Consider two classical binary codes $C = [n, k, d]$ and $C' = [n, k', d']$ with the following properties: (1) $C' \subset C$, and (2) C and $C'_\perp$ are both t-error correcting codes. C and C' can be used to construct a class of quantum error correcting codes known as Calderbank-Shor-Steane (CSS) codes [15, 16].*

(a) C' partitions C into N cosets:

$$C = C' \cup (c_1 + C') \cdots \cup (c_N + C') ,$$

where $c_1, \ldots, c_N \in C$. Show that $N = 2^{k-k'}$.

(b) CSS codes identify the codestates with the cosets $\{c_i + C' : i = 1, \ldots, N\}$:

$$|\bar{c}_i\rangle = \frac{1}{\sqrt{2^{k'}}} \sum_{c' \in C'} |c_i + c'\rangle .$$

Show that if the classical codewords $e, f \in C$ belong to the same coset $e - f \in C'$, then $|\bar{e}\rangle = |\bar{f}\rangle$. Argue that this causes CSS codes to map $k - k'$ qubits into n qubits. We will see in Chapter 3 that the resulting quantum code corrects t errors.

(c) Let C be the $[7, 4, 3]$ binary Hamming code (Problem 1.4) and C' the $[7, 3, 4]$ binary simplex code (Problem 1.6). Thus $C'_\perp$ is the $[7, 4, 3]$ binary Hamming code. Write out the cosets of C' in C and then construct the $2^{(4-3)}$ encoded computational basis states $|\bar{0}\rangle$ and $|\bar{1}\rangle$. This CSS code maps 1 qubit into 7 qubits.

(d) We will see in Chapter 3 that CSS codes are examples of quantum stabilizer codes. The generators for these codes can be constructed from the parity check matrices $H(C)$ and $H(C'_\perp)$ as follows. We construct $n - k$ generators by identifying a generator with each of the $n - k$ rows of $H(C)$ and replacing 0 with the identity operator I and 1 with σ_x^i, where i is the column label that now labels qubit i. The remaining k' generators are identified with the k' rows of $H(C'_\perp)$ only now $1 \to \sigma_z^i$. Show that the six generators of the CSS code in part (c) are:

$$g_1 = \sigma_x^4 \sigma_x^5 \sigma_x^6 \sigma_x^7$$
$$g_2 = \sigma_x^2 \sigma_x^3 \sigma_x^6 \sigma_x^7$$
$$g_3 = \sigma_x^1 \sigma_x^3 \sigma_x^5 \sigma_x^7$$
$$g_4 = \sigma_z^4 \sigma_z^5 \sigma_z^6 \sigma_z^7$$
$$g_5 = \sigma_z^2 \sigma_z^3 \sigma_z^6 \sigma_z^7$$
$$g_6 = \sigma_z^1 \sigma_z^3 \sigma_z^5 \sigma_z^7$$

(e) Show that the generators $g_1, \ldots, g_6$ in part (d) fix the encoded computational basis states $|\bar{0}\rangle$ and $|\bar{1}\rangle$ in part (c).

References

[1] Leakey, R., *The Origin of Humankind*, BasicBooks, New York, 1994.

[2] Mano, M. M. and Kime, C. R., *Logic and Computer Design Fundamentals*, Prentice Hall, Upper Saddle River, NJ, 1997.

[3] Benioff, P., The computer as a physical system: a microscopic quantum mechanical Hamiltonian model of computers as represented by Turing machines, *J. Stat. Phys.* **22**, 563, 1980.

[4] Feynman, R. P., Simulating physics with computers, *Int. J. Theor. Phys.* **21**, 467, 1982.

[5] Deutsch, D., Quantum theory, the Church-Turing principle, and the universal quantum computer, *Proc. R. Soc. Lond. A* **400**, 97, 1985.

[6] Deutsch, D., Quantum computational networks, *Proc. R. Soc. Lond. A* **425**, 73, 1989.

[7] Barenco, A. et al., Elementary gates for quantum computation, *Phys. Rev. A* **52**, 3457, 1995.

[8] Deutsch, D., Barenco, A., and Ekert, A., Universality in quantum computation, *Proc. R. Soc. Lond. A* **449**, 669 (1995).

[9] Lloyd, S., Almost any quantum gate is universal, *Phys. Rev. Lett.* **75**, 346, 1995.

[10] Shor, P. W., Algorithms for quantum computation: discrete logarithms and factoring, in *Proceedings, 35th Annual Symposium on the Foundations of Computer Science*, IEEE Press, Los Alamitos, CA, 1994.

[11] Shor, P. W., Polynomial-time algorithms for prime factorization and discrete logarithm on a quantum computer, *SIAM J. Comp.* **26**, 1484, 1997.

[12] Grover, L., Quantum mechanics helps in searching for a needle in a haystack, *Phys. Rev. Lett.* **79**, 325, 1997.

[13] Shor, P. W., Scheme for reducing decoherence in quantum computer memory, *Phys. Rev. A* **52**, 2493, 1995.

[14] Steane, A. M., Error correcting codes in quantum theory, *Phys. Rev. Lett.* **77**, 793, 1996.

[15] Calderbank, A. R. and Shor, P. W., Good quantum error correcting codes exist, *Phys. Rev. A* **54**, 1098, 1996.

[16] Steane, A. M., Multiple-particle interference and quantum error correction, *Proc. R. Soc. Lond. A* **452**, 2551, 1996.

[17] Steane, A. M., Simple quantum error-correcting codes, *Phys. Rev. A* **54**, 4741, 1996.

[18] Bennett, C. H., DiVincenzo, D. P., Smolin, J. A., and Wootters, W. K., Mixed state entanglement and quantum error correction, *Phys. Rev. A* **54**, 3824, 1996.

[19] Ekert, A. and Macchiavello, C., Error correction in quantum communication, *Phys. Rev. Lett.* **77**, 2585, 1996.

[20] Laflamme, R., Miquel, C., Paz, J.-P., and Zurek, W. H., Perfect quantum error correction code, *Phys. Rev. Lett.* **77**, 198, 1996.

[21] Knill, E. and Laflamme, R., Theory of quantum error-correcting codes, *Phys. Rev. A* **55**, 900, 1997.

[22] Gottesman, D., Class of quantum error correcting codes saturating the quantum Hamming bound, *Phys. Rev. A* **54**, 1862, 1996.

[23] Calderbank, A. R. et al., Quantum error correction and orthogonal geometry, *Phys. Rev. Lett.* **78**, 405, 1997.

[24] Calderbank, A. R. et al., Z_4 Kerdock codes, orthogonal spreads, and extremal Euclidean line-sets, *Proc. London Math. Soc.* **75**, 436, 1997.

[25] Calderbank, A. R. et al., Quantum error correction via codes over GF(4), *IEEE Trans. Inf. Theor.* **44**, 1369, 1998.

[26] Shor, P. W., Fault-tolerant quantum computation, in *Proceedings, 37th Annual Symposium on Fundamentals of Computer Science*, IEEE Press, Los Alamitos, CA, 1996, pp. 56–65.

[27] DiVincenzo, D. P. and Shor, P. W., Fault-tolerant error correction with efficient quantum codes, *Phys. Rev. Lett.* **77**, 3260, 1996.

[28] Gottesman, D., Theory of fault-tolerant quantum computation, *Phys. Rev. A* **57**, 127, 1998.

[29] Plenio, M. B., Vedral, V., and Knight, P. L., Conditional generation of error syndromes in fault-tolerant error correction, *Phys. Rev. A* **55**, 4593, 1997.

[30] Steane, A. M., Active stabilization, quantum computation, and quantum state synthesis, *Phys. Rev. Lett.* **78**, 2252, 1997.

[31] Steane, A. M., Efficient fault-tolerant quantum computing, *Nature* **399**, 124, 1999.

[32] Gottesman, D., Stabilizer codes and quantum error correction, Ph. D. thesis, California Institute of Technology, Pasadena, CA, 1997.

[33] Preskill, J., Fault-tolerant quantum computation, in *Introduction to Quantum Computation and Information*, H.-K. Lo, S. Popescu, and T. Spiller, Eds., World Scientific, Singapore, 1998.

[34] Preskill, J., Reliable quantum computers, *Proc. R. Soc. Lond. A* **454**, 385, 1998.

[35] Knill, E. and Laflamme, R., Concatenated Quantum Codes, download at http://arXiv.org/abs/quant-ph/9608012, 1996.

[36] Knill, E., Laflamme, R., and Zurek, W. H., Resilient quantum computation, *Science* **279**, 342, 1998.

[37] Knill, E., Laflamme, R., and Zurek, W.H., Resilient quantum computation: error models and thresholds, *Proc. R. Soc. Lond. A* **454**, 365, 1998.

[38] Aharonov, D. and Ben-Or, M., Fault-tolerant computation with constant error, in *Proceedings of the Twenty-Ninth ACM Symposium on the Theory of Computing*, 1997, pp. 176–188.

[39] Zalka, C., Threshold Estimate for Fault-Tolerant Quantum Computing, download at http://arXiv.org/abs/quant-ph/9612028, 1996.

[40] Kitaev, A. Y., Quantum computation: algorithms and error correction, *Russ. Math. Surv.* **52**, 1191, 1997.

[41] Kitaev, A. Y., Quantum error correction with imperfect gates, in *Quantum Communication, Computing, and Measurement*, Plenum Press, New York, 1997, pp. 181–188.

[42] Shannon, C. E., A mathematical theory of communication, *Bell Sys. Tech. J.* **27**, 379–423 and 623–656, 1948.

[43] Shannon, C. E. and Weaver, W., *A Mathematical Theory of Communication*, University of Illinois Press, Urbana, IL, 1963.

[44] MacWilliams, F. J. and Sloane, N. J. A., *The Theory of Error Correcting Codes*, North-Holland, New York, 1977.

[45] Peterson, W. W. and Weldon, E. J., *Error-Correcting Codes*, MIT Press, Cambridge, MA, 1972.

[46] Blahut, R. E., *Theory and Practice of Error-Control Codes*, Addison-Wesley, Reading, MA, 1983.

[47] Roman, S., *Coding and Information Theory*, Springer-Verlag, New York, 1992.

[48] Dirac, P. A. M., *The Principles of Quantum Mechanics*, 4th ed., Oxford University Press, London, 1958.

[49] Nielsen, M. A. and Chuang, I. L., *Quantum Computation and Quantum Information*, Cambridge University Press, London, 2000.

[50] Wooters, W. K. and Zurek, W. H., A single quantum cannot be cloned, *Nature* **299**, 802 (1982).

[51] Dieks, D., Communication by EPR devices, *Phys Lett A* **92**, 271, 1982.

[52] See Reference [49], pp. 604–605.

[53] Wheeler, J. A. and Zurek, W. H., *Quantum Theory and Measurement*, Princeton University Press, Princeton, NJ, 1983.

[54] d'Espagnat, B., *Conceptual Foundations of Quantum Mechanics*, Addison-Wesley, New York, 1976.

[55] Jammer, M., *The Philosophy of Quantum Mechanics*, Wiley, New York, 1974.

[56] Zurek, W. H., Decoherence, einselection, and the quantum origins of the classical, *Rev. Mod. Phys.* **75**, 715, 2003.

[57] von Neumann, J., *Mathematical Foundations of Quantum Mechanics*, Princeton University Press, Princeton, NJ, 1955. See also Wigner, E., Interpretation of quantum mechanics, in Reference [53], pp. 260–314.

[58] Peres, A., *Quantum Theory: Concepts and Methods*, Kluwer Academic Publishers, Boston, 1995.

[59] Einstein, A., Podolsky, B., and Rosen, N., Can quantum-mechanical description of physical reality be considered complete?, *Physical Review* **47**, 777, 1935.

[60] Bell, J. S., On the Einstein Podolsky Rosen paradox, *Physics* **1**, 195, 1964.

[61] For a summary of recent efforts, see Garcia-Patron, R., Fiurasek, J., and Cerf, N. J., Loophole-free test of quantum nonlocality using high efficiency homodyne detectors, *Phys. Rev. A* **71**, 022105, 2005.

[62] Bennett, C. H., Brassard, G., Crepeau, C., Jozsa, R., Peres, A., and Wooters, W. K., Teleporting an unknown quantum state via dual classical and Einstein-Podolsky-Rosen channels, *Phys. Rev. Lett.* **70**, 1895, 1993.

[63] Bennett, C. H. and Brassard, G., Quantum cryptography: public key distribution and coin tossing, in *Proc. of IEEE Int. Conf. on Comps., Sys., and Sig. Proc.*, Bangalore, India, 175 (1984); and Ekert, A., Quantum cryptography based on Bell's theorem, *Phys. Rev. Lett.* **67**, 661, 1991.

[64] Bennett, C. H., DiVincenzo, D. P., Smolin, J. A., and Wooters, W. K., Mixed state entanglement and quantum error correction, *Phys. Rev. A* **54**, 3824, 1996.

[65] Schiff, L. I., *Quantum Mechanics*, 3rd ed., McGraw-Hill, New York, 1968, Sec. 42.

[66] Merzbacher, E., *Quantum Mechanics*, 2nd ed., Wiley, New York, 1970, Chap. 13.

[67] Feynman, R. P. and Hibbs, A. R., *Quantum Mechanics and Path Integrals*, McGraw-Hill, New York, 1965, Chap. 1.

[68] Halmos, P. R., *Finite-Dimensional Vector Spaces*, Springer, New York, 1987, p. 142.

2

Quantum Error Correcting Codes

Having surveyed the necessary background in Chapter 1, this chapter begins developing the theory of quantum error correcting codes and fault-tolerant quantum computing. Section 2.1 introduces the dynamical framework of quantum operations. In this approach a quantum computer is treated as an open quantum system and a superoperator describes the effective interaction it has with its environment. The operator-sum representation for a quantum operation is derived, as are some of its essential properties. This section also shows how the noise model introduced in Section 1.4.1 can be represented as a quantum operation, and briefly describes other important noise models. Section 2.2 introduces a number of fundamental notions associated with quantum error correcting codes, and the seven-qubit Calderbank-Shor-Steane code is used in Section 2.3 to show how these notions manifest in a specific quantum error correcting code.

2.1 Quantum Operations

As noted in Chapter 1, to properly assess the computational power of a quantum computer, it is essential to include the dynamical effects of its environment in the analysis. Quantum operations provide a powerful framework in which a quantum computer is treated as an open quantum system and a superoperator is used to describe the coupling it has to its environment. A book-length review of quantum operations is given by Kraus [1], together with references to the original papers. Nielsen and Chuang [2] give a comprehensive, highly accessible presentation. In this section we limit our discussion to those aspects of quantum operations that are most relevant to the purposes of this book. This discussion draws on Refs. [2–5]. We begin by examining the unitary dynamics of a closed system made up of a quantum computer and its environment. We show how tracing out the environment leads to the operator-sum representation for the superoperator that captures the dynamical consequences of the environment. Three important properties of the operator-sum representation are derived and this representation is used to describe the noise model introduced in Section 1.4.1. A brief discussion of other important noise models closes out the section.

2.1.1 Operator-Sum Representation

Consider the situation where a quantum register Q is acted on by an environment E. For our purposes, you might imagine E to be a noisy imperfect quantum gate that is applied to Q. The composite system of register plus environment is modeled as a closed quantum system. We assume that initially Q and E are not entangled. Denoting the initial states of Q and E by the density operators ρ and ϵ_0, respectively, the initial state of the composite system will be the direct product state $\rho \otimes \epsilon_0$. Q and E then interact for a period of time. Because the composite system is closed, its dynamics is unitary and its final state is specified through a unitary operator U: $U(\rho \otimes \epsilon_0) U^\dagger$. Usually, the object of interest for us will be the reduced density operator for Q at the end of the interaction. This operator is found by tracing out the environment E from the final state of the composite system:

$$\rho_f = \mathrm{tr}_E \left[U(\rho \otimes \epsilon_0) U^\dagger \right]$$
$$\equiv \xi(\rho) \ . \tag{2.1}$$

ξ is an example of a quantum operation ($\xi: \rho \to \rho_f$) that maps density operators to density operators. Being a map from operators to operators, quantum operations are also referred to as superoperators [3–5]. It follows immediately from eq. (2.1) that ξ is a linear (super)operator.

Let the initial state of the environment be $\epsilon_0 = \sum_l \lambda_l |\phi_l\rangle\langle\phi_l|$, where $|\phi_l\rangle$ are the eigenstates of ϵ_0, and let $\{|e_m\rangle\}$ be an orthonormal basis for the environment's Hilbert space $\mathcal{H}_E$. Inserting these expressions into eq. (2.1) gives

$$\rho_f = \sum_{l,m} \lambda_l \langle e_m | U \{\rho \otimes |\phi_l\rangle\langle\phi_l| \} U^\dagger | e_m \rangle$$
$$= \sum_{l,m} E_{lm} \rho E_{lm}^\dagger \ , \tag{2.2}$$

where $E_{lm} = \sqrt{\lambda_l} \langle e_m | U | \phi_l \rangle$. The operators E_{lm} are known as the operation elements for the quantum operation ξ and are seen to be linear operators on states in Q's Hilbert space $\mathcal{H}_Q$. Once they are known, eq. (2.2) completely specifies the effect of the environment E on the dynamics of Q. We will also refer to the operation elements as error operators, and will often replace the double subscript lm by a single subscript k. Eq. (2.2) is known as the operator-sum representation for the quantum operation ξ.

Exercise 2.1 *Show that each operation element E_{lm} is a linear operator on the states of $\mathcal{H}_Q$.*

Quantum operations fall into two classes: (1) trace-preserving for which $\mathrm{tr}\,\xi(\rho) = \mathrm{tr}\,\rho = 1$, and (2) non-trace-preserving for which $\mathrm{tr}\,\xi(\rho) < 1$. Note

Quantum Error Correcting Codes

that for trace-preserving quantum operations,

$$\begin{aligned}
\operatorname{tr}\rho &= \operatorname{tr}\xi(\rho) \\
&= \operatorname{tr}\left[\sum_k E_k\rho E_k^\dagger\right] \\
&= \operatorname{tr}\left[\rho \sum_k E_k^\dagger E_k\right] .
\end{aligned} \qquad (2.3)$$

Since eq. (2.3) must be true for arbitrary ρ, we must have

$$\sum_k E_k^\dagger E_k = I \qquad (2.4)$$

for a trace-preserving quantum operation. We will say more about non-trace-preserving quantum operations below when we discuss measurements.

Before moving on, a few remarks are in order concerning the derivation of eq. (2.2).

(1) The initial state for the composite system was assumed to be a direct product state $\rho \otimes \epsilon_0$. This is not the most general starting point, but will be sufficient for our needs. As seen in Section 1.3.5, the protocol for quantum error correction leaves the combined system of register/ancilla/environment in a direct product state in which correctable errors have been removed from the register's state. This direct product state becomes the input for the next phase of quantum computation at which point the analysis leading to eq. (2.2) can be applied. Having said this, note that it is possible to do better. It has been shown that the restriction to an initial direct product state can be removed from the derivation of the operator-sum representation [2,4,5].

(2) If the Hilbert space $\mathcal{H}_Q$ for the quantum register Q is n-dimensional, then a quantum operation ξ acting on Q will contain at most n^2 operation elements. To see why, recall that the operation elements E_k are linear operators on $\mathcal{H}_Q$ and thus are represented by $n \times n$ matrices with complex elements. The set of such $n \times n$ matrices form an n^2-dimensional vector space over the field of complex numbers. There are thus n^2 orthonormal basis operators $\{B_a\}$ that span the space, and the trace norm $\langle A|B\rangle = \operatorname{tr}(A^\dagger B)$ can be used to define the inner product. The operation elements E_k can thus be expanded in this basis,

$$E_k = \sum_{a=1}^{n^2} C_{ka} B_a ,$$

where the C_{ka} are complex numbers. Since the vector space of matrices is n^2-dimensional, there can be no more than n^2 linearly independent operation elements E_k. This then is the most that can appear in a quantum operation

ξ acting on Q.

(3) The derivation of eq. (2.2) can be extended to allow the final reduced density operator ρ_f to correspond to a different quantum system $Q' \neq Q$. This extension simply requires that the appropriate subsystem be traced out in eq. (2.1) where ρ_f is defined. For us, the quantum register will always be the quantum system of interest, both before and after the interaction with the environment E, and so we point out this additional flexibility in the formalism, though we will not need to make use of it.

(4) References [4, 5] give an axiomatic formulation of quantum operations. This approach is reviewed in Nielsen and Chuang [2]. For our purposes, the operator-sum representation will do just fine and so we will not discuss this complimentary approach to quantum operations. The interested reader is encouraged to examine these references for further details.

Physical Content

To bring out the physical content of the operator-sum representation we again consider the situation where the register Q and its environment E are initially non-entangled and then interact according to the unitary transformation U. The initial states of Q and E are again ρ and ϵ_0, respectively, though having seen above how to handle the case of a mixed state for ϵ_0, we simplify the following discussion by letting ϵ_0 correspond to a pure state $\epsilon_0 = |e_0\rangle\langle e_0|$. At the end of the interaction we measure the observable $M = \sum_m \mu_m P_m$ of E, where $\{\mu_m\}$ are the eigenvalues of M and P_m is the projection operator onto the subspace of states of E with eigenvalue μ_m. In the interests of keeping the discussion simple, we assume the eigenvalues μ_m are non-degenerate. The reader can easily modify the following argument to allow for degeneracy. If the measurement outcome is μ_k, then the final (normalized) state of the composite system is (see Appendix B):

$$\rho_{tot}^k = \frac{P_k U \{\rho \otimes |e_0\rangle\langle e_0|\} U^\dagger P_k}{\text{Tr}\left(P_k U \{\rho \otimes |e_0\rangle\langle e_0|\} U^\dagger P_k\right)},$$

where Tr is a trace over Q and E. The trace in the denominator can be reduced to a simpler form:

$$\text{Tr}\left(P_k U \{\rho \otimes |e_0\rangle\langle e_0|\} U^\dagger P_k\right) = \text{tr}_Q \text{tr}_E \left(P_k U \{\rho \otimes |e_0\rangle\langle e_0|\} U^\dagger P_k\right)$$
$$= \text{tr}_Q \left(E_k \rho E_k^\dagger\right), \quad (2.5)$$

where $E_k = \langle \mu_k | U | e_0 \rangle$ is an operation element, and $|\mu_k\rangle$ is the eigenstate associated with the measurement outcome μ_k. Thus we can write

$$\rho_{tot}^k = \frac{P_k U \{\rho \otimes |e_0\rangle\langle e_0|\} U^\dagger P_k}{\text{tr}_Q \left(E_k \rho E_k^\dagger\right)}. \quad (2.6)$$

The (normalized) reduced density operator ρ_k for Q is found by tracing out E in ρ_{tot}^k:

$$\rho_k = \frac{\text{tr}_E \left(P_k U \{ \rho \otimes |e_0\rangle\langle e_0| \} U^\dagger P_k \right)}{\text{tr}_Q \left(E_k \rho E_k^\dagger \right)}$$

$$= \frac{E_k \rho E_k^\dagger}{\text{tr}_Q \left(E_k \rho E_k^\dagger \right)} . \qquad (2.7)$$

The probability that μ_k is the measurement outcome is (see Appendix B):

$$P(k) = \text{Tr} \left(P_k U \{ \rho \otimes |e_0\rangle\langle e_0| \} U^\dagger \right)$$
$$= \text{Tr} \left(P_k U \{ \rho \otimes |e_0\rangle\langle e_0| \} U^\dagger P_k \right)$$
$$= \text{tr}_Q \left(E_k \rho E_k^\dagger \right) , \qquad (2.8)$$

where we used $P_k^2 = P_k$ and $\text{Tr} AB = \text{Tr} BA$ to go from the first line to the second, and eq. (2.5) to go from the second to the third line. If the measurement outcome is *not* observed, the composite system will be in the mixed state:

$$\rho_{tot} = \sum_k P(k) \rho_{tot}^k .$$

The reduced density operator for Q after the measurement is then

$$\rho_Q = \sum_k P(k) \, \text{tr}_E \left(\rho_{tot}^k \right)$$
$$= \sum_k P(k) \rho_k \qquad (2.9)$$
$$= \sum_k E_k \rho E_k^\dagger . \qquad (2.10)$$

We see that after the measurement, the reduced density operator for Q is given by the operator-sum representation (eq. (2.2)) for the quantum operation that E applies to Q. Note that the original derivation of the operator-sum representation did not implement a measurement on the environment, whereas the derivation of eq. (2.10) did, and yet the same reduced density operator was found in both cases. This is an example of the principle of implicit measurement [2]. It states that once a subsystem E of a composite quantum system C has finished interacting with the rest of the composite system $Q = C - E$, the reduced density operator for Q will not be affected by any operations that we carry out solely on E.

Exercise 2.2 (Principle of Implicit Measurement)

Consider two quantum systems Q and E whose composite state after their interactions have ceased is represented by the density operator ρ_c. The reduced density operator for Q is then $\rho_Q = \text{tr}_E[\rho_c]$. Let M be a physical observable of E as discussed in the derivation of eq. (2.10), and assume again that its eigenvalues are non-degenerate. With Q and E in the joint state ρ_c, imagine that we implement a unitary operation U_E on E and then measure M (without observing the outcome).

(a) Show that the state of the composite system after the pair of operations is

$$\rho'_c = \sum_m P_m U_E \rho_c U_E^\dagger P_m \ .$$

(b) Let ρ'_Q be the reduced density operator for Q after the operations on E have been carried out. Show that $\rho'_Q = \rho_Q$ (i. e., that the final reduced density operator for Q is unaffected by whether or not the pair of operations were applied to E).

Because of the principle of implicit measurement, we can think of the quantum operation that E applies to Q as producing the map $\rho \to \rho_k$ with probability $P(k)$ (see eq. (2.9)). This is the physical picture we will usually have in mind when discussing a quantum operation.

Measurement of E

We again consider the situation where Q and E are initially in the direct product state $\rho \otimes \epsilon_0$ and E is in the mixed state $\epsilon_0 = \sum_l \lambda_l |\phi_l\rangle\langle\phi_l|$. Without loss of generality, the eigenstates $\{|\phi_l\rangle\}$ of ϵ_0 can be assumed to form an orthonormal set, and $\text{tr}\,\epsilon_0 = \sum_l \lambda_l = 1$. Q and E then undergo an interaction of finite duration described by the unitary transformation U. At the end of the interaction we make a measurement of E, however this time we observe the outcome. The state of E after the measurement is assumed to lie in the subspace $\mathcal{S} \subset \mathcal{H}_E$. The projection operator associated with $\mathcal{S}$ is

$$P_E = \sum_m |g_m\rangle\langle g_m| \ ,$$

and the $\{|g_m\rangle\}$ are an orthonormal basis for $\mathcal{S}$. The dual space $\mathcal{S}_\perp$ is the set of vectors that are orthogonal to the elements of $\mathcal{S}$. Let $\{|h_b\rangle\}$ be an orthonormal basis for $\mathcal{S}_\perp$. Since $\mathcal{S} \oplus \mathcal{S}_\perp = \mathcal{H}_E$, the combined set of $\{|g_m\rangle\}$ and $\{|h_b\rangle\}$ forms an orthonormal basis for $\mathcal{H}_E$.

Adapting arguments made earlier in this section, we start with the unnormalized reduced density operator for Q after the measurement:

$$\xi(\rho) = \text{tr}_E \left(P_E U \{\rho \otimes \epsilon_0\} U^\dagger P_E \right)$$
$$= \sum_{lm} E_{lm} \rho E_{lm}^\dagger \ ,$$

where $E_{lm} = \sqrt{\lambda_l} \langle g_m|U|\phi_l\rangle$. The normalized reduced density operator is

$$\rho_f = \frac{\xi(\rho)}{\text{tr}_Q(\xi(\rho))} \ ,$$

Quantum Error Correcting Codes 63

and the probability $P(\mathcal{S})$ that the state of E is found in the subspace $\mathcal{S}$ after the measurement is

$$P(\mathcal{S}) = \text{tr}_Q\left(\xi(\rho)\right) . \tag{2.11}$$

Since $P(\mathcal{S})$ is a probability, it follows that

$$0 \leq \text{tr}_Q\left(\xi(\rho)\right) \leq 1 .$$

We previously defined a quantum operation for which $\text{tr}_Q\left(\xi(\rho)\right) = 1$ as being trace-preserving, and one for which $0 \leq \text{tr}_Q\left(\xi(\rho)\right) < 1$ as being non-trace-preserving. It is possible to establish physical settings that lead to these two classes of quantum operations. They correspond to non-selective and selective dynamics [1,5].

(1) *Non-Selective Dynamics:* Here the outcome of the measurement of E is not observed. In this case, $P_E = I_E$, the $\{|g_m\rangle\}$ span $\mathcal{H}_E$, and

$$\sum_{lm} E_{lm}^\dagger E_{lm} = \sum_{lm} \lambda_l \langle \phi_l | U^\dagger | g_m \rangle \langle g_m | U | \phi_l \rangle$$
$$= \sum_l \lambda_l \langle \phi_l | U^\dagger U | \phi_l \rangle .$$

Note that $U^\dagger U = I_Q \otimes I_E$ since U is unitary and acts on the composite system. Thus,

$$\sum_{lm} E_{lm}^\dagger E_{lm} = \sum_l \lambda_l \langle \phi_l | \phi_l \rangle I_Q$$
$$= I_Q \sum_l \lambda_l$$
$$= I_Q .$$

Here we used that $\sum_l \lambda_l = \text{tr}\,\epsilon_0 = 1$. Thus trace-preserving quantum operations correspond to non-selective dynamics where nothing is learned about the final state of the environment E.

(2) *Selective Dynamics:* Here we observe the measurement outcome so that $P_E \neq I_E$ and $\mathcal{S}$ is a proper subset of $\mathcal{H}_E$. Since $\mathcal{S} \oplus \mathcal{S}_\perp = \mathcal{H}_E$, it is certain that the state of E after the measurement will lie either in $\mathcal{S}$ or $\mathcal{S}_\perp$. Stated in terms of probabilities, we have

$$P(\mathcal{S}) + P(\mathcal{S}_\perp) = 1 .$$

Using eq. (2.11), this becomes

$$\text{tr}\left(\xi(\rho)\right) = P(\mathcal{S})$$
$$= 1 - P(\mathcal{S}_\perp) < 1 .$$

Thus non-trace-preserving quantum operations correspond to selective dynamics where we gain some knowledge about the final state of E.

Recall that the $\{|h_b\rangle\}$ form an orthonormal basis for $\mathcal{S}_\perp$. Define the set of operators $\{H_{lb} = \sqrt{\lambda_l}\,\langle h_b|U|\phi_l\rangle\}$, one for each pairing of states $|h_b\rangle$ and $|\phi_l\rangle$. Then

$$\sum_{lm} E_{lm}^\dagger E_{lm} + \sum_{lb} H_{lb}^\dagger H_{lb}$$

$$= \sum_{lm} \lambda_l \langle\phi_l|U^\dagger|g_m\rangle\langle g_m|U|\phi_l\rangle + \sum_{lb} \lambda_l \langle\phi_l|U^\dagger|h_b\rangle\langle h_b|U|\phi_l\rangle$$

$$= \sum_{l} \lambda_l \langle\phi_l|U^\dagger \left\{ \sum_m |g_m\rangle\langle g_m| + \sum_b |h_b\rangle\langle h_b| \right\} U|\phi_l\rangle$$

$$= \sum_{l} \lambda_l \langle\phi_l|U^\dagger U|\phi_l\rangle$$

$$= I_Q \;. \tag{2.12}$$

We used that the $\{|g_m\rangle\}$ and $\{|h_b\rangle\}$ span $\mathcal{H}_E$ in going from the second to the third line. For non-trace-preserving quantum operations one sometimes sees the content of eq. (2.12) expressed informally as $\sum_{lm} E_{lm}^\dagger E_{lm} < I$.

Non-uniqueness

Here we examine the non-uniqueness of the operator-sum representation associated with a quantum operation. One background topic needs to be discussed first: the non-uniqueness of the ensemble associated with a given density operator. We then work out the connection between two operator-sum representations that give rise to the same quantum operation.

(1) Non-uniqueness of Ensemble: Consider two ensembles E_ψ and E_ϕ that are both described by the density operator ρ. Let E_ψ correspond to the probabilities and states $\{p_i, |\psi_i\rangle\}$ with $i = 1, \ldots, L_\psi$, and similarly, let E_ϕ correspond to the probabilities and states $\{p_\alpha, |\phi_\alpha\rangle\}$ with $\alpha = 1, \ldots, L_\phi$. By assumption,

$$\sum_i p_i |\psi_i\rangle\langle\psi_i| = \rho = \sum_\alpha p_\alpha |\phi_\alpha\rangle\langle\phi_\alpha| \;. \tag{2.13}$$

The set of states $\{|\psi_i\rangle\}$ need not be mutually orthogonal, nor even linearly independent, though we do assume that they are normalized. Similar remarks apply to the states $\{|\phi_\alpha\rangle\}$.

Since ρ is an Hermitian operator (acting on a Hilbert space $\mathcal{H}$), it can be diagonalized. We denote its eigenstates by $\{|r\rangle\}$, and its eigenvalues by $\{p_r\}$, where $0 \leq p_r \leq 1$ and $\sum_r p_r = 1$. It follows that

$$\rho = \sum_r p_r |r\rangle\langle r| \;.$$

Since ρ is diagonal in the r-basis, its rank R_ρ is equal to the number of non-zero eigenvalues, $p_n \neq 0$. The corresponding eigenstates $\{|n\rangle\}$ span an R_ρ-dimensional subspace $\mathcal{S}_n \subset \mathcal{H}$. Eigenstates $|z\rangle$, corresponding to zero eigenvalue $p_z = 0$, span a subspace $\mathcal{S}_z$ that is the orthogonal complement of $\mathcal{S}_n$: $\mathcal{H} = \mathcal{S}_n \oplus \mathcal{S}_z$ and $\langle z|n\rangle = 0$ for all $|n\rangle \in \mathcal{S}_n$ and $|z\rangle \in \mathcal{S}_z$. The probability p_r can be written as

$$p_r = \langle r|\rho|r\rangle$$
$$= \sum_i p_i |\langle r|\psi_i\rangle|^2 = \sum_\alpha p_\alpha |\langle r|\phi_\alpha\rangle|^2 \ . \quad (2.14)$$

If $|r\rangle \in \mathcal{S}_z$, then $p_r = 0$, and eq. (2.14) indicates that the states $\{|\psi_i\rangle\}$ and $\{|\phi_\alpha\rangle\}$ are orthogonal to an arbitrary state in $\mathcal{S}_z$. They thus belong to $\mathcal{S}_n$, the orthogonal complement of $\mathcal{S}_z$, and can be expanded in the $\{|n\rangle\}$:

$$|\psi_i\rangle = \sum_n a_{in} |n\rangle \quad (2.15)$$

and

$$|\phi_\alpha\rangle = \sum_n b_{\alpha n} |n\rangle \ . \quad (2.16)$$

Here $a_{in} = \langle n|\psi_i\rangle$ and $b_{\alpha n} = \langle n|\phi_\alpha\rangle$ are the matrix elements of the linear operators a and b that carry out the mappings in eqs. (2.15) and (2.16). Now consider an eigenvalue $p_n \neq 0$ for which we can write

$$p_n|n\rangle = \rho|n\rangle = \sum_i p_i |\psi_i\rangle\langle\psi_i|n\rangle \ , \quad (2.17)$$

so that

$$|n\rangle = \sum_i \frac{p_i}{p_n} \langle\psi_i|n\rangle|\psi_i\rangle = \sum_i c_{ni}|\psi_i\rangle \ . \quad (2.18)$$

If we instead write ρ in terms of the $\{|\phi_\alpha\rangle\}$ and repeat the steps leading from eq. (2.17) to (2.18), we arrive at

$$|n\rangle = \sum_\alpha \frac{p_\alpha}{p_n} \langle\phi_\alpha|n\rangle|\phi_\alpha\rangle = \sum_\alpha d_{n\alpha}|\phi_\alpha\rangle \ . \quad (2.19)$$

The c_{ni} and $d_{n\alpha}$ are the matrix elements of the linear operators c and d that carry out the mappings in eqs. (2.18) and (2.19). We see that the set of states $\{|n\rangle\}$ can be expanded in either the set of states $\{|\psi_i\rangle\}$ or $\{|\phi_\alpha\rangle\}$. Since the $\{|n\rangle\}$ span $\mathcal{S}_n$, the $\{|\psi_i\rangle\}$ and $\{|\phi_\alpha\rangle\}$ must also span $\mathcal{S}_n$, not just a subspace of it. Thus L_ψ and L_ϕ must be greater than or equal to R_ρ, the dimension of $\mathcal{S}_n$. Strict inequality for L_ψ (L_ϕ) implies the states $\{|\psi_i\rangle\}$ ($\{|\phi_\alpha\rangle\}$) are linearly dependent. Combining eqs. (2.15) and (2.19) gives

$$|\psi_i\rangle = \sum_\alpha U_{i\alpha} |\phi_\alpha\rangle \ , \quad (2.20)$$

where
$$U_{i\alpha} = \sum_n a_{in} d_{n\alpha} \ . \tag{2.21}$$

Similarly, using eqs. (2.16) and (2.18) gives
$$|\phi_\alpha\rangle = \sum_i V_{\alpha i} |\psi_i\rangle \ , \tag{2.22}$$

where
$$V_{\alpha i} = \sum_n b_{\alpha n} c_{ni} \ . \tag{2.23}$$

Consistency of eqs. (2.20) and (2.22) requires
$$\sum_\alpha U_{i\alpha} V_{\alpha j} = \delta_{ij} \tag{2.24}$$

and
$$\sum_i V_{\alpha i} U_{i\beta} = \delta_{\alpha\beta} \ . \tag{2.25}$$

Thus if the ensembles E_ψ and E_ϕ are to have the same density operator, the two sets of states $\{|\psi_i\rangle\}$ and $\{|\phi_\alpha\rangle\}$ must be linearly related (eqs. (2.20) and (2.22)), and the linear transformations U and V must satisfy eqs. (2.24) and (2.25).

To see that our expressions for U and V satisfy eqs. (2.24) and (2.25), we note that eqs. (2.15) and (2.18) indicate that a and c are left-inverses of each other:
$$a = c^{-1_l} \iff c = a^{-1_l} \ . \tag{2.26}$$

The subscript l on -1_l reminds us that we are dealing with a left-inverse. If $L_\psi = R_\rho$, then unique two-sided inverses exist [6], and $a = c^{-1}$. Similarly, eqs. (2.16) and (2.19) indicate that b and d are left-inverses of each other:
$$b = d^{-1_l} \iff d = b^{-1_l} \ , \tag{2.27}$$

and if $L_\phi = R_\rho$, then $b = d^{-1}$. From eqs. (2.21) and (2.23) we have
$$U = ad$$
$$V = bc \ .$$

Using eqs. (2.26) and (2.27) gives
$$UV = \left(ab^{-1_l}\right)(bc)$$
$$= ac$$
$$= c^{-1_l} c$$
$$= I_\psi \ ,$$

where I_ψ is the $L_\psi \times L_\psi$ identity matrix. Similar manipulations show that $VU = I_\phi$, where I_ϕ is the $L_\phi \times L_\phi$ identity matrix. Note that when $L_\psi = L_\phi = R_\rho$, we have that $UV = VU = I_R$, where I_R is the $R_\rho \times R_\rho$ identity matrix. In this case, U and V are the unique two-sided inverses of one another: $U = V^{-1}$. Otherwise, a unique two-sided inverse does not exist [6], and eqs. (2.24) and (2.25) only imply that U and V are left-inverses of one another.

We now show that if the set of states $\{|\psi_i\rangle\}$ and $\{|\phi_\alpha\rangle\}$ are related through eqs. (2.20) and (2.21), then the ensembles E_ψ and E_ϕ define the same density operator ρ. From eq. (2.20) we can write

$$\sum_i p_i |\psi_i\rangle\langle\psi_i| = \sum_{i\alpha\beta} p_i U_{i\alpha} U_{i\beta}^* |\phi_\alpha\rangle\langle\phi_\beta| \ .$$

Using eq. (2.21) as well as $a_{in} = \langle n|\psi_i\rangle$ and $d_{n\alpha} = (p_\alpha/p_n)\langle\phi_\alpha|n\rangle$ gives

$$\sum_i p_i |\psi_i\rangle\langle\psi_i| = \sum_{i\alpha\beta nm} \frac{p_i p_\alpha p_\beta}{p_n p_m} |\phi_\alpha\rangle\langle\phi_\alpha|n\rangle\langle n|\psi_i\rangle\langle\psi_i|m\rangle\langle m|\phi_\beta\rangle\langle\phi_\beta|$$

$$= \sum_{\alpha nm} \frac{p_\alpha}{p_n p_m} |\phi_\alpha\rangle\langle\phi_\alpha|n\rangle\langle n|\rho|m\rangle\langle m|\rho \ .$$

Using $\rho|m\rangle = p_m|m\rangle$ and the orthonormality of the $\{|n\rangle\}$ gives

$$\sum_i p_i |\psi_i\rangle\langle\psi_i| = \sum_{\alpha n} p_\alpha |\phi_\alpha\rangle\langle\phi_\alpha|n\rangle\langle n|$$

$$= \sum_\alpha p_\alpha |\phi_\alpha\rangle\langle\phi_\alpha| \ .$$

This establishes the following.

THEOREM 2.1

Let E_ψ and E_ϕ be two ensembles with probabilities and states $\{p_i, |\psi_i\rangle\}$ and $\{p_\alpha, |\phi_\alpha\rangle\}$, respectively. Here $i = 1, \ldots, L_\psi$; $\alpha = 1, \ldots, L_\phi$; and L_ψ and L_ϕ need not be equal. E_ψ and E_ϕ define the same density operator ρ if and only if

$$|\psi_i\rangle = \sum_\alpha U_{i\alpha} |\phi_\alpha\rangle \ ,$$

where $U_{i\alpha} = \sum_n a_{in} d_{n\alpha}$, $a_{in} = \langle n|\psi_i\rangle$, $d_{n\alpha} = (p_\alpha/p_n)\langle\phi_\alpha|n\rangle$, and the $\{|n\rangle\}$ are an orthonormal basis for a subspace S_n of dimension $R_\rho = Rank(\rho)$.

(2) Non-uniqueness of Operation-Sum Representations: It is now possible to state necessary and sufficient conditions for when two operator-sum representations describe the same quantum operation ξ.

THEOREM 2.2

Let $A = \{ A_i : i = 1, \ldots, L_A\}$ and $B = \{ B_j : j = 1, \ldots, L_B\}$ be two sets of operation elements. The operator-sum representations associated with A and

B describe the same quantum operation ξ if and only if

$$A_i = \sum_j \tilde{U}_{ij} B_j ,$$

where $\tilde{U}^\dagger = \tilde{U}^{-1_l}$ is the left-inverse of $\tilde{U}$, and $\tilde{U}$ is unitary when $L_A = L_B$.

PROOF ($\Longrightarrow$) By assumption,

$$\sum_i A_i \rho A_i^\dagger = \xi(\rho) = \sum_j B_j \rho B_j^\dagger , \qquad (2.28)$$

for all ρ. We treat the case of a pure state first, then extend the analysis to cover mixed states.

Let $|\psi\rangle$ be an arbitrary state in $\mathcal{H}_Q$ and $\rho = |\psi\rangle\langle\psi|$ the input density operator in eq. (2.28). Define the states

$$|\tilde{\psi}_i\rangle = A_i |\psi\rangle \qquad (i = 1, \ldots, L_A) \qquad (2.29)$$

and

$$|\tilde{\phi}_j\rangle = B_j |\psi\rangle \qquad (j = 1, \ldots, L_B) . \qquad (2.30)$$

Notice that

$$\langle \tilde{\psi}_i | \tilde{\psi}_i \rangle = tr_Q \left(A_i \rho A_i^\dagger \right) \qquad (2.31)$$

$$\langle \tilde{\phi}_j | \tilde{\phi}_j \rangle = tr_Q \left(B_j \rho B_j^\dagger \right) . \qquad (2.32)$$

Previously, it was shown that the probability that ξ puts Q in the state $A_i \rho A_i^\dagger / tr_Q(A_i \rho A_i^\dagger)$ is $p_i = tr_Q(A_i \rho A_i^\dagger)$. A parallel remark applies to the probability $p_j = tr_Q(B_j \rho B_j^\dagger)$. Thus $\langle \tilde{\psi}_i | \tilde{\psi}_i \rangle = p_i$ and $\langle \tilde{\phi}_j | \tilde{\phi}_j \rangle = p_j$, and the states $\{|\tilde{\psi}_i\rangle\}$ and $\{|\tilde{\phi}_j\rangle\}$ are non-normalized states that produce the same density operator $\xi(\rho)$. Using the results of Problem 2.1, it follows that

$$|\tilde{\psi}_i\rangle = \sum_j \tilde{U}_{ij} |\tilde{\phi}_j\rangle ,$$

or

$$A_i |\psi\rangle = \sum_j \tilde{U}_{ij} B_j |\psi\rangle . \qquad (2.33)$$

Since $|\psi\rangle$ is arbitrary, eq. (2.33) implies

$$A_i = \sum_j \tilde{U}_{ij} B_j . \qquad (2.34)$$

It is important to note that $\tilde{U}$ is independent of the initial state $|\psi\rangle$. Problem 2.1 also established that $\tilde{U}^\dagger = \tilde{U}^{-1_l}$, and that $\tilde{U}$ is unitary when $L_A = L_B$.

For the mixed state case, we note that ξ is a linear superoperator so that for $\rho = \sum_k p_k |k\rangle\langle k|$,

$$\xi\left(\sum_k p_k |k\rangle\langle k|\right) = \sum_k p_k \xi(|k\rangle\langle k|) \ .$$

Thus the full ensemble E of N identical quantum systems breaks up into a mixture of subensembles E_k containing $N_k = p_k N$ quantum systems in the state $\xi(|k\rangle\langle k|)$. The pure state analysis just presented is applicable to each subensemble, where the state before ξ acts is the pure state $|k\rangle$. Since the operator $\tilde{U}$ appearing in eq. (2.34) is independent of the initial pure state $|\psi\rangle$, analysis of each subensemble leads to eq. (2.34) with the *same* operator $\tilde{U}$ since each subensemble only differs in the initial pure state, and $\tilde{U}$ is not sensitive to this difference. Thus we find that

$$A_i = \sum_j \tilde{U}_{ij} B_j$$

for each subensemble E_k, and hence for the full ensemble E. The pure state analysis also establishes that for each E_k: (1) $\tilde{U}^\dagger = \tilde{U}^{-1_l}$ and (2) $\tilde{U}$ is unitary when $L_A = L_B$. Being true for each subensemble, these properties will be true for the full ensemble as well.

($\Longleftarrow$) Here we are given that

$$A_i = \sum_j \tilde{U}_{ij} B_j \ ,$$

where $\tilde{U}^\dagger = \tilde{U}^{-1_l}$. A straightforward calculation establishes the desired result:

$$\begin{aligned}
\sum_i A_i \rho A_i^\dagger &= \sum_i \sum_{j,j'} \tilde{U}_{ij} B_j \rho \tilde{U}^*_{ij'} B_{j'}^\dagger \\
&= \sum_{j,j'} \left[\sum_i \tilde{U}^\dagger_{j'i} \tilde{U}_{ij}\right] B_j \rho B_{j'}^\dagger \\
&= \sum_{j,j'} \left[\sum_i \tilde{U}^{-1_l}_{j'i} \tilde{U}_{ij}\right] B_j \rho B_{j'}^\dagger \\
&= \sum_{j,j'} \delta_{j'j} B_j \rho B_{j'}^\dagger \\
&= \sum_j B_j \rho B_j^\dagger \ .
\end{aligned}$$

2.1.2 Depolarizing Channel

In Section 1.4.1 we introduced the error model that will be used throughout this book. Here we represent it as a quantum operation and show that it is physically equivalent to the depolarizing channel [7]. This channel, with probability $1-p$, passes a qubit without altering its state, while with probability p, leaves it in the state $\rho_f = I/2$ that maximizes the von-Neumann entropy $S(\rho) = -Tr\rho\log\rho = 1$.

For ease of reference we restate the properties that define the model: (1) errors on different qubits occur independently; (2) single-qubit errors σ_x, σ_y, and σ_z are equally likely; and (3) all qubits have the same single-error probability ϵ. The essential error process is thus the occurrence/non-occurrence of a single-qubit error. For this model, the probability that t errors occur is ϵ^t. For later convenience we write $\epsilon = p/4$.

The total probability that a qubit experiences a single-qubit error σ_x, σ_y, or σ_z is $3p/4$, and the probability that no error occurs is $1 - 3p/4$. The operation elements for our model are then:

$$E_0 = \sqrt{1-3p/4}\, I \;\; ; \;\; E_1 = \sqrt{p/4}\,\sigma_x \;\; ; \;\; E_2 = \sqrt{p/4}\,\sigma_y \;\; ; \;\; E_3 = \sqrt{p/4}\,\sigma_z \;.$$

The resulting quantum operation is easily shown to be trace-preserving and its operator-sum representation is

$$\xi(\rho) = \left(1 - \frac{3p}{4}\right)\rho + \frac{p}{4}\left(\sigma_x\rho\sigma_x + \sigma_y\rho\sigma_y + \sigma_z\rho\sigma_z\right) \;. \tag{2.35}$$

To make the connection with the depolarizing channel we need the following identity:

$$2I = \rho + \sigma_x\rho\sigma_x + \sigma_y\rho\sigma_y + \sigma_z\rho\sigma_z \;. \tag{2.36}$$

To prove this identity, note that the RHS of eq. (2.36) is a 2×2 matrix and so can be expanded in the complete set of 2×2 matrices:

$$e_0 = I \;\; ; \;\; e_1 = \sigma_x \;\; ; \;\; e_2 = \sigma_y \;\; ; \;\; e_3 = \sigma_z \;.$$

We can thus write

$$\rho + \sigma_x\rho\sigma_x + \sigma_y\rho\sigma_y + \sigma_z\rho\sigma_z = \sum_{i=0}^{3} a_i\, e_i \;. \tag{2.37}$$

Using that $tr(e_i e_j) = 2\delta_{ij}$, it is straightforward to show that $a_0 = 2$ and $a_i = 0$ for $i = 1, 2, 3$. We derive the result for a_0 and leave evaluation of the remaining a_i to the following exercise. Taking the trace of eq. (2.37) gives

$$tr\left(\rho + \sigma_x\rho\sigma_x + \sigma_y\rho\sigma_y + \sigma_z\rho\sigma_z\right) = a_0\, tr I \;.$$

Quantum Error Correcting Codes

Interchanging the two sides of this equation, using $trI = 2$ and $tr AB = tr BA$, gives

$$2a_0 = tr\left(\rho + \sigma_x^2\rho + \sigma_y^2\rho + \sigma_z^2\rho\right)$$
$$= tr(4\rho)$$
$$= 4 ,$$

which yields $a_0 = 2$ as promised.

Exercise 2.3

(a) Use the following well-known property of the Pauli matrices $\sigma_j\sigma_k = \delta_{jk}I + i\sum_l \epsilon_{jkl}\sigma_l$ ($j,k,l = x,y,z$ and ϵ_{jkl} is the Levi-Civita completely antisymmetric tensor), and that the Pauli matrices are traceless, to show that $tr\,\sigma_j\sigma_k = 2\delta_{jk}$ and that $\sigma_k\sigma_j\sigma_k = (2\delta_{jk} - 1)\sigma_j$.

(b) Use the result of (a) and that $tr\rho = 1$ to show that $a_1 = a_2 = a_3 = 0$ in eq. (2.37), thus completing the proof of eq. (2.36).

Using eq. (2.36) in eq. (2.35) gives

$$\xi(\rho) = \left(1 - \frac{3p}{4}\right)\rho + \frac{p}{4}(2I - \rho)$$
$$= (1-p)\rho + p\left(\frac{I}{2}\right) , \quad (2.38)$$

which is the quantum operation for the depolarizing channel.

Exercise 2.4 Use the Bloch/polarization vector representation for the density operator ρ (eq.(1.68)) to show that the polarization vector $\mathbf{R}$ vanishes when $\rho = I/2$. Show that the von-Neumann entropy $S(\rho) = -Tr\,\rho\log_2\rho$ takes its maximum value of 1 when $\rho = I/2$.

2.1.3 Other Error Models

Before leaving the subject of quantum operations we point out some ways in which our error model can be modified.

First, the errors σ_x, σ_y, and σ_z need not be equally likely. This is the case for the amplitude damping channel. The operation elements for this channel are

$$E_0 = \begin{pmatrix} 1 & 0 \\ 0 & \sqrt{1-\epsilon^2} \end{pmatrix} , \quad E_1 = \begin{pmatrix} 0 & \epsilon \\ 0 & 0 \end{pmatrix},$$

which act on a single qubit. If $|\psi\rangle = a|0\rangle + b|1\rangle$ is the initial (normalized) qubit state ($\rho = |\psi\rangle\langle\psi|$), then the effect of the amplitude damping channel is to map $\rho \to \xi(\rho)$ with

$$\xi(\rho) = \begin{pmatrix} |a|^2 + \epsilon^2|b|^2 & ab^*\sqrt{1-\epsilon^2} \\ a^*b\sqrt{1-\epsilon^2} & |b|^2(1-\epsilon^2) \end{pmatrix} . \quad (2.39)$$

The probabilities $P(0)$ and $P(1)$ that E_0 and E_1 occur, respectively, are

$$P(0) = \text{tr}\left(E_0 \rho E_0^\dagger\right) = 1 - \epsilon^2 |b|^2$$
$$P(1) = \text{tr}\left(E_1 \rho E_1^\dagger\right) = \epsilon^2 |b|^2 \quad . \tag{2.40}$$

The Bloch vectors $\mathbf{R}_\rho$ and $\mathbf{R}_\xi$ for ρ and $\xi(\rho)$, respectively, are

$$R_{\rho,x} = 2\,\text{Re}\,(a^*b)$$
$$R_{\rho,y} = 2\,\text{Im}\,(a^*b)$$
$$R_{\rho,z} = |a|^2 - |b|^2 \tag{2.41}$$

and

$$R_{\xi,x} = R_{\rho,x}\sqrt{1-\epsilon^2}$$
$$R_{\xi,y} = R_{\rho,y}\sqrt{1-\epsilon^2}$$
$$R_{\xi,z} = R_{\rho,z}\left(1-\epsilon^2\right) + \epsilon^2 \quad . \tag{2.42}$$

It follows from eq. (2.39) that when $\rho = |1\rangle\langle 1|$, and so $\mathbf{R}_\rho = -\hat{\mathbf{z}}$, the amplitude damping channel causes the transition $|1\rangle \to |0\rangle$ to occur with probability ϵ^2. In this case the channel causes the error $\sigma_x + i\sigma_y$ to occur. The final Bloch vector is $\mathbf{R}_\xi = \left(-1 + 2\epsilon^2\right)\hat{\mathbf{z}}$. Its reduced length relative to $\mathbf{R}_\rho$ is due to the non-vanishing probability for occupation of the $|0\rangle$-state after the quantum operation has completed. For $\rho = |0\rangle\langle 0|$, $\xi(\rho) = |0\rangle\langle 0|$ and the channel causes no error. Spontaneous emission is an example of a physical process that can be modeled as an amplitude damping channel [2].

The error model we will use also assumes that errors occurring on different qubits are uncorrelated. This is another convenient assumption that need not be true for a particular environment. However, if the error probability for correlated errors drops off sufficiently rapidly with the number of errors k so that correlated errors with more than, say, t errors are unlikely, one can protect the computational data using a quantum error correcting code that fixes up to t errors. Such a quantum code will be adequate to diagnose and remove the most probable correlated errors for which $k \leq t$. A decoding error will only occur when $k > t$, and ideally t is chosen so this occurs with a suitably small probability.

Finally, we have been tacitly assuming that the state of a qubit will always lie within a particular two-dimensional Hilbert space and that errors due to entanglement with the environment or to imperfectly applied quantum gates leave the qubit state in this space. Errors that knock the qubit state out of this two-dimensional Hilbert space are known as leakage errors. For example, a qubit might be stored in the two-dimensional subspace spanned by the ground state ($|0\rangle$) of an ion and one of its metastable excited states ($|1\rangle$). An error that moves the state vector out of this subspace would be an example of a leakage error. This type of error can also be handle using the quantum error correction protocols described in this book [8,9]. A measurement that determines whether the ion state lies in the computational subspace or not can be used to determine whether a leakage error has occured. If so, the leaked qubit can be discarded and replaced with a new qubit prepared in some standard state (say $|0\rangle$). At that point, the qubit state has been returned to the computational subspace, although most likely, it will contain an error. The usual quantum error correction protocol can then be applied to diagnose the error and remove it.

2.2 Quantum Error Correcting Codes: Definitions

Recall that in Section 1.4.2 we defined a quantum error correcting code that encodes k qubits into n qubits as a 2^k-dimensional subspace $\mathcal{C}_q$ of the n-qubit Hilbert space H_2^n, together with a recovery operation R. The subspace $\mathcal{C}_q$ is called the code space; the states belonging to $\mathcal{C}_q$, the codewords; and the encoded computational basis states (eq. (1.76)), the basis codewords. The recovery operation $R = \{R_r\}$ is relative to a set of correctable errors $E = \{E_a\}$. It is important to note that a linear combination of correctable errors is also a correctable error. This follows from Theorem 1.12, which showed (*inter alia*) that the maximal set of correctable errors $E(\mathcal{C}_q, R)$ is linearly closed. Thus we can focus our attention on correcting a linearly independent set of errors. To generate such a set, we recall that the errors I, σ_x, σ_y, and σ_z form a basis for single-qubit errors, and that a basis for n-qubit errors is produced from them by forming all possible n-fold direct products:

$$e(j_1,\ldots,j_n) = \sigma_{j_1}^1 \otimes \cdots \otimes \sigma_{j_n}^n \ . \tag{2.43}$$

Here superscripts on the RHS label the qubits $1,\ldots,n$; $j_i = 0, x, y, z$, and σ_0^i is the identity operator I^i for the i-th qubit. This n-qubit error-basis can be transformed into a multiplicative group known as the Pauli group $\mathcal{G}_n$ if we allow the elements $e(j_1,\ldots,j_n)$ to be multiplied by -1 and $\pm i$. This is necessary since products such as $\sigma_x \sigma_y = i\sigma_z$ generate factors of i and closure under multiplication requires $i\sigma_z \in \mathcal{G}_n$. The weight of an error operator $e(j_1,\ldots,j_n) \in \mathcal{G}_n$ is defined to be the number of qubits on which $\sigma_{j_i}^i \neq I^i$. Thus, for example, the error operator $I^1 \otimes \sigma_x^2 \otimes \sigma_z^3 \otimes I^4$ has weight 2.

Theorem 1.13 established necessary and sufficient conditions for a quantum code to correct a set of errors $E = \{E_a\}$. Denoting the basis codewords by $|\bar{i}\rangle$, these conditions were shown to be

$$\langle \bar{i}|E_a^\dagger E_b|\bar{i}\rangle = \langle \bar{j}|E_a^\dagger E_b|\bar{j}\rangle$$

and

$$\langle \bar{i}|E_a^\dagger E_b|\bar{j}\rangle = 0 \ ,$$

for all basis codewords $|\bar{i}\rangle$, $|\bar{j}\rangle$, and for all errors $E_a, E_b \in E$. These conditions can be consolidated into the single matrix equation:

$$\langle\bar{i}|E_a^\dagger E_b|\bar{j}\rangle = C_{ab}\,\delta_{ij} \ . \tag{2.44}$$

It is a simple matter to show that the matrix elements C_{ab} satisfy $C_{ab} = C_{ba}^*$ so that the square matrix C is Hermitian. It can thus be diagonalized and its rank will be given by the number of its non-zero eigenvalues c_n.

Exercise 2.5 *Show that C is an Hermitian matrix.*

If U is the operator that diagonalizes C; c_n its eigenvalues; and we define the set of operators $\{F_l\}$ via

$$F_l = \sum_a U_{la} E_a \ ,$$

then eq. (2.44) can be transformed into

$$\langle\bar{i}|F_l^\dagger F_m|\bar{j}\rangle = c_l\,\delta_{lm}\,\delta_{ij} \ . \tag{2.45}$$

By Theorem 1.12, the $\{F_l\}$ are correctable errors; however if $c_l = 0$, it follows from eq. (2.45) that

$$\langle\bar{i}|F_l^\dagger F_l|\bar{i}\rangle = 0 \ ,$$

and so

$$F_l|\bar{i}\rangle = 0 \ .$$

Thus an error operator F_l corresponding to a zero eigenvalue of C annihilates all basis codewords, and hence all codewords. Such errors can be ignored since they can never be present in a non-zero quantum state and thus can never appear as an observable error. Note that when C has vanishing eigenvalues, it is a singular matrix since its determinant will be zero. A quantum error correcting code for which C is singular is said to be a degenerate code, and codes with non-singular C are called non-degenerate codes. It is clear from eq. (2.44) that C is determined by the set of errors $\{E_a\}$ that one wants to correct. Hence the question of whether a quantum code is degenerate or not depends on what errors are to be corrected.

The condition for quantum error correction, eq. (2.44), allows us to define the distance of a quantum error correcting code. Writing $E = E_a^\dagger E_b$, eq. (2.44) becomes

$$\langle\bar{i}|E|\bar{j}\rangle = C_E\,\delta_{ij} \ , \tag{2.46}$$

where all the dependence on the basis codewords on the RHS resides in the Kronecker delta δ_{ij}. A quantum error correcting code has distance d if all errors of weight less than d satisfy eq. (2.46), and at least one error of weight d exists that violates it. The notation $[n, k, d]$ is usually used to denote a quantum code that maps k qubits into n qubits and which has distance d. This notation is identical with that used for classical codes; however this should not cause confusion as the context of the discussion or explicit statements will allow the reader to identify which class of code is being considered.

Suppose we want a quantum code that corrects up to t arbitrary errors. This requires that all error operators E_a, E_b with weight less than or equal to t satisfy eq. (2.44). Thus for correctable errors, the maximum weight of the product $E = E_a^\dagger E_b$ will be $2t$. The smallest weight possible for an error operator E that violates eq. (2.46) is then $2t+1$, and our quantum code must have distance $d = 2t+1$. Although this is the same condition on the distance as found for a classical code, the underlying arguments for the two cases are (not surprisingly) completely different. Next, suppose we want a quantum code that can *detect* (but not correct) up to s errors. In this case, all that is required is the ability to distinguish errors E_b of weight less than or equal to s from the identity operator $E_a = I$. Here $E = E_a^\dagger E_b = E_b$ in eq. (2.46), and by assumption, $s + 1$ is the smallest possible weight for an error E that violates this condition. Thus the quantum code appropriate for this situation must have distance $d = s + 1$. This second example gives us an alternative way to think about the distance of a quantum error correcting code: it is the weight of the smallest error E that *cannot* be detected by the code. Finally, suppose we want a quantum code that can correct errors that occur on, at most, r known qubits. If E_a and E_b are two such errors, then the weight of the product $E = E_a^\dagger E_b$ will be less than or equal to r and so the smallest weight for an error that violates eq. (2.46) is $r + 1$. The distance for the appropriate quantum code must then be $d = r + 1$.

2.3 Example: Calderbank-Shor-Steane [7, 1, 3] Code

Calderbank and Shor [10] and Steane [11] discovered an important class of quantum error correcting codes whose structure makes them convenient for use in fault-tolerant quantum computing. The Calderbank-Shor-Steane (CSS) codes are constructed from two classical binary codes C and C' that have the following properties:

1. C and C' are $[n, k, d]$ and $[n, k', d']$ codes, respectively;
2. $C' \subset C$;
3. C and $C'_\perp$ are both t-error correcting codes.

The code construction first partitions C into the cosets of C':

$$C = C' \cup (c_1 + C') \cdots \cup (c_N + C') \quad,$$

where $c_1, \ldots, c_N \in C$, and N is the number of cosets. From Lagrange's theorem (Appendix A), $N = 2^k/2^{k'}$. Next, $N = 2^{k-k'}$ basis codewords are defined by identifying each one with a coset: $|\bar{c}_i\rangle \Leftrightarrow c_i + C'$. Specifically,

$$|\bar{c}_i\rangle = \frac{1}{\sqrt{2^{k'}}} \sum_{c' \in C'} |c_i + c'\rangle \qquad (i = 1, \ldots, N) \ . \qquad (2.47)$$

Note that if e and f are codewords in C that belong to the same coset of C', they define the same basis codeword: $|\bar{e}\rangle = |\bar{f}\rangle$. This follows since, if e and f belong to the same coset, there exists an $h \in C'$ such that $e = f + h$, and from eq. (2.47):

$$|\bar{e}\rangle = \frac{1}{\sqrt{2^{k'}}} \sum_{c' \in C'} |e + c'\rangle$$

$$= \frac{1}{\sqrt{2^{k'}}} \sum_{c' \in C'} |f + h + c'\rangle \ . \qquad (2.48)$$

Since C' is a subspace of C, it is closed under addition and so $h' = h + c' \in C'$. Note that as c' ranges over C', so does h'. Thus eq. (2.48) can be rewritten as:

$$|\bar{e}\rangle = \frac{1}{\sqrt{2^{k'}}} \sum_{h' \in C'} |f + h'\rangle$$

$$= |\bar{f}\rangle \ .$$

The code space $\mathcal{C}_q$ is the subspace of H_2^n spanned by the $|\bar{c}_i\rangle$ and so is $N = 2^{k-k'}$ dimensional. Thus the CSS construction causes $k - k'$ qubits to be encoded in n qubits. We will see in Chapter 3 that a CSS code will correct as many errors as do C and $C'_\perp$.

To give an example of this construction, we work out the CSS code that maps 1 qubit into 7. We show that it has distance $d = 3$ and so corrects arbitrary 1-qubit errors, and is non-degenerate. Here the two classical codes are the [7, 4, 3] binary Hamming code (C) and the [7, 3, 4] binary simplex code (C'). As shown in Problem 1.6, $C'_\perp$ is the [7, 4, 3] binary Hamming code. Thus C and $C'_\perp$ are both 1-error correcting codes. In this case, $n = 7$, $k = 4$, $k' = 3$, and $k - k' = 1$ so that 1 qubit is mapped into 7. From Problem 1.7 we have the basis codewords

$$|\bar{0}\rangle = \frac{1}{\sqrt{2^3}} \left[\begin{array}{c} |0000000\rangle + |0110011\rangle + |1010101\rangle + |1100110\rangle \\ +|0001111\rangle + |0111100\rangle + |1011010\rangle + |1101001\rangle \end{array} \right]$$

$$|\bar{1}\rangle = \frac{1}{\sqrt{2^3}} \left[\begin{array}{c} |1111111\rangle + |1001100\rangle + |0101010\rangle + |0011001\rangle \\ +|1110000\rangle + |1000011\rangle + |0100101\rangle + |0010110\rangle \end{array} \right] , \quad (2.49)$$

and the generators

$$\begin{aligned} g_1 &= \sigma_x^4 \sigma_x^5 \sigma_x^6 \sigma_x^7 & ; \quad g_4 &= \sigma_z^4 \sigma_z^5 \sigma_z^6 \sigma_z^7 \\ g_2 &= \sigma_x^2 \sigma_x^3 \sigma_x^6 \sigma_x^7 & ; \quad g_5 &= \sigma_z^2 \sigma_z^3 \sigma_z^6 \sigma_z^7 \\ g_3 &= \sigma_x^1 \sigma_x^3 \sigma_x^5 \sigma_x^7 & ; \quad g_6 &= \sigma_z^1 \sigma_z^3 \sigma_z^5 \sigma_z^7 \ . \end{aligned} \qquad (2.50)$$

Quantum Error Correcting Codes

Direct calculation shows that the generators $g_1, \ldots, g_6$ fix the basis codewords $|\bar{0}\rangle$ and $|\bar{1}\rangle$. By definition, the distance d of this code is the weight of the smallest error E that violates eq. (2.46), which we rewrite here for easy reference:

$$\langle \bar{i}|E|\bar{j}\rangle = C_E\, \delta_{ij} . \tag{2.51}$$

It is straightforward, though tedious, to check that all 1- and 2-qubit errors satisfy this equation, while the 3-qubit error $\sigma_x^1\sigma_x^2\sigma_x^3$ does not. In the latter case, direct calculation gives

$$\langle \bar{0}|\sigma_x^1\sigma_x^2\sigma_x^3|\bar{1}\rangle = 1 ,$$

though eq. (2.51) requires this matrix element to vanish. Thus the distance $d = 3$ and the $[7,1,3]$ CSS code corrects arbitrary 1-qubit errors. From Problem 2.4, this code will be degenerate if the stabilizer S has an element whose weight is less than 3 (excluding the identity element). Direct calculation using eq. (2.50) shows that the elements of S have weights that are greater than or equal to 4. Thus the $[7,1,3]$ CSS code is non-degenerate. Finally, note that the operators $X = \sigma_x^1\sigma_x^2\sigma_x^3$ and $Z = \sigma_z^1\sigma_z^2\sigma_z^3$ (i) act as the encoded σ_x and σ_z operators on the basis codewords, respectively; (ii) are mutually anticommuting: $\{X, Z\} = 0$; and (iii) commute with the generators of S, and hence with all of S.

Exercise 2.6

(a) Show that for $\bar{i} = 0, 1$,

$$X|\bar{i}\rangle = |\bar{i} \oplus 1\rangle$$
$$Z|\bar{i}\rangle = (-1)^{\bar{i}}|\bar{i}\rangle ,$$

where $\oplus$ is addition modulo 2.

(b) Show that $\{X, Z\} = 0$.

(c) Show that X and Z commute with the generators of S.

Problems

2.1 Let us reconsider the question of two ensembles described by the same density matrix (Section 2.1.1). Using the notation introduced there, consider the non-normalized states $|\tilde{\psi}_i\rangle = \sqrt{p_i}\,|\psi_i\rangle$, $|\tilde{\phi}_\alpha\rangle = \sqrt{p_\alpha}\,|\phi_\alpha\rangle$, and $|\tilde{n}\rangle = \sqrt{p_n}\,|n\rangle$, where $p_i, p_\alpha, p_n \neq 0$.

(a) Repeating the arguments given in Section 2.1.1, show that

$$|\tilde{\psi}_i\rangle = \sum_n \tilde{a}_{in} |\tilde{n}\rangle$$

$$|\tilde{\phi}_\alpha\rangle = \sum_n \tilde{b}_{\alpha n} |\tilde{n}\rangle \ ,$$

with $\tilde{a}_{in} = \langle \tilde{n}|\tilde{\psi}_i\rangle/p_n$ and $\tilde{b}_{\alpha n} = \langle \tilde{n}|\tilde{\phi}_\alpha\rangle/p_n$.

(b) Starting from the eigenvalue problem for ρ, derive the following two expressions for $|\tilde{n}\rangle$:

$$|\tilde{n}\rangle = \sum_i \tilde{c}_{ni} |\tilde{\psi}_i\rangle$$

$$|\tilde{n}\rangle = \sum_\alpha \tilde{d}_{n\alpha} |\tilde{\phi}_\alpha\rangle \ ,$$

where $\tilde{c}_{ni} = \left(\langle \tilde{\psi}_i|\tilde{n}\rangle\right)/p_n$ and $\tilde{d}_{n\alpha} = \left(\langle \tilde{\phi}_\alpha|\tilde{n}\rangle\right)/p_n$. Show that $\tilde{c} = \tilde{a}^\dagger$ and $\tilde{d} = \tilde{b}^\dagger$.

(c) Show that

$$|\tilde{\psi}_i\rangle = \sum_\alpha \tilde{U}_{i\alpha} |\tilde{\phi}_\alpha\rangle$$

$$|\tilde{\phi}_\alpha\rangle = \sum_i \tilde{V}_{\alpha i} |\tilde{\psi}_i\rangle \ ,$$

where $\tilde{U} = \tilde{a}\tilde{d}$, $\tilde{V} = \tilde{b}\tilde{c}$, $\tilde{a} = \tilde{c}^{-1_l}$, and $\tilde{b} = \tilde{d}^{-1_l}$. Show that $\tilde{U}^\dagger = \tilde{V} = \tilde{U}^{-1_l}$.

(d) Argue that if $L_\psi = L_\phi = R_\rho$, then a unique two-sided inverse exists, and the operators $\tilde{a}, \tilde{b}, \tilde{c}$, and $\tilde{d}$ are now unitary. Use this to show that, in this case, $\tilde{U}$ is unitary.

The other half of Theorem 2.1 now follows easily. We leave it to the reader to work out the details.

2.2 Consider a qubit, initially in the normalized state $|\psi\rangle = a|0\rangle + b|1\rangle$, that interacts with an amplitude damping channel.

(a) Show that $\xi(\rho)$ is given by eq. (2.39).

(b) Verify that the probabilities $(P(0), P(1))$ and the Bloch vectors $\mathbf{R}_\rho$ and $\mathbf{R}_\xi$ discussed in Section 2.1.3 are given by eqs. (2.40), (2.41), and (2.42), respectively.

(c) Let $\rho = (1/2)[\,|0\rangle + |1\rangle\,][\,\langle 0| + \langle 1|\,]$, corresponding to Bloch vector $\mathbf{R}_\rho = \hat{\mathbf{x}}$. Show that $P(0) = 1 - \epsilon^2/2$, $P(1) = \epsilon^2/2$, and

$$\mathbf{R}_\xi = \sqrt{1 - \epsilon^2}\,\hat{\mathbf{x}} + \epsilon^2 \hat{\mathbf{z}} \ .$$

Note that the action of the channel has generated a z-component in $\mathbf{R}_f$ due to a shift of the occupation probabiltites for the states $|0\rangle$ and $|1\rangle$.

Quantum Error Correcting Codes

2.3 Consider a qubit in the initial state $|\psi\rangle = a|0\rangle + b|1\rangle$ that interacts with a generalized amplitude damping channel. The operation elements for this channel are

$$E_0 = \sqrt{p}\begin{pmatrix} 1 & 0 \\ 0 & \sqrt{1-\gamma} \end{pmatrix} \quad ; \quad E_1 = \sqrt{p}\begin{pmatrix} 0 & \sqrt{\gamma} \\ 0 & 0 \end{pmatrix}$$

$$E_2 = \sqrt{1-p}\begin{pmatrix} \sqrt{1-\gamma} & 0 \\ 0 & 1 \end{pmatrix} \quad ; \quad E_3 = \sqrt{1-p}\begin{pmatrix} 0 & 0 \\ \sqrt{\gamma} & 0 \end{pmatrix}.$$

This channel generalizes the amplitude damping channel in that it allows transitions from $|0\rangle \to |1\rangle$ as well as from $|1\rangle \to |0\rangle$. The Bloch vector $\mathbf{R}_\rho$ for the initial state $\rho = |\psi\rangle\langle\psi|$ is given by eq. (2.41).

(a) Show that the quantum operation associated with this channel is trace-preserving.

(b) Evaluate $\xi(\rho)$ for this channel and show that the final Bloch vector $\mathbf{R}_\xi$ is given by

$$\mathbf{R}_\xi = R_{\rho,x}\sqrt{1-\gamma}\,\hat{\mathbf{x}} + R_{\rho,y}\sqrt{1-\gamma}\,\hat{\mathbf{y}} + [R_{\rho,z}(1-\gamma) + \gamma(2p-1)]\,\hat{\mathbf{z}}\ .$$

(c) Show that the probabilities $P(i)$ that processes E_i occur ($i = 0,\ldots,3$) are

$$P(0) = p\left[1 - \frac{\gamma}{2}(1-R_{\rho,z})\right] \quad ; \quad P(1) = \frac{\gamma p}{2}(1-R_{\rho,z})$$

$$P(2) = (1-p)\left[1 - \frac{\gamma}{2}(1+R_{\rho,z})\right] \quad ; \quad P(3) = \frac{\gamma(1-p)}{2}(1+R_{\rho,z}).$$

(d) Show that the density matrix $\rho_* = (1/2)(I + \mathbf{R}_* \cdot \boldsymbol{\sigma})$ with $\mathbf{R}_* = (2p-1)\hat{\mathbf{z}}$ is a fixed point of this channel: $\xi(\rho_*) = \rho_*$.

2.4 We shall see in Chapter 3 that all elements of the Pauli group are unitary operators. For now, simply assume that this is true. Let (C_q, R) be a quantum error correcting code with distance d. Prove that (C_q, R) is a degenerate code if and only if its stabilizer S has an element with weight less than d (excluding the identity element).

2.5 Here we introduce Shor's [9, 1, 3] quantum code [12], which gave birth to the subject of quantum error correcting codes. The basis codewords are

$$|\bar{0}\rangle = \frac{1}{\sqrt{2^3}}[|000\rangle + |111\rangle][|000\rangle + |111\rangle][|000\rangle + |111\rangle]$$

$$|\bar{1}\rangle = \frac{1}{\sqrt{2^3}}[|000\rangle - |111\rangle][|000\rangle - |111\rangle][|000\rangle - |111\rangle]\ .$$

(a) Here $n - k = 8$ so that the stabilizer group S has eight generators. In Chapter 3 we will see that Shor's code is an example of a concatenated code and will show how to obtain its generators. Here we simply write them down:

$$g_1 = \sigma_z^1 \sigma_z^2$$
$$g_2 = \sigma_z^1 \sigma_z^3$$
$$g_3 = \sigma_z^4 \sigma_z^5$$
$$g_4 = \sigma_z^4 \sigma_z^6$$
$$g_5 = \sigma_z^7 \sigma_z^8$$
$$g_6 = \sigma_z^7 \sigma_z^9$$
$$g_7 = \sigma_x^1 \sigma_x^2 \sigma_x^3 \sigma_x^4 \sigma_x^5 \sigma_x^6$$
$$g_8 = \sigma_x^1 \sigma_x^2 \sigma_x^3 \sigma_x^7 \sigma_x^8 \sigma_x^9$$

Show that these generators fix the basis codewords $|\bar{0}\rangle$ and $|\bar{1}\rangle$.

(b) Show that this code has a distance $d = 3$.

(c) Show that Shor's code is degenerate.

References

[1] Kraus, K., *States, Effects, and Operations: Fundamental Notions of Quantum Theory*, Lecture Notes in Physics, Vol. 190, Springer-Verlag, Berlin, 1983.

[2] Nielsen, M. A. and Chuang, I. L., *Quantum Computation and Quantum Information*, Cambridge University Press, London, 2000.

[3] Ernst, R. R., Bodenhausen, G., and Wokaun, A., *Principles of Nuclear Magnetic Resonance in One and Two Dimensions*, Oxford University Press, Oxford, 1987.

[4] Schumacher, B., Sending entanglement through a noisy quantum channel, *Phys. Rev. A* **54**, 2614, 1996.

[5] Caves, C. C., Quantum error correction and reversible operations, *J. Superconductivity* **12**, 707, 1999.

[6] Noble, B., *Applied Linear Algebra*, Prentice-Hall, Inc., Englewood Cliffs, NJ, 1969, pp. 134–136.

[7] Bennett, C. H., DiVincenzo, D. P., Smolin, J. A., and Wooters, W. K., Mixed-state entanglement and quantum error correction, *Phys. Rev. A* **54**, 3824, 1996.

[8] Preskill, J., Fault-tolerant quantum computation, in *Introduction to Quantum Computation and Information*, H.-K. Lo, S. Popescu, and T. Spiller, Eds., World Scientific, Singapore, 1998.

[9] Plenio, M. B. and Knight, P. L., *Proc. R. Soc. Lond. A* **453**, 2017, 1997.

[10] Calderbank, A. R. and Shor, P. W., Good quantum error correcting codes exist, *Phys. Rev. A* **54**, 1098, 1996.

[11] Steane, A. M., Multiple-particle interference and quantum error correction, *Proc. R. Soc. Lond. A* **452**, 2551, 1996.

[12] Shor, P. W., Scheme for reducing decoherence in quantum computer memory, *Phys. Rev. A* **52**, 2493, 1995.

3

Quantum Stabilizer Codes

In this chapter we begin to examine quantum stabilizer codes in detail. Section 3.1 introduces the theoretical framework that underlies this class of quantum error correcting codes, and some examples are given in Section 3.2. Section 3.3 shows how these codes can be described in terms of a binary vector space with symplectic inner product, and finally, concatenated quantum error correcting codes are introduced in Section 3.4. The latter discussion will focus on concatenated codes that are put together using quantum stabilizer codes. We will see in Chapter 6 that concatenated codes play an important role in establishing the accuracy threshold theorem for fault-tolerant quantum computing.

3.1 General Framework

The discussion in this section breaks up into four parts. Section 3.1.1 summarizes key points presented in the first two chapters connected with quantum error correcting codes, with the discussion quickly zeroing in on quantum stabilizer codes. A number of useful properties of the Pauli group are also worked out. Within the context of quantum stabilizer codes, (i) Section 3.1.2 examines the conditions under which an error will be detectable, correctable, or non-detectable by the code; (ii) Section 3.1.3 updates our earlier discussion of the quantum error correction protocol; and finally, (iii) Section 3.1.4 discusses encoded quantum operations.

3.1.1 Summary

Recall (Sections 1.4 and 2.2) that a quantum error correcting code (QECC) that encodes k qubits into n qubits is defined through a map ξ from the k-qubit Hilbert space H_2^k onto a 2^k-dimensional subspace $\mathcal{C}_q$ of the n-qubit Hilbert space H_2^n. The image space $\mathcal{C}_q$ is called the code space; its elements $|c\rangle \in \mathcal{C}_q$ are the codewords; and the encoded computational basis states (eq. (1.76)) are the basis codewords. A QECC also includes a recovery operation R, though its presence will be tacit in this chapter.

Quantum stabilizer codes [1–3] were introduced in Section 1.4. These codes are also known as additive codes [3]. In a quantum stabilizer code, the code space $\mathcal{C}_q$ is identified with the unique subspace of H_2^n that is fixed by the elements of an Abelian group $\mathcal{S}$ known as the stabilizer of $\mathcal{C}_q$. The stabilizer $\mathcal{S}$ is constructed from $n-k$ generators $g_1, \ldots, g_{n-k}$ so that any element $s \in \mathcal{S}$ can be written as a unique product of powers of the generators:

$$s = g_1^{p_1} \cdots g_{n-k}^{p_{n-k}} .$$

The generators were shown to have order 2 (Section 1.4.1), making $\mathcal{S}$ isomorphic to F_2^{n-k}, the vector space of bit strings of length $n-k$. There are 2^{n-k} bit strings in F_2^{n-k}, and so the order of $\mathcal{S}$ is $|\mathcal{S}| = 2^{n-k}$. Note that $\mathcal{S}$ must not include the elements $-I$ or iI if the code space $\mathcal{C}_q$ is to be non-trivial. To see this, note that if $-I \in \mathcal{S}$, it must fix all codewords $|c\rangle$: $-I|c\rangle = |c\rangle$. But clearly, $-I|c\rangle = -|c\rangle$. The null vector $|c\rangle = 0$ is the only solution to both equations. All other elements of $\mathcal{S}$ fix the null vector; and so when $-I \in \mathcal{S}$, the code space $\mathcal{C}_q$ is trivial, containing only the null vector 0. If $iI \in \mathcal{S}$, closure of $\mathcal{S}$ under multiplication demands that $(iI)^2 = -I$ also belong to $\mathcal{S}$, and this again forces $\mathcal{C}_q$ to be trivial.

The Pauli group $\mathcal{G}_n$ of n-qubit errors was introduced in Section 2.2. We saw that any error $e \in \mathcal{G}_n$ can be written as

$$e = i^\lambda \sigma_{j_1}^1 \otimes \cdots \otimes \sigma_{j_n}^n , \qquad (3.1)$$

where (i) $\lambda = 0, 1, 2, 3$; (ii) superscripts on the $\sigma_{j_k}^k$ label the qubits $k = 1, \ldots, n$; (iii) the subscripts take values $j_k = 0, x, y, z$; and (iv) $\sigma_0^k = I^k$ is the identity operation on the kth qubit. Note that since $\sigma_y^k = -i\sigma_x^k \sigma_z^k$, eq. (3.1) can be rewritten as

$$e = i^{\lambda'} \sigma_x(a) \sigma_z(b) , \qquad (3.2)$$

where $a = a_1 \cdots a_n$ and $b = b_1 \cdots b_n$ are bit strings of length n, and

$$\sigma_x(a) \equiv \left(\sigma_x^1\right)^{a_1} \otimes \cdots \otimes \left(\sigma_x^n\right)^{a_n}$$
$$\sigma_z(b) \equiv \left(\sigma_z^1\right)^{b_1} \otimes \cdots \otimes \left(\sigma_z^n\right)^{b_n} . \qquad (3.3)$$

Although the factor $i^{\lambda'}$ in eq. (3.2) is needed to insure that $\mathcal{G}_n$ is a group (Section 2.2), in many discussions it is only necessary to work with the quotient group $\mathcal{G}_n/\mathcal{C}$, where $\mathcal{C} = \{\pm I, \pm iI\}$ is the center of $\mathcal{G}_n$. For such discussions, each coset $e\mathcal{C} = \{\pm e, \pm ie\}$ is treated as a single error e, and one can ignore the factor $i^{\lambda'}$ in eq. (3.2). When this is appropriate, there is a 1-1 correspondence between errors and bit strings of length $2n$. This correspondence will be examined further in Section 3.3. The following theorem establishes a number of useful properties of the Pauli group $\mathcal{G}_n$.

THEOREM 3.1
(1) The orders of $\mathcal{G}_n$ and $\mathcal{G}_n/\mathcal{C}$ are $|\mathcal{G}_n| = 2^{2n+2}$ and $|\mathcal{G}_n/\mathcal{C}| = 2^{2n}$.
(2) For all $e \in \mathcal{G}_n$, (a) $e^2 = \pm I$; (b) $e^\dagger = \pm e$; and (c) $e^{-1} = e^\dagger$.
(3) For all $e, f \in \mathcal{G}_n$, either $[e, f] = 0$ or $\{e, f\} = 0$.

Quantum Stabilizer Codes

PROOF (1) Our starting point is eq. (3.2). Since $\lambda = 0, 1, 2, 3$, $i^{\lambda'}$ can only take four values. As a and b are bit strings of length n, there are 2^n possible assignments for each. The total number of errors in $\mathcal{G}_n$ is thus $4 \times 2^n \times 2^n$ and so $|\mathcal{G}_n| = 2^{2n+2}$. Since quotienting out the center $\mathcal{C}$ of $\mathcal{G}_n$ removes the factors of i from eq. (3.2), it follows that the order of $\mathcal{G}_n/\mathcal{C}$ is given by the number of ways that a and b can be assigned values and so $|\mathcal{G}_n/\mathcal{C}| = 2^{2n}$.

(2) (a) Recall that: σ_x^i and σ_z^j anti-commute when $i = j$ and commute when $i \neq j$; and $\sigma_x^2 = \sigma_z^2 = I$. Then,

$$\begin{aligned} e^2 &= i^{2\lambda} \, \sigma_x(a)\sigma_z(b)\sigma_x(a)\sigma_z(b) \\ &= (-1)^\lambda (-1)^{a \cdot b} [\sigma_x(a)]^2 [\sigma_z(b)]^2 \\ &= (-1)^{\lambda + a \cdot b} \, I \; . \end{aligned} \qquad (3.4)$$

Here the factor $(-1)^{a \cdot b}$ in the second line arises from commuting the right-most $\sigma_x(a)$ in the first line through the left-most $\sigma_z(b)$, and the scalar product $a \cdot b$ has the usual form $a \cdot b = a_1 b_1 + \cdots + a_n b_n$. We also used that $(\sigma_x(a))^2 = (\sigma_z(b))^2 = I$ in going from the second to the third line. Since $\lambda + a \cdot b$ is an integer, $(-1)^{\lambda + a \cdot b} = \pm 1$, establishing that $e^2 = \pm I$.

(b) Since the Pauli matrices are Hermitian, $\sigma_x^\dagger(a) = \sigma_x(a)$ and $\sigma_z^\dagger(b) = \sigma_z(b)$. Thus,

$$\begin{aligned} e^\dagger &= (-i)^\lambda \, \sigma_z^\dagger(b) \, \sigma_x^\dagger(a) \\ &= (-1)^\lambda (-1)^{a \cdot b} \, i^\lambda \sigma_x(a) \, \sigma_z(b) \\ &= \pm e \; . \end{aligned} \qquad (3.5)$$

Thus the elements of $\mathcal{G}_n$ are either Hermitian or anti-Hermitian operators.

(c) From eqs. (3.4) and (3.5) we see that when $\lambda + a \cdot b$ is an even integer, $e^{-1} = e = e^\dagger$; and when it is odd, $e^{-1} = -e = e^\dagger$. Either way, $e^{-1} = e^\dagger$. Thus the elements of $\mathcal{G}_n$ are unitary operators.

(3) Let $e = i^{\lambda_e} \sigma_x(a_e) \sigma_z(b_e)$ and $f = i^{\lambda_f} \sigma_x(a_f) \sigma_z(b_f)$. Then,

$$\begin{aligned} ef &= (i)^{\lambda_e + \lambda_f} \, \sigma_x(a_e)\sigma_z(b_e)\sigma_x(a_f)\sigma_z(b_f) \\ &= (i)^{\lambda_e + \lambda_f} (-1)^{b_e \cdot a_f} \, \sigma_x(a_e)\sigma_x(a_f)\sigma_z(b_e)\sigma_z(b_f) \\ &= (i)^{\lambda_e + \lambda_f} (-1)^{b_e \cdot a_f} (-1)^{a_e \cdot b_f} \, \sigma_x(a_f)\sigma_z(b_f)\sigma_x(a_e)\sigma_z(b_e) \\ &= (-1)^{b_e \cdot a_f + a_e \cdot b_f} \, fe \; . \end{aligned} \qquad (3.6)$$

Since $b_e \cdot a_f + a_e \cdot b_f$ is an integer, it follows that $[e, f] = 0$ whenever this integer is even, and $\{e, f\} = 0$ whenever it is odd. ∎

3.1.2 Errors

The error syndrome was introduced during our preliminary discussion of quantum stabilizer codes (Section 1.4.1). For convenience, its definition is restated here.

DEFINITION 3.1 Let C_q be a quantum stabilizer code with generators $g_1, \ldots, g_{n-k}$ and let e be an error in the Pauli group: $e \in \mathcal{G}_n$. The error syndrome $S(e)$ for e is the bit string $l = l_1 \cdots l_{n-k}$, where the component bits l_i are determined by

$$l_i = \begin{cases} 0 & \text{if } [e, g_i] = 0 \\ 1 & \text{if } \{e, g_i\} = 0 \end{cases} \quad (i = 1, \cdots, n-k) \ . \tag{3.7}$$

The error syndrome allows a number of useful results for C_q to be established.

(1) It follows from Definition 3.1 that any error $e \in \mathcal{G}_n$ that has a non-vanishing error syndrome must anticommute with a subset of the generators of S. For such errors, Theorem 1.10 insures that

$$\langle i|e|j\rangle = 0 \tag{3.8}$$

for all basis codewords $|i\rangle$ and $|j\rangle$. Recall (Section 2.2) that C_q can detect an error e if

$$\langle i|e|j\rangle = C_e \, \delta_{ij} \ . \tag{3.9}$$

Since eq. (3.8) is eq. (3.9) with $C_e = 0$, errors with non-vanishing error syndrome are detectable by C_q.

(2) Let $E = \{E_a\}$ be a set of errors belonging to $\mathcal{G}_n$ for which $S(E_a^\dagger E_b) \neq 0$ for all $E_a, E_b \in E$. Theorem 1.10 again gaurantees that

$$\langle i|E_a^\dagger E_b|j\rangle = 0 \ , \tag{3.10}$$

for all basis codewords $|i\rangle$ and $|j\rangle$. This is eq. (2.44) with $C_{ab} = 0$ so that the set of errors E is correctable by C_q.

(3) Errors having a vanishing syndrome $S(e) = 0$ commute with all generators of the stabilizer S. The centralizer $C(S)$ of S is the set of errors $e \in \mathcal{G}_n$ that commute with all elements of S.

Exercise 3.1 Show that the centralizer $C(S)$ is a subgroup of $\mathcal{G}_n$.

Since the stabilizer S is Abelian, $S \subset C(S)$. Let $e \in C(S)$. Either (i) $e \in S$, or (ii) $e \in C(S) - S$, the complement of S in $C(S)$. In the first case, e is detectable and trivially correctable since it fixes codewords and so requires no correction. Formally, $\langle i|e|j\rangle = \langle i|j\rangle = \delta_{ij}$. This is eq. (3.9) with $C_e = 1$ so that e is detectable. For the second case, $e \in C(S) - S$. Let $s \in S$ and $|c\rangle \in C_q$; then

$$se|c\rangle = es|c\rangle = e|c\rangle \ . \tag{3.11}$$

Quantum Stabilizer Codes

Thus $e|c\rangle$ is fixed by all $s \in \mathcal{S}$ and so is an element of $\mathcal{C}_q$. Since $e \notin \mathcal{S}$, it must non-trivially map at least one codeword $|c'\rangle$: $e|c'\rangle \neq |c'\rangle$. Without loss of generality we can suppose that the basis codewords were chosen so that $|c'\rangle$ is a basis codeword, say $|c'\rangle = |i\rangle$. Then since $e|i\rangle \in \mathcal{C}_q$, it can be expanded in the basis codewords $|k\rangle$:

$$e|i\rangle = \sum_k a_k |k\rangle \ .$$

Since $e|i\rangle \neq |i\rangle$, there must be at least one value of k (say $k_* \neq i$) with $a_{k_*} \neq 0$. Then

$$\langle k_*|e|i\rangle = a_{k_*} \neq 0 \ . \tag{3.12}$$

But eq. (3.9) requires that $\langle k_*|e|i\rangle = 0$ when $k_* \neq i$ if e is to be a detectable error. Thus errors $e \in \mathcal{C}(\mathcal{S}) - \mathcal{S}$ violate eq. (3.9) and so are not detectable by $\mathcal{C}_q$.

(4) Since $\mathcal{C}(\mathcal{S})$ is a subgroup of $\mathcal{G}_n$, its cosets can be used to partition $\mathcal{G}_n$. Because they form a partition, the cosets do not overlap, and each error will belong to a unique coset $e\,\mathcal{C}(\mathcal{S})$ (Appendix A):

$$e\,\mathcal{C}(\mathcal{S}) = \{\, ec : c \in \mathcal{C}(\mathcal{S})\,\} \ . \tag{3.13}$$

THEOREM 3.2
Two errors $e_1, e_2 \in \mathcal{G}_n$ have the same error syndrome $l = l_1 \cdots l_{n-k}$ if and only if they belong to the same coset of $\mathcal{C}(\mathcal{S})$.

PROOF ($\Longrightarrow$) Here $S(e_1) = S(e_2) = l_1 \cdots l_{n-k}$. Then for each generator g_i

$$e_1 e_2 g_i = (-1)^{l_i} \, e_1 g_i e_2 = (-1)^{2l_i} \, g_i e_1 e_2 = g_i e_1 e_2 \ . \tag{3.14}$$

Thus $e_1 e_2$ commutes with all the generators of $\mathcal{S}$ and so $e_1 e_2 \in \mathcal{C}(\mathcal{S})$. Since $e_1^\dagger = \pm e_1$ (Theorem 3.1), $e_1^\dagger e_2$ also belongs to the centralizer $\mathcal{C}(\mathcal{S})$. Consequently, there exists a $c \in \mathcal{C}(\mathcal{S})$ such that $e_1^\dagger e_2 = c$ and so $e_2 = e_1 c$. Thus $e_2 \in e_1 \mathcal{C}(\mathcal{S})$. By definition, $e_2 \in e_2 \mathcal{C}(\mathcal{S})$ so that e_2 is a common element of $e_1 \mathcal{C}(\mathcal{S})$ and $e_2 \mathcal{C}(\mathcal{S})$. Since cosets do not overlap, two cosets must be identical if they share a common element, and so $e_1 \mathcal{C}(\mathcal{S}) = e_2 \mathcal{C}(\mathcal{S})$. Since $e_1 \in e_1 \mathcal{C}(\mathcal{S})$, we have shown that e_1 and e_2 belong to the same coset of $\mathcal{C}(\mathcal{S})$.

($\Longleftarrow$) Assume e_1 and e_2 belong to the same coset of $\mathcal{C}(\mathcal{S})$: $e_1, e_2 \in e_1 \mathcal{C}(\mathcal{S})$. By definition (eq. (3.13)), there exists a $c \in \mathcal{C}(\mathcal{S})$ such that $e_2 = e_1 c$ and so $e_1^\dagger e_2 = c \in \mathcal{C}(\mathcal{S})$. Thus $e_1^\dagger e_2$ commutes with all generators of $\mathcal{S}$ and since $e_1^\dagger = \pm e_1$, so does $e_1 e_2$. Let $S(e_1) = l_1 \cdots l_{n-k}$, $S(e_2) = l'_1 \cdots l'_{n-k}$, and g_i be a generator of $\mathcal{S}$. Then,

$$e_1 e_2 g_i = (-1)^{l'_i} \, e_1 g_i e_2 = (-1)^{l'_i + l_i} \, g_i e_1 e_2 \ . \tag{3.15}$$

But $e_1 e_2$ commutes with g_i, so $l'_i + l_i = 0 \pmod{2}$, and so $l'_i = l_i \pmod{2}$. Since g_i was arbitrary, $l'_i = l_i$ for all $i = 1, \ldots, n-k$, and so $S(e_1) = S(e_2)$. ∎

(5) Because of Theorem 3.1-(3), every error $e \in \mathcal{G}_n$ has a well-defined error syndrome; and since the cosets of $\mathcal{C}(\mathcal{S})$ partition $\mathcal{G}_n$, every error belongs to one of these cosets. Theorem 3.2 established that all elements of a coset of $\mathcal{C}(\mathcal{S})$ have the same error syndrome and different cosets have different syndrome values. Thus each coset has a unique syndrome value and so the number of cosets cannot exceed the total number of syndrome values 2^{n-k}. The index $[\mathcal{G}_n : \mathcal{C}(\mathcal{S})]$ is the number of cosets of $\mathcal{C}(\mathcal{S})$ in $\mathcal{G}_n$, so we have shown that $[\mathcal{G}_n : \mathcal{C}(\mathcal{S})] \leq 2^{n-k}$. On the other hand, recall (p. 34) that the n-qubit Hilbert space H_2^n partitions into 2^{n-k} subspaces $\mathcal{C}(l)$ of dimension 2^k, where $l = l_1 \cdots l_{n-k}$ is a bit string of length $n - k$. Theorem 1.11 proved that an error E maps the code space $\mathcal{C}_q$ onto the subspace $\mathcal{C}(l)$, where l is given by eq. (1.86) (a. k. a. eq.(3.7)). This connection between an error E and $\mathcal{C}(l)$ was in fact the original motivation for the definition of the error syndrome given on page 36: $S(E) \equiv l$. Each subspace $\mathcal{C}(l)$ thus identifies an error $E_l : \mathcal{C}_q \to \mathcal{C}(l)$, where

$$E_l = \sum_{ij} (E_l)_{ij} \, |l; \bar{\delta}_i\rangle\langle l = 0; \bar{\delta}_j| \ .$$

Here $(E_l)_{ij} = \langle l; \bar{\delta}_i | E_l | l = 0; \bar{\delta}_j \rangle$; $\mathcal{C}_q = \mathcal{C}(l=0)$; and the basis vectors $|l; \bar{\delta}\rangle$ were defined in eq. (1.80). By construction, there are 2^{n-k} errors E_l and each has a different error syndrome $S(E_l) \neq S(E_{l'})$ for $l \neq l'$. Thus, by Theorem 3.2, the errors E_l must belong to different cosets of $\mathcal{C}(\mathcal{S})$. Thus the number of cosets of $\mathcal{C}(\mathcal{S})$ cannot be less than 2^{n-k}: $[\mathcal{G}_n : \mathcal{C}(\mathcal{S})] \geq 2^{n-k}$. Consistency between these two inequalities is only possible if

$$[\mathcal{G}_n : \mathcal{C}(\mathcal{S})] = 2^{n-k} \ . \tag{3.16}$$

Theorem 3.1 established that the order of $\mathcal{G}_n$ is 2^{2n+2}. From Lagrange's theorem (Appendix A), $|\mathcal{G}_n| = |\mathcal{C}(\mathcal{S})| \, [\mathcal{G}_n : \mathcal{C}(\mathcal{S})]$, so that

$$|\mathcal{C}(\mathcal{S})| = 2^{n+k+2} \ . \tag{3.17}$$

If we quotient out the center $\mathcal{C} = \{\pm I, \pm iI\}$ of $\mathcal{G}_n$, Theorem 3.1 and Lagrange's theorem give

$$|\mathcal{C}(\mathcal{S})/\mathcal{C}| = 2^{n+k} \ . \tag{3.18}$$

This will be useful later.

(6) In Section 2.2 we defined the distance of a QECC to be d if all errors $e \in \mathcal{G}_n$ of weight less than d are detectable, and at least one error of weight d is non-detectable. Since non-detectable errors belong to $\mathcal{C}(\mathcal{S}) - \mathcal{S}$, a QECC will have distance d if and only if $\mathcal{C}(\mathcal{S}) - \mathcal{S}$ has an element of weight d and none of weight less than d.

Quantum Stabilizer Codes

The following theorem establishes a simple criterion for determining when a quantum stablizer code is degenerate.

THEOREM 3.3
A quantum stabilizer code $\mathcal{C}_q$ with distance d is a degenerate code if and only if its stabilizer $\mathcal{S}$ has an element of weight less than d (excluding the identity element).

PROOF ($\Longrightarrow$) If $\mathcal{C}_q$ is degenerate, then there exists at least one linear combination F of correctable errors $\{E_a\}$ that annihalates all basis codewords (Section 2.2):

$$F|i\rangle = \sum_a U_{la}\, E_a\, |i\rangle = 0 \ . \tag{3.19}$$

To simplify the proof, assume $F = e_1 - e_2$. Then eq. (3.19) implies

$$e_1|i\rangle = e_2|i\rangle$$

or

$$e_1^\dagger e_2|i\rangle = |i\rangle \ .$$

Thus $e_1^\dagger e_2$ fixes all basis codewords so that $e_1^\dagger e_2 \in \mathcal{S}$. Since e_1 and e_2 are correctable errors, Theorem 1.13 tells us that

$$\langle i|e_1^\dagger e_2|j\rangle = C_{12}\,\delta_{ij} \ .$$

Comparing with eq. (3.9) we see that $e_1^\dagger e_2$ is a detectable error. Since $\mathcal{C}_q$ has distance d, the weight of $e_1^\dagger e_2$ must be less than d (Section 2.2). Since $e_1^\dagger e_2 \in \mathcal{S}$, the stabilizer $\mathcal{S}$ contains an element of weight less than d other than the identity element.

($\Longleftarrow$) By assumption, $\mathcal{S}$ has an element s whose weight is less than d. Let s_a be an element of $\mathcal{S}$ distinct from s. Closure of $\mathcal{S}$ under multiplication insures that $s_a s = s_b \in \mathcal{S}$. Solving for s gives $s = s_a^\dagger s_b$. Since $s \in \mathcal{S}$, it fixes all basis codewords $|i\rangle$:

$$|i\rangle = s|i\rangle = s_a^\dagger s_b|i\rangle \ ,$$

or

$$s_a|i\rangle = s_b|i\rangle \quad\Longrightarrow\quad (s_a - s_b)|i\rangle = 0 \ ,$$

for all basis codewords $|i\rangle$. Since $s_a, s_b \in \mathcal{S}$, they are correctable and so $F = s_a - s_b$ is also correctable (Theorem 1.12), and it annihalates all codewords. Following the discussion on page 74, under this condition, the matrix C_{ab} will have a zero eigenvalue. This makes C_{ab} singular, and the code $\mathcal{C}_q$ degenerate. Since the matrix C_{ab} must be either singular or non-singular, the code $\mathcal{C}_q$ must be either degenerate or non-degenerate, respectively. ∎

Errors e_1 and e_2 are said to be degenerate errors if $e_1 e_2 \in \mathcal{S}$.

(7) Finally, let C_q be a non-degenerate QECC. Since C_{ab} is non-singular for such a code, there is no linear combination of correctable errors that annihalates all codewords. Let E_a and E_b be two linearly independent correctable errors with syndromes $S(E_a)$ and $S(E_b)$, respectively. Since E_a and E_b are correctable, $E_a^\dagger E_b$ has weight less than d. Furthermore, since C_q is non-degenerate, $E_a^\dagger E_b$ cannot belong to $\mathcal{S}$ (Theorem 3.3), nor can it belong to $\mathcal{C}(\mathcal{S}) - \mathcal{S}$ since C_q has distance d and so all errors in this set have weight greater than or equal to d. Thus $E_a^\dagger E_b \in \mathcal{G}_n - \mathcal{C}(\mathcal{S})$ and so it must anticommute with at least one generator g_* of $\mathcal{S}$. Consequently, if E_a anticommutes with g_*, E_b must commute with it, or vice versa. Either way, it follows that $S(E_a) \neq S(E_b)$. Thus, for a non-degenerate quantum stabilizer code, linearly independent correctable errors will have unequal error syndromes. It is worth noting that since $E_a^\dagger E_b \in \mathcal{G}_n - \mathcal{C}(\mathcal{S})$ for all correctable errors, Theorem 1.10 insures that $\langle i|E_a^\dagger E_b|j\rangle = 0$ for all basis codewords $|i\rangle$, $|j\rangle$. This is eq. (2.44) with $C_{ab} = 0$. Thus for quantum stabilizer codes, only degenerate codes can have $\langle i|E_a^\dagger E_b|i\rangle \neq 0$.

Exercise 3.2 *The normalizer $N_\mathcal{G}(\mathcal{S})$ of the stabilizer $\mathcal{S}$ is the set of errors $e \in \mathcal{G}_n$ that fix $\mathcal{S}$ under conjugation: $e\mathcal{S}e^\dagger = \mathcal{S}$ for all $e \in N_\mathcal{G}(\mathcal{S})$. Show that $N_\mathcal{G}(\mathcal{S}) = \mathcal{C}(\mathcal{S})$. Some presentations of quantum stabilizer codes work with the normalizer $N_\mathcal{G}(\mathcal{S})$ instead of the centralizer $\mathcal{C}(\mathcal{S})$. When the Pauli group $\mathcal{G}_n$ is the relevant parent group ($\mathcal{G}_n \supset \mathcal{S}$), the two approaches are clearly equivalent.*

3.1.3 Quantum Error Correction Reprise

The discussion given on page 37 of the quantum error correction protocol is re-expressed here using language developed in Section 3.1.2. As noted there, the quantum error correction protocol uses ancilla qubits to extract the syndrome of the error E that has corrupted the initial codeword $|c\rangle$. Chapter 5 will explain how syndrome extraction can be done fault-tolerantly. The measured syndrome value S determines a set of errors $\{E_i\}$ whose error syndromes are equal to the measured value: $S(E_i) = S$. For a quantum stabilizer code C_q, this set of errors is one of the cosets of the centralizer $\mathcal{C}(\mathcal{S})$. Denoting the most probable error in this coset by E_S, and using it as the coset representative, the remarks so far can be summarized by saying that the actual error $E \in E_S \mathcal{C}(\mathcal{S})$. If more than one error in this coset qualifies as most probable, then one of these errors is arbitrarily chosen to be the most probable error E_S. Applying E_S^{-1} to the corrupted state is the quantum analog of maximum likelihood decoding discussed in Section 1.2.2. Since elements $e \in \mathcal{G}_n$ are unitary (Theorem 3.1), $E_S^{-1} = E_S^\dagger$, and so the quantum error correction protocol applies $E_S^\dagger$ to the corrupted state $E|c\rangle$ to produce the recovered state $E_S^\dagger E|c\rangle$. If the actual error E is the expected error E_S, then $E_S^\dagger E = I$ and the recovered state is the original codeword $|c\rangle$ as desired. More generally, if $E_S \neq E$, syndrome extraction insures that E_S and E have the same error

Quantum Stabilizer Codes

syndrome so that $S(E_S^\dagger E) = 0$, and $E_S^\dagger E \in C(S)$. If $E_S^\dagger E \in S \subset C(S)$, then the recovered state $E_S^\dagger E|c\rangle = |c\rangle$, and the protocol recovers the original codeword. Otherwise, if $E_S^\dagger E \in C(S) - S$, then $E_S^\dagger E$ is a non-detectable error for the code C_q. In this case, the recovered state $E_S^\dagger E|c\rangle \in C_q$, but generally $E_S^\dagger E|c\rangle \neq |c\rangle$. (See the discussion following Exercise 3.1.) This latter case might appear to be a blemish on the protocol, but we will see in Section 3.1.4 that such non-detectable errors in fact implement operations on encoded qubits and so are a good thing.

3.1.4 Encoded Operations

We have seen that a QECC introduces an encoding map ξ that sends (i) unencoded k-qubit states $|u\rangle \in H_2^k$ to n-qubit codewords $|c\rangle = \xi|u\rangle$ and (ii) unencoded operators $\mathcal{O} \in \mathcal{G}_k$ to encoded operators $\overline{\mathcal{O}} = \xi \mathcal{O} \xi^\dagger$ that map codewords to codewords $C_q \to C_q$. In particular, the unencoded Pauli operators σ_x^i and σ_z^i are mapped to

$$X_i = \xi \sigma_x^i \xi^\dagger$$
$$Z_i = \xi \sigma_z^i \xi^\dagger \quad , \qquad (3.20)$$

where $i = 1, \ldots, k$ labels the qubits. Using eqs. (3.2) and (3.3), the unencoded operator $\mathcal{O} \in \mathcal{G}_k$ can be written as

$$\mathcal{O} = i^\lambda \sigma_x(a) \sigma_z(b) \quad , \qquad (3.21)$$

where $a = a_1 \cdots a_k$ and $b = b_1 \cdots b_k$ are bit strings of length k, and $\lambda = 0, 1, 2, 3$. Encoding sends $\mathcal{O}$ to

$$\overline{\mathcal{O}} = \xi \left[i^\lambda \sigma_x(a) \sigma_z(b) \right] \xi^\dagger$$
$$= i^\lambda X(a) Z(b) \quad , \qquad (3.22)$$

where

$$X(a) = (X_1)^{a_1} \cdots (X_k)^{a_k}$$
$$Z(b) = (Z_1)^{b_1} \cdots (Z_k)^{b_k} \quad . \qquad (3.23)$$

It is straightforward to show that

$$[X_i, X_j] = [Z_i, Z_j] = 0 \qquad (3.24)$$
$$[X_i, Z_j] = 0 \quad (i \neq j) \qquad (3.25)$$
$$\{X_i, Z_i\} = 0 \quad . \qquad (3.26)$$

Exercise 3.3 *Prove eqs. (3.24)–(3.26).*

We will see in Section 3.3 that the X_i and Z_i ($i = 1, \ldots, k$) can always be chosen to satisfy eqs. (3.25) and (3.26) and to commute with the generators

$\{g_l : l = 1, \ldots, n-k\}$ of the stabilizer $\mathcal{S}$. They thus belong to the centralizer $\mathcal{C}(\mathcal{S})$. It follows from eqs. (3.22) and (3.23), and that $\mathcal{C}(\mathcal{S})$ is closed under multiplication, that all encoded operators $\overline{\mathcal{O}}$ also belong to $\mathcal{C}(\mathcal{S})$. Now consider the set $\mathcal{P}$ of 2^{n+k+2} operators generated by all possible products of powers of the g_l, X_i, Z_j:

$$\hat{\mathcal{O}}(a,b,c) = i^\lambda (X_1)^{a_1} \cdots (X_k)^{a_k} (Z_1)^{b_1} \cdots (Z_k)^{b_k} (g_1)^{c_1} \cdots (g_{n-k})^{c_{n-k}}$$
$$= i^\lambda X(a) Z(b) g(c) \ . \tag{3.27}$$

Here $\lambda = 0, 1, 2, 3$; $a = a_1 \cdots a_k$ and $b = b_1 \cdots b_k$ are bit strings of length k; $c = c_1 \cdots c_{n-k}$ is a bit string of length $n-k$; and $g(c) \equiv (g_1)^{c_1} \cdots (g_{n-k})^{c_{n-k}}$. By appropriately adapting the argument above showing $\overline{\mathcal{O}} \in \mathcal{C}(\mathcal{S})$, it is possible to show that $\hat{\mathcal{O}}(a,b,c) \in \mathcal{C}(\mathcal{S})$, and so the set $\mathcal{P} = \{\hat{\mathcal{O}}(a,b,c)\}$ is a subset of $\mathcal{C}(\mathcal{S})$. Recall that the order of the centralizer is $|\mathcal{C}(\mathcal{S})| = 2^{n+k+2}$. Since $\mathcal{P}$ and $\mathcal{C}(\mathcal{S})$ have the same number of elements, $\mathcal{P}$ cannot be a proper subset of $\mathcal{C}(\mathcal{S})$, and so $\mathcal{P} = \mathcal{C}(\mathcal{S})$. We arrive at the result that the centralizer $\mathcal{C}(\mathcal{S})$ is generated by the encoded Pauli operators $\{X_i, Z_i : i = 1, \ldots, k\}$ and the generators $\{g_l : l = 1, \ldots, n-k\}$ of $\mathcal{S}$.

The decoding map $\xi^\dagger$ pulls back codewords to unencoded states: $\xi^\dagger |c\rangle = |u\rangle$, where $|c\rangle \in \mathcal{C}_q$ and $|u\rangle \in H_2^k$. By definition, elements of the stabilizer $s \in \mathcal{S}$ fix codewords $|c\rangle$:

$$s|c\rangle = |c\rangle \ .$$

Applying $\xi^\dagger$ to both sides and appropriately inserting the identity $\xi \xi^\dagger$ gives

$$\xi^\dagger s \left(\xi \xi^\dagger\right) |c\rangle = \xi^\dagger |c\rangle \ ,$$

which reduces to

$$\left(\xi^\dagger s \xi\right) |u\rangle = |u\rangle \ . \tag{3.28}$$

Since $\xi^\dagger$ maps $\mathcal{C}_q$ onto H_2^k, eq. (3.28) is true for all $|u\rangle \in H_2^k$ and so $\xi^\dagger s \xi$ is the identity I_k on H_2^k. Thus all elements of the stabilizer $\mathcal{S}$ are pulled back by $\xi^\dagger$ to the identity operator $I_k \in \mathcal{G}_k$. Notice that the decoding operation $\xi^\dagger$ is a homomorphism from $\mathcal{C}(\mathcal{S}) \to \mathcal{G}_k$. It maps $e \to \xi^\dagger e \xi$, where

$$\xi^\dagger e \xi = \xi^\dagger \left[i^\lambda X(a) Z(b) g(c)\right] \xi$$
$$= i^\lambda \sigma_x(a) \sigma_z(b) \ , \tag{3.29}$$

and we have used that $g(c) \in \mathcal{S}$ is pulled back to $I_k \in \mathcal{G}_k$. The image of the homomorphism is clearly $\mathcal{G}_k$, and eq. (3.28) shows that $\mathcal{S}$ is its kernel: $\text{Ker}\left(\xi^\dagger\right) = \mathcal{S}$.

Exercise 3.4 *Show that $\xi^\dagger$ is a homomorphism from $\mathcal{C}(\mathcal{S})$ onto $\mathcal{G}_k$.*

We saw in Section 3.1.2-(3) that an error $e \in \mathcal{C}(\mathcal{S}) - \mathcal{S}$ is non-detectable by $\mathcal{C}_q$, and that $e|c\rangle \in \mathcal{C}_q$ for all codewords $|c\rangle$. Such errors map codewords to codewords and so implement encoded operations. Because elements s of the

Quantum Stabilizer Codes

stabilizer $\mathcal{S}$ fix codewords, the errors e and es will produce the same encoded operation since

$$es|c\rangle = e|c\rangle$$

for all $|c\rangle$ in $\mathcal{C}_q$. This suggests that encoded operations should be identified with the cosets of $\mathcal{S}$ in $\mathcal{C}(\mathcal{S})$. To see that this is so, apply the Fundamental Homomorphism Theorem (Appendix A) to the decoding homomorphism $\xi^\dagger$: $\mathcal{C}(\mathcal{S}) \to \mathcal{G}_k$. It follows from this theorem that

$$\mathcal{C}(\mathcal{S})/\mathcal{S} \cong \mathcal{G}_k \ . \tag{3.30}$$

The quotient group $\mathcal{C}(\mathcal{S})/\mathcal{S}$ (whose elements are the cosets of $\mathcal{S}$ in $\mathcal{C}(\mathcal{S})$) is thus *isomorphic* to the group of unencoded operations $\mathcal{G}_k$. Because of the isomorphism, the orders of the two groups are equal. Using Theorem 3.1 gives

$$|\mathcal{C}(\mathcal{S})/\mathcal{S}| = |\mathcal{G}_k| = 2^{2k+2} \ . \tag{3.31}$$

This isomorphism is the link that allows us to unambiguously identify encoded operations $\overline{\mathcal{O}}$ with the cosets of $\mathcal{C}(\mathcal{S})/\mathcal{S}$. To see this, note that eqs. (3.21) and (3.22) establish a 1-1 correspondence between $\mathcal{O}$ and $\overline{\mathcal{O}}$ since the same bit strings a and b determine both $\mathcal{O}$ and $\overline{\mathcal{O}}$. The isomorphism of eq. (3.30) in turn sets up a 1-1 correspondence between cosets and unencoded operations $\mathcal{O}$. Thus the encoding ξ sets up a 1-1 correspondence between $\overline{\mathcal{O}}$ and $\mathcal{O}$, and the isomorphism a 1-1 correspondence between $\mathcal{O}$ and the cosets in $\mathcal{C}(\mathcal{S})/\mathcal{S}$. Together we have a 1-1 correspondence between the encoded operations $\overline{\mathcal{O}}$ and the cosets of $\mathcal{C}(\mathcal{S})/\mathcal{S}$: $\overline{\mathcal{O}} \Leftrightarrow \overline{\mathcal{O}}\mathcal{S}$. It is an immediate consequence of this 1-1 correspondence that each of the encoded Pauli operators $\{\overline{X}_i, \overline{Z}_j : i, j = 1, \ldots, k\}$ must belong to a different coset of $\mathcal{S}$. We leave it to the reader to show that the cosets $\{\overline{X}_i\mathcal{S}, \overline{Z}_j\mathcal{S} : i, j = 1, \ldots, k\}$ generate the quotient group $\mathcal{C}(\mathcal{S})/\mathcal{S}$.

Exercise 3.5 *The elements of the quotient group $\mathcal{C}(\mathcal{S})/\mathcal{S}$ are the cosets of $\mathcal{S}$ in $\mathcal{C}(\mathcal{S})$ with coset multiplication defined as $(a\mathcal{S})(b\mathcal{S}) = (ab)\mathcal{S}$. Show that the $2k$ cosets $\{\overline{X}_i\mathcal{S}, \overline{Z}_i\mathcal{S} : i = 1, \ldots, k\}$ and $i^\lambda : i = 0, 1, 2, 3$ generate $\mathcal{C}(\mathcal{S})/\mathcal{S}$.*

3.2 Examples

This section presents three new examples of quantum stabilizer codes. Two of these are multi-qubit encodings. Many more examples of QECCs can be found in Refs. [3, 4].

3.2.1 [5,1,3] Code

Of all QECCs that encode 1 qubit of data and correct all single-qubit errors, the [5,1,3] code is the most efficient [5, 6], saturating the quantum Hamming bound (see Chapter 7). It encodes $k = 1$ qubit in $n = 5$ qubits, and so constructs $\mathcal{S}$ from $n - k = 4$ generators. To unclutter the notation, the qubit label i in the unencoded Pauli operators σ_k^i will be suppressed, and it is understood that the left-most operator in any 5-qubit operator acts on qubit 1, the operator to its right acts on qubit 2, etc. We also suppress factors of i^λ in operators, and normalization constants in codewords. The [5,1,3] code can be constructed from the following set of generators:

$$\begin{aligned} g_1 &= \sigma_x \, \sigma_z \, \sigma_z \, \sigma_x \, I \\ g_2 &= I \, \sigma_x \, \sigma_z \, \sigma_z \, \sigma_x \\ g_3 &= \sigma_x \, I \, \sigma_x \, \sigma_z \, \sigma_z \\ g_4 &= \sigma_z \, \sigma_x \, I \, \sigma_x \, \sigma_z \end{aligned} \qquad (3.32)$$

It is a simple matter to check that the generators are mutually commuting. The [5,1,3] code is a cyclic code: its set of generators is invariant under a cyclic permutation of the qubits. Each element $s \in \mathcal{S}$ is a possible product of the generators:

$$s = (g_1)^{p_1} \cdots (g_4)^{p_4} \; , \qquad (3.33)$$

where $p_i = 0, 1$ and $i = 1, \ldots, 4$. The encoded Pauli operators X and Z can be represented by:

$$\begin{aligned} X &= \sigma_x \, \sigma_x \, \sigma_x \, \sigma_x \, \sigma_x \\ Z &= \sigma_z \, \sigma_z \, \sigma_z \, \sigma_z \, \sigma_z \end{aligned} \qquad (3.34)$$

The reader can verify that X and Z anticommute, and that each commutes with all of the generators. Note that the product $g_1 X$ is $I \, \sigma_y \, \sigma_y \, I \, \sigma_x$. It has weight 3 and is an element of $\mathcal{C}(\mathcal{S}) - \mathcal{S}$ (why?). The reader can check by direct calculation that $\mathcal{C}(\mathcal{S}) - \mathcal{S}$ has no elements of weight less than 3. It follows from the discussion in Section 3.1.2-(6) that the distance of this code is $d = 3$, justifying the '3' in the [5,1,3] label for this code. Direct calculation also shows that the smallest weight for elements of $\mathcal{S}$ (other than the identity) is 4. Since this is greater than the distance $d = 3$, Theorem 3.3 establishes that the [5,1,3] code is non-degenerate. An easily constructed choice for the basis codewords is

$$|\bar{0}\rangle = \sum_{s \in \mathcal{S}} s \, |00000\rangle \qquad (3.35)$$

and

$$|\bar{1}\rangle = X \, |\bar{0}\rangle \; . \qquad (3.36)$$

Here the state $|00000\rangle \equiv |0\rangle_1 \otimes \cdots \otimes |0\rangle_5$ is the direct product of the single qubit CB states $|0\rangle_i$ fixed by the σ_z^i: $\sigma_z^i |0\rangle_i = |0\rangle_i$ (see eq. (1.74), though we have slightly modified the notation). The reader can easily check that these states are fixed by the elements of $\mathcal{S}$, and that $Z|\bar{i}\rangle = (-1)^{\bar{i}}|\bar{i}\rangle$.

Quantum Stabilizer Codes

3.2.2 [4,2,2] Code

The [4,2,2] code is the simplest example of a class of $[n-1, k+1, d-1]$ codes that are derivable from an $[n,k,d]$ code [3, 7]. It is our first explicit example of a multi-qubit encoding. It is derivable from the [5,1,3] code (see Problem 3.2), and can detect a single qubit error. The presentation in this and the following subsection parallels that of the [5,1,3] code and so more abbreviated discussions will be given.

With $n=4$ and $k=2$, this code uses $n-k=2$ generators:

$$g_1 = \sigma_x\,\sigma_z\,\sigma_z\,\sigma_x$$
$$g_2 = \sigma_y\,\sigma_x\,\sigma_x\,\sigma_y \quad . \tag{3.37}$$

Each encoded qubit i has its own X_i and Z_i. Convenient choices are:

$$X_1 = \sigma_x\,I\,\sigma_y\,\sigma_y \quad ; \quad Z_1 = \sigma_y\,\sigma_z\,\sigma_y\,I$$
$$X_2 = \sigma_x\,I\,\sigma_x\,\sigma_z \quad ; \quad Z_2 = I\,\sigma_x\,\sigma_z\,\sigma_z \quad . \tag{3.38}$$

The reader can check that the X_i and Z_i commute with the generators g_1 and g_2, as well as satisfy eqs. (3.24)–(3.26). The element $X_1 Z_2 g_1 = I\,\sigma_y\,\sigma_y\,I$ belongs to $\mathcal{C}(\mathcal{S}) - \mathcal{S}$ and clearly has weight 2. Direct calculation shows that there is no element in $\mathcal{C}(\mathcal{S}) - \mathcal{S}$ that has weight 1 and so the code has distance $d=2$ (see Section 3.1.2-(6)). Direct calculation shows that the smallest weight for elements of $\mathcal{S}$ (other than the identity) is $4 > d$ so that the [4,2,2] code is non-degenerate. An easily constructed set of basis codewords $|\overline{ij}\rangle$ is

$$|\overline{ij}\rangle = (X_1)^{\overline{i}}(X_2)^{\overline{j}} \sum_{s \in \mathcal{S}} s\,|0000\rangle \quad . \tag{3.39}$$

3.2.3 [8,3,3] Code

The [8,3,3] code is a special case of a class of $[2^j, 2^j - j - 2, 3]$ codes [1, 2, 4, 8]. Here, three qubits are encoded in eight, and this code corrects all single-qubit errors. It has $n-k=5$ generators:

$$g_1 = \sigma_x\,\sigma_x\,\sigma_x\,\sigma_x\,\sigma_x\,\sigma_x\,\sigma_x\,\sigma_x$$
$$g_2 = \sigma_z\,\sigma_z\,\sigma_z\,\sigma_z\,\sigma_z\,\sigma_z\,\sigma_z\,\sigma_z$$
$$g_3 = I\,\sigma_x\,I\,\sigma_x\,\sigma_y\,\sigma_z\,\sigma_y\,\sigma_z$$
$$g_4 = I\,\sigma_x\,\sigma_z\,\sigma_y\,I\,\sigma_x\,\sigma_z\,\sigma_y$$
$$g_5 = I\,\sigma_y\,\sigma_x\,\sigma_z\,\sigma_x\,\sigma_z\,I\,\sigma_y \quad , \tag{3.40}$$

and the X_i and Z_i can be chosen to be:

$$X_1 = \sigma_x\,\sigma_x\,I\,I\,I\,\sigma_z\,I\,\sigma_z \quad ; \quad Z_1 = I\,\sigma_z\,I\,\sigma_z\,I\,\sigma_z\,I\,\sigma_z$$
$$X_2 = \sigma_x\,I\,\sigma_x\,\sigma_z\,I\,I\,\sigma_z\,I \quad ; \quad Z_2 = I\,I\,\sigma_z\,\sigma_z\,I\,I\,\sigma_z\,\sigma_z$$
$$X_3 = \sigma_x\,I\,I\,\sigma_z\,\sigma_x\,\sigma_z\,I\,I \quad ; \quad Z_3 = I\,I\,I\,I\,\sigma_z\,\sigma_z\,\sigma_z\,\sigma_z \quad . \tag{3.41}$$

It is straightforward to check that the X_i and Z_i commute with the g_j and satisfy eqs. (3.24)–(3.26). The element $X_1 X_3 Z_1 = I\ \sigma_y\ I\ I\ \sigma_x\ \sigma_z\ I\ I$ has weight 3 and belongs to $\mathcal{C}(\mathcal{S}) - \mathcal{S}$. The reader can check that no element of $\mathcal{C}(\mathcal{S}) - \mathcal{S}$ has smaller weight and so the code distance $d = 3$. The reader can further check that all elements of $\mathcal{S}$ (except the identity) have weight greater than 3 so that this code is non-degenerate. A convenient choice for the basis codewords is

$$|\overline{ijk}\rangle = (X_1)^{\bar{i}} (X_2)^{\bar{j}} (X_3)^{\bar{k}} \sum_{s \in \mathcal{S}} s |0000\ 0000\rangle \ . \tag{3.42}$$

3.3 Alternate Formulation: Finite Geometry

Recall that $\mathcal{C} = \{\pm I, \pm iI\}$ is the center of $\mathcal{G}_n$. Being a normal subgroup of $\mathcal{G}_n$, the quotient group $\mathcal{G}_n/\mathcal{C}$ can be constructed from its cosets, with coset multiplication defined as

$$(e\mathcal{C})(f\mathcal{C}) = (ef)\mathcal{C} \ .$$

Calderbank et al. [3] showed that there is a 1-1 correspondence between $\mathcal{G}_n/\mathcal{C}$ and the $2n$-dimensional binary vector space F_2^{2n} whose elements are bit strings of length $2n$. A vector $v \in F_2^{2n}$ is denoted $v = (a|b)$, where $a = a_1 \cdots a_n$ and $b = b_1 \cdots b_n$ are bit strings of length n. Scalars take values in the Galois field $F_2 = \{0,1\}$ and vector addition adds components modulo 2. For example, let $e = i^\lambda \sigma_x(a_e)\sigma_z(b_e) \in \mathcal{G}_n$. It belongs to the coset $e\mathcal{C} = \{\pm e, \pm ie\}$ whose coset representative is chosen to be $\sigma_x(a_e)\sigma_z(b_e)$. It is clear that the bit strings a_e and b_e uniquely determine the coset representative and hence the coset $e\mathcal{C}$, and so are used to define the image vector $v_e = (a_e|b_e)$ in F_2^{2n}. The reader can check that this defines a 1-1 correspondence between $\mathcal{G}_n/\mathcal{C}$ and F_2^{2n}. At times it proves convenient to speak in terms of a coset member $i^\lambda \sigma_x(a_e)\sigma_z(b_e)$ even though the object of discussion is actually the coset $e\mathcal{C}$. This will only be done in cases where confusion seems unlikely.

An inner product $\langle v_e, v_f \rangle$ can be introduced that maps pairs of vectors $v_e, v_f \in F_2^{2n}$ onto F_2. If $v_e = (a_e|b_e)$ and $v_f = (a_f|b_f)$, then

$$\langle v_e, v_f \rangle \equiv a_e \cdot b_f + b_e \cdot a_f \ , \tag{3.43}$$

where

$$a_e \cdot b_f = a_{e,1} b_{f,1} + \cdots + a_{e,n} b_{f,n}$$
$$b_e \cdot a_f = b_{e,1} a_{f,1} + \cdots + b_{e,n} a_{f,n} \ ,$$

and all addition operations are modulo 2. The form of the inner product is inspired by eq. (3.6), which can be rewritten as

$$ef = (-1)^{\langle v_e, v_f \rangle} fe \ , \tag{3.44}$$

Quantum Stabilizer Codes

where the operators $e = i^{\lambda_e} \sigma_x(a_e) \sigma_z(b_e)$ and $f = i^{\lambda_f} \sigma_x(a_f) \sigma_z(b_f)$ correspond, respectively, to the vectors v_e and v_f defined above eq. (3.43). The inner product thus encodes whether the operators e and f commute or anticommute:

$$\langle v_e, v_f \rangle = \begin{cases} 0 & \text{if and only if} \quad [e, f] = 0 \\ 1 & \text{if and only if} \quad \{e, f\} = 0 \end{cases} \quad . \tag{3.45}$$

Representing vectors as column vectors, eq. (3.43) can be rewritten as

$$\langle v_e, v_f \rangle = v_e^T J v_f \quad , \tag{3.46}$$

where v_e^T is the transpose of v_e and J is the $2n \times 2n$ matrix

$$J = \begin{pmatrix} 0_n & I_n \\ I_n & 0_n \end{pmatrix} \quad , \tag{3.47}$$

with I_n the $n \times n$ identity matrix and 0_n the $n \times n$ matrix whose elements are all 0. Notice that $\langle v_e, v_e \rangle = 0$ for all $v_e \in F_2^{2n}$. This self-orthogonality follows from eq. (3.43) and that the range of the inner product is F_2 where $a + a = 0$ for all $a \in F_2$. In light of eq. (3.45), this self-orthogonality is simply another way of expressing that $[e, e] = 0$ for all $e \in \mathcal{G}_n$. The inner product is also symmetric: $\langle v_e, v_f \rangle = \langle v_f, v_e \rangle$. This follows immediately from eq. (3.43). It is left to the reader to show that the inner product is bilinear.

Exercise 3.6 *An inner product is said to be bilinear if for all vectors v_e, v_f, v_g:*

$$<v_e + v_f, v_g> = <v_e, v_g> + <v_f, v_g>$$
$$<v_e, v_f + v_g> = <v_e, v_f> + <v_e, v_g> \quad .$$

Show that the inner product defined by eq. (3.43) is bilinear.

DEFINITION 3.2 *An inner product $<,>$ that maps $F_2^{2n} \otimes F_2^{2n} \to F_2$ is symplectic [9] if for all $u, v \in F_2^{2n}$ it is (1) self-orthogonal: $<u, u> = 0$; (2) symmetric: $<u, v> = <v, u>$; and (3) bilinear.*

We see that eq. (3.43) defines a symplectic inner product. This was first pointed out in Ref. [3].

DEFINITION 3.3 *The weight of a vector $v = (a_1 \cdots a_n | b_1 \cdots b_n)$ is equal to the number of coordinates i for which $a_i = 1$ and/or $b_i = 1$.*

DEFINITION 3.4 *The distance between two vectors $v_e = (a_e | b_e)$ and $v_f = (a_f | b_f)$ is equal to the weight of their sum $v_e + v_f$.*

Example 3.1
The operators $X = \sigma_x \sigma_x \sigma_x \sigma_x \sigma_x$ and $Z = \sigma_z \sigma_z \sigma_z \sigma_z \sigma_z$ map to the vectors $v(X) = (11111|00000)$ and $v(Z) = (00000|11111)$, respectively. The weight of

both $v(X)$ and $v(Z)$ is 5. The product XZ maps to $v(XZ) = (11111|11111)$, which is clearly the vector sum of $v(X)$ and $v(Z)$. The weight of $v(XZ)$ is seen to be 5 and thus the distance between $v(X)$ and $v(Z)$ is also 5. □

An element s of the stabilizer $\mathcal{S}$ maps to the vector $v(s)$ in F_2^{2n}. In particular, the generator $g_j = i^{\lambda_j} \sigma_x(a_j)\sigma_z(b_j)$ maps to the vector $v_j = (a_j|b_j)$. The resulting set of vectors $\{v_j : j = 1,\ldots,n-k\}$ is linearly independent since no collection of generators with less than $n-k$ elements can generate all of $\mathcal{S}$. We leave it to the reader to verify that if the $\{v_j\}$ are linearly dependent, then one of the generators can be written as a product of the remaining generators. Let S be the image of $\mathcal{S}$ in F_2^{2n}. Because of the linear independence of the $\{v_j\}$, it is an $n-k$ dimensional subspace of F_2^{2n}. The dual space $S_\perp$ is defined to be the set of vectors $v_d \in F_2^{2n}$ that is orthogonal to all vectors $v(s) \in S$:

$$\langle v_d, v(s)\rangle = 0 \ .$$

It follows from eqs. (3.44) and (3.45) that $S \subset S_\perp$ and that $S_\perp$ is the image of the centralizer $\mathcal{C}(\mathcal{S})$.

Let $v_e = (a_e|b_e)$ be the image in F_2^{2n} of the error $e \in \mathcal{G}_n$, and the set $\{v_j = (a_j|b_j)\}$ the respective images of the generators $\{g_j\}$. It follows from eqs. (3.44), (3.45), and Definition 3.1 that the integers

$$l_j = \langle v_j, e\rangle \qquad j = 1,\ldots,n-k \tag{3.48}$$

specify the error syndrome $S(e) = l_1 \cdots l_{n-k}$. Eq. (3.48) can be rewritten as

$$l_j = v_j^T J v_e$$
$$= (b_j|a_j)\begin{pmatrix} a_e \\ b_e \end{pmatrix}$$
$$= H^T(j) \cdot v_e \ . \tag{3.49}$$

Thus, as noted in Section 1.4.1, to each generator g_j there corresponds a row vector

$$H^T(j) = v_j^T J = (b_j|a_j) \ ,$$

which acts like a parity check on the vector v_e. The presence of J in the inner product is the reason the bit strings a_j and b_j in $v_j^T = (a_j|b_j)$ appear in reverse order in $H^T(j)$. The parity check (row) vectors $H^T(j)$ can be collected into a parity check matrix H in which $H^T(j)$ appears as the jth row:

$$H = \begin{pmatrix} \text{———} & H^T(1) & \text{———} \\ & \vdots & \\ \text{———} & H^T(n-k) & \text{———} \end{pmatrix} \ .$$

Eq. (3.49) allows the error syndrome $S(e)$ to be found from H just as in a classical linear error correcting code:

$$S(e) = Hv_e \ , \tag{3.50}$$

Quantum Stabilizer Codes

where the product on the RHS is matrix-vector multiplication with addition done modulo 2, and $S(e)$ is output as an n-component column vector.

The encoded operators X_i and Z_i map to vectors $v(X_i) = (a(X_i)|b(X_i))$ and $v(Z_i) = (a(Z_i)|b(Z_i))$, respectively. Since these operators belong to the centralizer $\mathcal{C}(\mathcal{S})$, the vectors $v(X_i)$ and $v(Z_i)$ belong to the dual space $S_\perp$. The following remarks are worked out for X_i, though a parallel set of remarks can be made for Z_i. We leave it to the interested reader to make the appropriate modifications. The vector $v(X_i)$ has $2n$ components whose values must satisfy $n + k$ linear constraints. These constraints arise from the requirements that X_i: (i) commute with all the generators g_j so that

$$\langle v(X_i), v_j \rangle = 0 \qquad j = 1, \cdots, n - k \; ; \tag{3.51}$$

(ii) commute with all X_j and with all the Z_j for which $j \neq i$:

$$\langle v(X_i), v(X_j) \rangle = 0 \qquad j = 1, \cdots, k \tag{3.52}$$
$$\langle v(X_i), v(Z_j) \rangle = 0 \qquad j \neq i \; ; \tag{3.53}$$

and (iii) anticommute with Z_i:

$$\langle v(X_i), v(Z_i) \rangle = 1 \; . \tag{3.54}$$

Because the contraints are linear, $v(X_i)$ has $n - k$ degrees of freedom. There are thus 2^{n-k} possible ways of choosing the X_i so that they commute with the generators of $\mathcal{S}$ and also satisfy eqs. (3.24)–(3.26). That $v(X_i)$ has $n - k$ degrees of freedom is an alternative way of expressing our earlier result that encoded operations are represented by cosets of $\mathcal{S}$ in $\mathcal{C}(\mathcal{S})$. To see this more clearly, let

$$X_i = i^\lambda \sigma_x(a) \sigma_z(b)$$
$$g_j = i^{\lambda_j} \sigma_x(a_j) \sigma_z(b_j) \qquad j = 1, \cdots, n - k \; , \tag{3.55}$$

and

$$s = (g_1)^{c_1} \cdots (g_{n-k})^{c_{n-k}}$$
$$= i^{\bar{\lambda}} \sigma_x \left(\sum_{j=1}^{n-k} c_j a_j \right) \sigma_z \left(\sum_{j=1}^{n-k} c_j b_j \right) \; . \qquad r \tag{3.56}$$

Defining $X_i' = X_i s$, we have

$$X_i' = i^{\lambda + \bar{\lambda}} \sigma_x \left(a + \sum_{j=1}^{n-k} c_j a_j \right) \sigma_z \left(b + \sum_{j=1}^{n-k} c_j b_j \right) \; . \tag{3.57}$$

Thus, from eq. (3.55), X_i maps to $v(X_i) = (a|b)$, and from eq. (3.57), X_i' maps to

$$v(X_i') = v(X_i) + \sum_{j=1}^{n-k} c_j v_j \; . \tag{3.58}$$

The bilinear character of the inner product, together with $v_j \in S$, insures that for any assignment of the $n-k$ parameters $\{c_j\}$, $v(X'_i)$ will satisfy eqs. (3.51)–(3.54) whenever $v(X_i)$ does. From eq. (3.56), each assignment of the $\{c_j\}$ specifies a particular $s \in \mathcal{S}$, which in turn specifies a particular $X'_i = X_i s$ whose effect on codewords is the same as X_i. We find again that an encoded operation such as X_i is only determined to within multiplication by an element of the stabilizer $\mathcal{S}$, and we see that the $n-k$ degrees of freedom in $v(X_i)$ are a manifestation of this.

Example 3.2
The generators g_1 and g_2 of the [4,2,2] code give rise to the vectors

$$v(g_1) = (1001|0110)$$
$$v(g_2) = (1111|1001) \ .$$

The parity check matrix H is then

$$H = \begin{pmatrix} 0110 \ | \ 1001 \\ 1001 \ | \ 1111 \end{pmatrix},$$

and the vectors associated with the encoded operators X_i and Z_i are

$$v(X_1) = (1011|0011)$$
$$v(X_2) = (1010|0001)$$
$$v(Z_1) = (1010|1110)$$
$$v(Z_2) = (0100|0011) \ .$$

The sum of $v(g_1)$ and $v(Z_2)$ is $(1101|0101)$ and so the distance between $v(g_1)$ and $v(Z_2)$ is 3. □

We close with a reminder that the 1-1 correspondence that underlies the alternate formulation given in this Section is between $\mathcal{G}_n/\mathcal{C}$ and F_2^{2n}. Thus a vector $v \in F_2^{2n}$ pulls back to a coset $e\mathcal{C}$ and so the error operator e is only determined to within a factor of i^λ.

3.4 Concatenated Codes

We have seen that an $[n_1, k_1, d_1]$ quantum error correcting code uses entangled states of n_1 qubits to protect k_1 qubits of data. If the recovery operation R can be applied perfectly, encoding will reduce the effective error probability for the data from $\mathcal{O}(p)$ to $\mathcal{O}(p^2)$. Section 1.4.1 demonstrated this reduction for the

Quantum Stabilizer Codes

phase-flip channel. The question arises: if using an $[n_1, k_1, d_1]$ code improves the reliability of the k_1 data qubits, can a further improvement be obtained by adding in another layer of encoding? Specifically, would splitting up the n_1 qubits into blocks of k_2 qubits and encoding each block using an $[n_2, k_2, d_2]$ code further reduce the effective error probability on the k_1 data qubits? As will be seen when proving the accuracy threshold theorem in Chapter 6, under appropriate circumstances the answer is yes, and this layering in of entanglement by recursively encoding blocks of qubits is the essence of what concatenated QECCs do [10, 11]. These codes will play an essential role in establishing the accuracy threshold theorem [10–18].

This section explains how to construct concatenated QECCs out of quantum stabilizer codes. The discussion will be limited to two layers of concatenation, though it should be clear from what is said how to add in further layers. The case where the second code encodes a single qubit will be explained in Section 3.4.1, and the Shor [9,1,3] code will be obtained by concatenating two [3,1,1] codes. Section 3.4.2 explains how to put in the second layer of entanglement using a multi-qubit encoding, and constructs an [8,2,2] code by layering two [4,2,2] codes.

3.4.1 Single Qubit Encoding

Here we construct a concatenated quantum error correcting code $\mathcal{C}$ using two layers of encoding. The first layer is an $[n_1, k_1, d_1]$ code $\mathcal{C}_1$ with generators $G_1 = \{g_i^1 : i = 1, \ldots, n_1 - k_1\}$, and the second is an $[n_2, 1, d_2]$ code $\mathcal{C}_2$ with generators $G_2 = \{g_j^2 : j = 1, \ldots, n_2 - 1\}$. The concatenated code $\mathcal{C}$ is to map k_1 qubits into $n = n_1 n_2$ qubits, with code construction parsing the n qubits into n_1 blocks $B(b)$ ($b = 1, \ldots, n_1$), each containing n_2 qubits. The generators of $\mathcal{C}$ are constructed as follows. (1) To each block $B(b)$ we associate a copy of the generators of $\mathcal{C}_2$: $G_2(b) = \{g_j^2(b) : j = 1, \ldots, n_2 - 1\}$. Each generator $g_j^2(b)$ is a block operator that only acts on qubits belonging to $B(b)$. With $n_2 - 1$ generators coming from each of the n_1 blocks, this step introduces a total of $n_1(n_2 - 1)$ generators. (2) Finally, the encoded images of the generators of $\mathcal{C}_1$ are added to the generators of $\mathcal{C}$. Specifically, if ξ_2 is the encoding operator for $\mathcal{C}_2$ and g_i^1 is a generator of $\mathcal{C}_1$, then $\overline{g_i^1} = \xi_2 \, g_i^1 \, \xi_2^\dagger$ is a generator of $\mathcal{C}$. This adds $n_1 - k_1$ generators to $\mathcal{C}$ for a total of $n_1(n_2 - 1) + n_1 - k_1 = n_1 n_2 - k_1$ generators. Since an $[n_1 n_2, k, d]$ code will have $n_1 n_2 - k$ generators, we see that the above construction produces an $[n_1 n_2, k_1, d]$ code, as desired. A general formula for the code distance d is not known, though a lower bound can be established.

THEOREM 3.4
Let $\mathcal{C}$ be the $[n_1 n_2, k_1, d]$ concatenated code constructed from the $[n_1, k_1, d_1]$ and $[n_2, 1, d_2]$ quantum stabilizer codes $\mathcal{C}_1$ and $\mathcal{C}_2$, respectively. The code distances satisfy $d \geq d_1 d_2$.

PROOF Let S, S_1, and S_2 be the respective stabilizers and $C(S)$, $C(S_1)$, and $C(S_2)$ the respective centralizers of the codes C, C_1, and C_2. Choose E_* to be a smallest weight element of $C(S) - S$. From Section 3.1.2-(6), its weight is equal to the code distance d: $\text{wt}(E_*) = d$. Let $e_* \in \mathcal{G}_{n_1}$ be mapped to E_* under the encoding of C_2: $E_* = \xi_2 e_* \xi_2^\dagger$. We first show that e_* commutes with all generators of C_1. Thus, let g_i^1 be an arbitrary generator of C_1. Then,

$$[e_*, g_i^1] = \xi_2^\dagger \left[E_*, \overline{g_i^1}\right] \xi_2$$
$$= \xi_2^\dagger [0] \xi_2$$
$$= 0 \ ,$$

where we used that $g_i^1 = \xi_2^\dagger \overline{g_i^1} \xi_2$, and that $E_* \in C(S) - S$ and so commutes with all generators of C. Thus $e_* \in C(S_1)$. In fact, $e_* \in C(S_1) - S_1$. To see this, assume the contrary so that $e_* \in S_1$ and thus is a product of the generators of S_1. Under step 2 of the concatenation procedure, the encoding operation ξ_2 maps generators of C_1 into generators of the final code C. This means that the image E_* must be an element of S, which is a contradiction, since by assumption, $E_* \in C(S) - S$. Thus $e_* \notin S_1$ and so $e_* \in C(S_1) - S_1$. It follows from this that $\text{wt}(e_*) \geq d_1$, where d_1 is the distance of C_1. Since $e_* \in \mathcal{G}_{n_1}$, we can write (see Section 3.1.1):

$$e_* = i^\lambda \left(\sigma_x^1\right)^{a_1} \cdots \left(\sigma_x^{n_1}\right)^{a_{n_1}} \left(\sigma_z^1\right)^{b_1} \cdots \left(\sigma_z^{n_1}\right)^{b_{n_1}} \ , \quad (3.59)$$

with $a_i, b_i = 0, 1$. Under ξ_2: $\sigma_x^i \to X_i$, $\sigma_z^i \to Z_i$, and $e_* \to E_*$. Combining this with eq. (3.59) gives

$$E_* = i^\lambda (X_1)^{a_1} \cdots (X_{n_1})^{a_{n_1}} (Z_1)^{b_1} \cdots (Z_{n_1})^{b_{n_1}} \ , \quad (3.60)$$

where the $\{a_i\}$ and $\{b_i\}$ in eqs. (3.59) and (3.60) are the same. Since X_i and Z_i belong to $C(S_2) - S_2$, the weight of each of these operators must be greater than or equal to the code distance d_2 of C_2. This, in combination with eqs. (3.59) and (3.60) and the results of Problems 3.4 and 3.5, gives that $\text{wt}(E_*) \geq \text{wt}(e_*) d_2$. Since $\text{wt}(e_*) \geq d_1$, we have that $\text{wt}(E_*) \geq d_1 d_2$. Finally, since $\text{wt}(E_*) = d$, we have that $d \geq d_1 d_2$, which is the desired lower bound. Note that this result can also be obtained as a special case of Theorem 3.5. The proof of Theorem 3.5 does not require Problems 3.4 and 3.5 and so is less cumbersome than the proof just given. ∎

Example 3.3 *Construction of Shor [9,1,3] Code*
Here we construct the Shor [9,1,3] code by concatenating two [3,1,1] codes. The generators for the codes C_1 and C_2 are, respectively,

$$g_1^1 = \sigma_x^1 \sigma_x^2 \ ; \quad g_2^1 = \sigma_x^1 \sigma_x^3 \quad (3.61)$$

and

$$g_1^2 = \sigma_z^1 \sigma_z^2 \ ; \quad g_2^2 = \sigma_z^1 \sigma_z^3 \ . \quad (3.62)$$

Quantum Stabilizer Codes

Code $\mathcal{C}_1$ was introduced in Section 1.4.1 when discussing the phase-flip channel. $\mathcal{C}_2$ is a modification of $\mathcal{C}_1$ designed for use with the bit-flip channel whose operation elements are $E_0 = \sqrt{1-p}\, I$ and $E_1 = \sqrt{p}\, \sigma_x$. Choosing the unencoded computational basis states to be eigenstates of σ_z,

$$\sigma_z |i\rangle = (-1)^i |i\rangle \ ,$$

the basis codewords $|\bar{i}\rangle_1$ for $\mathcal{C}_1$ are

$$|\bar{0}\rangle_1 = \xi_1 |0\rangle$$
$$= \frac{1}{\sqrt{2^3}} [\,|0\rangle + |1\rangle\,] \otimes [\,|0\rangle + |1\rangle\,] \otimes [\,|0\rangle + |1\rangle\,] \ ;$$

$$|\bar{1}\rangle_1 = \xi_1 |1\rangle$$
$$= \frac{1}{\sqrt{2^3}} [\,|0\rangle - |1\rangle\,] \otimes [\,|0\rangle - |1\rangle\,] \otimes [\,|0\rangle - |1\rangle\,] \ , \quad (3.63)$$

and for $\mathcal{C}_2$ are

$$|\bar{0}\rangle_2 = \xi_2 |0\rangle$$
$$= |0\rangle \otimes |0\rangle \otimes |0\rangle \equiv |000\rangle \ ;$$

$$|\bar{1}\rangle_2 = \xi_2 |1\rangle$$
$$= |1\rangle \otimes |1\rangle \otimes |1\rangle \equiv |111\rangle \ . \quad (3.64)$$

A subscript $i = 1, 2$ on an encoded ket indicates that it belongs to the code $\mathcal{C}_i$. From eqs. (3.61) and (3.63) it is clear that $\mathcal{C}_1$ maps $\sigma_x \to X = \sigma_z^1 \sigma_z^2 \sigma_z^3$ and $\sigma_z \to Z = \sigma_x^1 \sigma_x^2 \sigma_x^3$; and from eqs. (3.62) and (3.64), that $\mathcal{C}_2$ maps $\sigma_x \to X = \sigma_x^1 \sigma_x^2 \sigma_x^3$ and $\sigma_z \to Z = \sigma_z^1 \sigma_z^2 \sigma_z^3$.

We now construct the generators of the concatenated code $\mathcal{C}$. This code uses 9 qubits that parse into three blocks, each containing 3 qubits. Qubits 1–3 belong to block 1; 4–6 to block 2; and 7–9 to block 3. To each block we associate a copy of the generators of $\mathcal{C}_2$. This gives the following six generators:

$$g_1 = \sigma_z^1 \sigma_z^2$$
$$g_2 = \sigma_z^1 \sigma_z^3$$
$$g_3 = \sigma_z^4 \sigma_z^5$$
$$g_4 = \sigma_z^4 \sigma_z^6$$
$$g_5 = \sigma_z^7 \sigma_z^8$$
$$g_6 = \sigma_z^7 \sigma_z^9 \ . \quad (3.65)$$

The remaining two generators are the encoded versions of g_1^1 and g_2^1. Here the qubit label i in eq. (3.61) becomes a block label, and each σ_x^i is replaced by $X_i = \sigma_x^{a_i} \sigma_x^{b_i} \sigma_x^{c_i}$. The superscripts a_i, b_i, and c_i are determined by which block is being encoded. Thus,

$$g_7 = X_1 X_2$$
$$= \left(\sigma_x^1 \sigma_x^2 \sigma_x^3\right) \left(\sigma_x^4 \sigma_x^5 \sigma_x^6\right) \ ;$$

$$g_8 = X_1 X_3$$
$$= \left(\sigma_x^1 \sigma_x^2 \sigma_x^3\right) \left(\sigma_x^7 \sigma_x^8 \sigma_x^9\right) \ . \tag{3.66}$$

Eqs. (3.65) and (3.66) are the generators of the Shor [9, 1, 3] code introduced in Problem 2.5. As shown there, this code has distance $d = 3$, which is consistent with Theorem 3.4 since $d \geq d_1 d_2 = 1$. The basis codewords for $\mathcal{C}$ are

$$|\bar{0}\rangle = \xi_2 \xi_1 |0\rangle$$
$$= \frac{1}{\sqrt{2^3}} \left[|000\rangle + |111\rangle\right] \otimes \left[|000\rangle + |111\rangle\right] \otimes \left[|000\rangle + |111\rangle\right] \ ;$$

$$|\bar{1}\rangle = \xi_2 \xi_1 |1\rangle$$
$$= \frac{1}{\sqrt{2^3}} \left[|000\rangle - |111\rangle\right] \otimes \left[|000\rangle - |111\rangle\right] \otimes \left[|000\rangle - |111\rangle\right] \ , \tag{3.67}$$

and the encoded Pauli operators $X = (\xi_2 \xi_1) \sigma_x (\xi_2 \xi_1)^\dagger$ and $Z = (\xi_2 \xi_1) \sigma_z (\xi_2 \xi_1)^\dagger$ are

$$X = \sigma_z^1 \sigma_z^2 \sigma_z^3 \sigma_z^4 \sigma_z^5 \sigma_z^6 \sigma_z^7 \sigma_z^8 \sigma_z^9$$

$$Z = \sigma_x^1 \sigma_x^2 \sigma_x^3 \sigma_x^4 \sigma_x^5 \sigma_x^6 \sigma_x^7 \sigma_x^8 \sigma_x^9 \ . \tag{3.68}$$

It is straightforward to check that $Z|\bar{i}\rangle = (-1)^{\bar{i}} |\bar{i}\rangle$ and $X|\bar{i}\rangle = |\overline{i \oplus 1}\rangle$, where $\oplus$ is addition modulo 2. □

3.4.2 Multi-Qubit Encoding

Here again the aim is to construct a concatenated code $\mathcal{C}$ using two layers of encoding. As in Section 3.4.1, the first layer is an $[n_1, k_1, d_1]$ code $\mathcal{C}_1$ with generators $G_1 = \{g_i^1 : i = 1, \cdots, n_1 - k_1\}$. This time, however, we allow the second code $\mathcal{C}_2$ to be an $[n_2, k_2, d_2]$ code with $k_2 > 1$. The generators for $\mathcal{C}_2$ are $G_2 = \{g_j^2 : j = 1, \cdots, n_2 - k_2\}$. Two concatenation procedures are presented [15]: the first is applicable when k_2 divides n_1, and the second is applicable when it does not. Since one of these two conditions will always be true, these two procedures are sufficient to handle any code $\mathcal{C}_2$. Note that the code construction introduced in Section 3.4.1 where $k_2 = 1$ is a special case of the first procedure introduced below since 1 divides n_1.

Procedure 1: *When $k_2 \mid n_1$*

Code construction here follows the same recipe as in Section 3.4.1. Step 1 is to encode k_1 qubits into n_1 qubits using C_1. Since by assumption, k_2 divides n_1, we can parse the n_1 qubits into blocks $B(b)$ ($b = 1, \ldots, n_1/k_2$), each containing k_2 qubits. Step 2 takes each block $B(b)$ and encodes its k_2 qubits into n_2 qubits using C_2. The total number of qubits used is $n = (n_1/k_2)n_2$. The generators for the final code C are constructed as in Section 3.4.1. First, to each block $B(b)$ we associate a copy of the generators of C_2: $G_2(b) = \{g_j^2(b) : j = 1, \ldots, n_2 - k_2\}$. Summing over all blocks, this introduces a total of $(n_2 - k_2)(n_1/k_2)$ generators. Finally, the encoded images of the generators of C_1 are added to the generators of C. The stabilizer S of C then has $(n_2 - k_2)(n_1/k_2) + (n_1 - k_1) = (n_1 n_2/k_2) - k_1$ generators. Since C is an $[n_1 n_2/k_2, k, d]$ code with $(n_1 n_2/k_2) - k$ generators, we see that it encodes $k = k_1$ qubits. The following theorem establishes a lower bound on the code distance d.

THEOREM 3.5
Let C_1 and C_2 be $[n_1, k_1, d_1]$ and $[n_2, k_2, d_2]$ quantum stabilizer codes, respectively, and suppose that k_2 divides n_1. If C is the $[n_1 n_2/k_2, k_1, d]$ concatenated code constructed from C_1 and C_2, then $d \geq d_1 d_2 / k_2$.

PROOF As in Theorem 3.4, let S, S_1, and S_2 be the stabilizers and $C(S)$, $C(S_1)$, and $C(S_2)$ the centralizers of the codes C, C_1, and C_2, respectively. Let E_* be a smallest weight element of $C(S) - S$ so that $\text{wt}(E_*) = d$. Without loss of generality, E_* can be written as a product of block operators $E_*(i)$:

$$E_* = \prod_{i=1}^{n_1/k_2} E_*(i) \ . \tag{3.69}$$

Since $E_* \in C(S) - S$, it commutes with all generators of C. Consequently, the block operator $E_*(i)$ must commute with the generators $G_2(i)$ of C_2 that are associated with block $B(i)$ by the concatenation procedure. Thus $E_*(i) \in C(S_2)$. Now those $E_*(i)$ that are not equal to a block identity-operator I_i cannot be elements of S_2 as this would make $E_* \in S$ and this would contradict our assumption that $E_* \in C(S) - S$. Thus the block operators $E_*(i) \neq I_i$ belong to $C(S_2) - S_2$ and so have

$$\text{wt}(E_*(i)) \geq d_2 \ . \tag{3.70}$$

Let r be the number of blocks $B(i)$ for which $E_*(i) \neq I_i$, and let $i_1, \ldots, i_r$ be the labels for those blocks. It follows from eq. (3.69) that

$$\text{wt}(E_*) = \text{wt}(E_*(i_1)) + \cdots + \text{wt}(E_*(i_r))$$
$$\geq r \, d_2 \ , \tag{3.71}$$

where we used eq. (3.70) to bound the weights on the RHS of the first line.

As in Theorem 3.4, E_* pulls back to an element $e_* = \xi_2^\dagger E_* \xi_2$, where $\xi_2^\dagger$ is the decoding operator for C_2. As shown in the proof of Theorem 3.4, $e_* \in C(S_1) - S_1$, and so $\text{wt}(e_*) \geq d_1$. Because of how the concatenated code C is constructed, the n_2 qubits in block $B(i)$ pull back to a block i containing k_2 qubits, and $E_*(i)$ pulls back to a block operator $e_*(i)$:

$$e_*(i) = \xi_2^\dagger E_*(i) \xi_2 . \tag{3.72}$$

Combining eqs. (3.69) and (3.72) with $e_* = \xi_2^\dagger E_* \xi_2$ gives

$$e_* = \prod_{i=1}^{n_1/k_2} e_*(i) . \tag{3.73}$$

Since $e_*(i)$ acts on the k_2 qubits in block i, its weight cannot exceed k_2: $\text{wt}(e_*(i)) \leq k_2$. It follows from eq. (3.72) that only r blocks in eq. (3.73) have $e_*(i)$ differing from the identity operator (for that block), and these blocks are again labeled by $i_1, \ldots, i_r$. Thus, from eq. (3.73) we have

$$\text{wt}(e_*) = \text{wt}(e_*(i_1)) + \cdots + \text{wt}(e_*(i_r))$$
$$\leq r k_2 .$$

Since $d_1 \leq \text{wt}(e_*)$ we have that $d_1 \leq r k_2$. Solving for r, plugging the result into eq. (3.71), and using that $\text{wt}(E_*) = d$ gives

$$d \geq \frac{d_1 d_2}{k_2} , \tag{3.74}$$

which establishes the desired lower bound on d. Notice that when $k_2 = 1$, we recover the lower bound in Theorem 3.4, although the derivation here is much simpler. ∎

Example 3.4

To illustrate Procedure 1 we set C_1 and C_2 both equal to the [4,2,2] code introduced in Section 3.2.2. This makes $n_1 = 4$ and $k_2 = 2$, and so Procedure 1 can be used as $k_2 \mid n_1$. Step 1 encodes 2 qubits into 4 qubits, which are then partitioned into two blocks. Block 1 contains qubits I and II, and block 2 contains qubits III and IV. Step 2 encodes block i into block $B(i)$ using code C_2. The qubits in block $B(1)$ will be labeled 1–4, and those in block $B(2)$ will be labeled 5–8. The final code C thus uses $n = 8$ qubits, which agrees with our general result $n = n_1 n_2 / k_2$. Each block $B(i)$ receives a copy of the generators of C_2 so that the following four operators are generators of C. For block $B(1)$ we have

$$g_1 = \sigma_x^1 \sigma_z^2 \sigma_z^3 \sigma_x^4$$
$$g_2 = \sigma_y^1 \sigma_x^2 \sigma_x^3 \sigma_y^4 ,$$

Quantum Stabilizer Codes

and for block $B(2)$ we have

$$g_3 = \sigma_x^5 \sigma_z^6 \sigma_z^7 \sigma_x^8$$
$$g_4 = \sigma_y^5 \sigma_x^6 \sigma_x^7 \sigma_y^8 \ .$$

The two remaining generators are the images of the generators of $\mathcal{C}_1$ under $\mathcal{C}_2$. The generators for $\mathcal{C}_1$ are (see eq. (3.37)):

$$g_1^1 = \left(\sigma_x^I \sigma_z^{II}\right)\left(\sigma_z^{III} \sigma_x^{IV}\right)$$
$$g_2^1 = \left(\sigma_y^I \sigma_x^{II}\right)\left(\sigma_x^{III} \sigma_y^{IV}\right) \ . \tag{3.75}$$

Under $\mathcal{C}_2$ the Pauli operators for the first qubit in each block $B(i)$ map as follows (see eq. (3.38)):

$$\begin{aligned}
\sigma_x^I &\longrightarrow X_I = \sigma_x^1 I_2 \sigma_y^3 \sigma_y^4 \\
\sigma_x^{III} &\longrightarrow X_{III} = \sigma_x^5 I_6 \sigma_y^7 \sigma_y^8 \\
\sigma_z^I &\longrightarrow Z_I = \sigma_y^1 \sigma_z^2 \sigma_y^3 I_4 \\
\sigma_z^{III} &\longrightarrow Z_{III} = \sigma_y^5 \sigma_z^6 \sigma_y^7 I_8 \ .
\end{aligned} \tag{3.76}$$

From these operators we find (suppressing factors of i^λ):

$$\begin{aligned}
\sigma_y^I &\longrightarrow Y_I = Z_I X_I \\
&= \sigma_z^1 \sigma_z^2 I_3 \sigma_y^4 \\
\sigma_y^{III} &\longrightarrow Y_{III} = Z_{III} X_{III} \\
&= \sigma_z^5 \sigma_z^6 I_7 \sigma_y^8 \ .
\end{aligned} \tag{3.77}$$

Similarly, $\mathcal{C}_2$ maps the Pauli operators on the second qubit in each block $B(i)$ as follows (see eq. (3.38)):

$$\begin{aligned}
\sigma_x^{II} &\longrightarrow X_{II} = \sigma_x^1 I_2 \sigma_x^3 \sigma_z^4 \\
\sigma_x^{IV} &\longrightarrow X_{IV} = \sigma_x^5 I_6 \sigma_x^7 \sigma_z^8 \\
\sigma_z^{II} &\longrightarrow Z_{II} = I_1 \sigma_x^2 \sigma_z^3 \sigma_z^4 \\
\sigma_z^{IV} &\longrightarrow Z_{IV} = I_5 \sigma_x^6 \sigma_z^7 \sigma_z^8 \ .
\end{aligned} \tag{3.78}$$

and

$$\begin{aligned}
Y_{II} &= \sigma_x^1 \sigma_x^2 \sigma_y^3 I_4 \\
Y_{IV} &= \sigma_x^5 \sigma_x^6 \sigma_y^7 I_8 \ .
\end{aligned} \tag{3.79}$$

Using eqs. (3.76)–(3.79) we see that $\mathcal{C}_2$ maps the generators of $\mathcal{C}_1$ (eq. (3.75)) to

$$\begin{aligned}
g_5 &= \overline{g_1^1} \\
&= \xi_2 \, g_1^1 \, \xi_2^\dagger \\
&= X_I Z_{II} Z_{III} X_{IV} \\
&= \left(\sigma_x^1 \sigma_x^2 \sigma_x^3 \sigma_x^4\right)\left(\sigma_z^5 \sigma_z^6 \sigma_z^7 \sigma_z^8\right)
\end{aligned}$$

and

$$\begin{aligned}g_6 &= \overline{g_2^1} \\ &= \xi_2 \, g_2^1 \, \xi_2^\dagger \\ &= Y_I \, X_{II} \, X_{III} \, Y_{IV} \\ &= \left(\sigma_y^1 \, \sigma_z^2 \, \sigma_x^3 \, \sigma_x^4\right) \left(I_5 \, \sigma_x^6 \, I_7 \, \sigma_y^8\right) \quad .\end{aligned}$$

By inspection we see that the generators $g_1, \ldots, g_6$ mutually commute. Clearly the number of generators $n-k$ is 6, which gives $k=2$ as expected. The reader can check that all one-qubit operators anticommute with at least one of the generators of $\mathcal{C}$. On the other hand, the distance 2 operator $\sigma_z^1 \sigma_y^2$ commutes with all of these generators. The reader can verify that $\sigma_z^1 \sigma_y^2 = X_{II} Z_I Z_{II}$ so that this error belongs to $\mathcal{C}(\mathcal{S})-\mathcal{S}$. Thus the code distance for $\mathcal{C}$ is $d=2$, which saturates the lower bound $d_1 d_2 / k_2$ of Theorem 3.5. We have thus constructed an [8,2,2] code. Finally, we construct the encoded Pauli operators X_i and Z_i ($i=1,2$) for $\mathcal{C}$. Starting with X_1, we have

$$\begin{aligned}X_1 &= \xi_2 \xi_1 \left(\sigma_x^1\right) \xi_1^\dagger \xi_2^\dagger \\ &= \xi_2 \left(\sigma_x^I \, I_{II} \, \sigma_y^{III} \, \sigma_y^{IV}\right) \xi_2^\dagger \\ &= X_I \, I_{II} \, Y_{III} \, Y_{IV} \\ &= \left(\sigma_x^1 \, I_2 \, \sigma_y^3 \, \sigma_y^4\right)\left(\sigma_y^5 \, \sigma_y^6 \, \sigma_y^7 \, \sigma_y^8\right) \quad .\end{aligned}$$

Similar calculations give

$$\begin{aligned}X_2 &= \left(\sigma_x^1 \, I_2 \, \sigma_y^3 \, \sigma_y^4\right)\left(\sigma_x^5 \, \sigma_x^6 \, \sigma_x^7 \, \sigma_x^8\right) \\ Z_1 &= \left(\sigma_z^1 \, \sigma_y^2 \, \sigma_z^3 \, \sigma_x^4\right)\left(\sigma_z^5 \, \sigma_z^6 \, I_7 \, \sigma_y^8\right) \\ Z_2 &= \left(\sigma_x^1 \, I_2 \, \sigma_x^3 \, \sigma_z^4\right)\left(\sigma_y^5 \, \sigma_y^6 \, \sigma_x^7 \, \sigma_z^8\right) \quad .\end{aligned}$$

☐

Procedure 2: When $k_2 \nmid n_1$

Here again, codes $\mathcal{C}_1$ and $\mathcal{C}_2$ are, respectively, $[n_1, k_1, d_1]$ and $[n_2, k_2, d_2]$ codes, however this time k_2 does not divide n_1. Procedure 2 begins by introducing k_2 sets of unencoded qubits $q(c)$ ($c = 1, \ldots, k_2$), each containing k_1 qubits. (i) The first step is to use $\mathcal{C}_1$ to map each set of k_1 qubits $q(c)$ into a bundle of n_1 qubits $Q(c)$. Associated with each bundle $Q(c)$ is a copy of the generators of $\mathcal{C}_1$: $G_1(c) = \{g_l^1(c) : l = 1, \ldots, n_1 - k_1\}$. Later, when forming generators for the final code $\mathcal{C}$, each of the $G_1(c)$ will be further encoded using $\mathcal{C}_2$. (ii) The next step takes the ith qubit from each bundle $Q(c)$ ($c = 1, \ldots, k_2$) to form a block $b(i)$ that contains k_2 qubits. Since the $Q(c)$ contain n_1 qubits, this produces a total of n_1 such blocks. (iii) Finally, $\mathcal{C}_2$ is used to map the k_2 qubits in block $b(i)$ into a block of n_2 qubits $B(i)$, with $i = 1, \ldots, n_1$. At the end of this construction we have n_1 blocks $B(i)$ each containing n_2 qubits so that the

Quantum Stabilizer Codes

final code $\mathcal{C}$ uses $n = n_1 n_2$ qubits. The generators of $\mathcal{C}$ are constructed in the same manner as in the two previous constructions. (1) To each block $B(i)$ we associate a copy of the generators of $\mathcal{C}_2$: $G_2(i) = \{g_m^2(i) : m = 1, \ldots, n_2 - k_2\}$. This introduces $n_1(n_2 - k_2)$ generators. (2) Finally, the image of the generator sets $G_1(c)$ under $\mathcal{C}_2$ are added to the generators of $\mathcal{C}$: $\overline{G_1}(c) = \{\overline{g_n^1}(c) : n = 1, \ldots, n_1 - k_1\}$. Here $\overline{g_n^1}(c) = \xi_2 \, g_n^1(c) \, \xi_2^\dagger$; ξ_2 is the encoding operator for $\mathcal{C}_2$; and $c = 1, \ldots, k_2$. This introduces $k_2(n_1 - k_1)$ generators. All together, $\mathcal{C}$ has $n_1(n_2 - k_2) + k_2(n_1 - k_1) = n_1 n_2 - k_1 k_2$ generators so that $\mathcal{C}$ encodes $k_1 k_2$ qubits. These are simply the $k_1 k_2$ unencoded qubits that we had at the beginning of the construction. Procedure 2 thus produces an $[n_1 n_2, k_1 k_2, d]$ code $\mathcal{C}$. The final code distance d satisfies the following lower bound.

THEOREM 3.6
Let $\mathcal{C}_1$ and $\mathcal{C}_2$ be $[n_1, k_1, d_1]$ and $[n_2, k_2, d_2]$ quantum stabilizer codes, respectively, and suppose that k_2 does not divide n_1. If $\mathcal{C}$ is the $[n_1 n_2, k_1 k_2, d]$ concatenated code constructed from $\mathcal{C}_1$ and $\mathcal{C}_2$, then $d \geq d_1 d_2$.

PROOF Here again, $\mathcal{S}$, $\mathcal{S}_1$, and $\mathcal{S}_2$ are the stabilizers, and $\mathcal{C}(\mathcal{S})$, $\mathcal{C}(\mathcal{S}_1)$, and $\mathcal{C}(\mathcal{S}_2)$ are the centralizers of $\mathcal{C}$, $\mathcal{C}_1$, and $\mathcal{C}_2$, respectively. We introduce an element $E_* \in \mathcal{C}(\mathcal{S}) - \mathcal{S}$ of smallest weight so that $\mathrm{wt}(E_*) = d$. As in eq. (3.69), E_* can be written as a product of block operators $E_*(i)$ over the blocks $B(i)$. As in Theorem 3.5, the $E_*(i) \neq I_i$ belong to $\mathcal{C}(\mathcal{S}_2) - \mathcal{S}_2$ and so

$$\mathrm{wt}(E_*(i)) \geq d_2 \ .$$

Let r be the number of blocks $B(i)$ for which $E_*(i) \neq I_i$ and $i_1, \ldots, i_r$ the labels for these blocks. As in eq. (3.71), one finds that

$$\mathrm{wt}(E_*) \geq r \, d_2 \ . \tag{3.80}$$

To learn about r, we pull back the operator E_* to an operator e_*. This is done by pulling back the blocks $B(i)$ to the blocks $b(i)$. Recall that $b(i)$ contains the ith qubit from each of the bundles $Q(c)$. Thus $E_*(i)$ pulls back to an operator $\epsilon(i)$ that contains k_2 factors, one for each bundle $Q(c)$:

$$\epsilon(i) = \prod_{c=1}^{k_2} \epsilon_c(i) \ .$$

Pulling back all the $E_*(i)$ thus produces the map $E_* \to e_*$, where

$$e_* = \prod_{c=1}^{k_2} e_*(c) \tag{3.81}$$

and

$$e_*(c) = \prod_{i=1}^{n_1} \epsilon_c(i) \ . \tag{3.82}$$

Since $E_*(i)$ contributes, at most, one non-identity element to $e_*(c)$, and only r blocks $B(i_1), \ldots, B(i_r)$ have $E_*(i) \neq I_i$, eq. (3.82) gives that

$$\text{wt}(e_*(c)) \leq r \ . \tag{3.83}$$

As in Theorem 3.4, one finds that $[e_*, g_j^1(c)] = 0$ for all $g_j^1(c)$. This together with eq. (3.81) gives that $[e_*(c), g_j^1(c)] = 0$ so that $e_*(c) \in \mathcal{C}(\mathcal{S}_1)$. As in Theorems 3.4 and 3.5, for those $e_*(c)$ that are not equal to the identity operator, $e_*(c) \in \mathcal{C}(\mathcal{S}_1) - \mathcal{S}_1$ and for them,

$$d_1 \leq \text{wt}(e_*(c)) \ . \tag{3.84}$$

Combining eqs. (3.83) and (3.84) gives $d_1 \leq r$. This together with eq. (3.80) and $d = \text{wt}(E_*)$ finally gives

$$d \geq d_1 d_2 \ ,$$

which proves the theorem. ∎

The reader is encouraged to solve Problem 3.6, which illustrates Procedure 2 by concatenating the [5,1,3] code with the [4,2,2] code.

Problems

3.1 Determine the basis codewords for the following codes: (a) [5,1,3], and (b) [4,2,2].

3.2 This problem shows how to obtain the [4,2,2] code from the [5,1,3] code. Ultimately the fifth qubit is discarded and two generators of the [5,1,3] code are converted into the X and Z operators of the second qubit in the [4,2,2] code. Some manipulations are needed before the fifth qubit is discarded to insure that the operators of the new code commute and anticommute properly.

(a) Choose two generators from the [5,1,3] code such that one ends in σ_x and the other ends in σ_z. If necessary, generators can be multiplied together to produce the appropriate ending. Strip off the last/fifth qubit operator and label the two new four-qubit operators X_2 and Z_2. Show that choosing g_2 and g_3 from eq. (3.32) yields the operators X_2 and Z_2 in eq. (3.38).

(b) Take the remaining two generators of the [5,1,3] code and multiply them as needed by the other generators so that the last/fifth qubit operator in each becomes the identity operator I. Strip off this identity operator from each and let the resulting four-qubit operators be the generators of the [4,2,2] code. Following the choice made in (a), show that this procedure yields the generators in eq. (3.37).

(c) The final step is to convert the X and Z operators of the five-qubit code into X_1 and Z_1 for the [4,2,2] code. To do this, multiply each of them as needed by the generators of the [5,1,3] code so that the last operator of each becomes the identity. Strip off this final identity operator and label the new four-qubit operators X_1 and Z_1. Specifically, use g_2 and g_3 to put X and Z into appropriate form and show that these manipulations yield the X_1 and Z_1 operators appearing in eq. (3.38) upon discarding the fifth qubit.

(d) Verify that the generators of the new code commute among themselves and also with the X_i and Z_i, and that the X_i and Z_i satisfy eqs. (3.24)-(3.26).

3.3 Section 3.3 introduced a 1-1 correspondence between the quotient group $\mathcal{G}_n/\mathcal{C}$ and the $2n$-dimensional binary vector space F_2^{2n}.

(a) Show that F_2^{2n} is an abelian group in which the group operation is vector addition.

(b) Show that $\mathcal{G}_n/\mathcal{C}$ is an abelian group.

(c) Show that the 1-1 correspondence between $\mathcal{G}_n/\mathcal{C}$ and F_2^{2n} is an isomorphism.

(d) With $X_i' = X_i s$ and X_i and s as defined in eqs. (3.55) and (3.56), use this isomorphism to give a simple derivation of eq. (3.58).

The following two problems establish some intermediate results needed to prove Theorem 3.4. In both these problems, factors of i^λ (such as those appearing in eqs. (3.59) and (3.60)) can be safely ignored.

3.4 Consider an $[n_1 n_2, k_1, d]$ concatenated code $\mathcal{C}$ as in Theorem 3.4. Let X_i and Z_i be the encoded σ_x^i and σ_z^i operators for block $B(i)$:

$$X_i = \left(\sigma_{x,1}^i\right)^{c_1^i} \cdots \left(\sigma_{x,n_2}^i\right)^{c_{n_2}^i} \left(\sigma_{z,1}^i\right)^{d_1^i} \cdots \left(\sigma_{z,n_2}^i\right)^{d_{n_2}^i}$$
$$Z_i = \left(\sigma_{x,1}^i\right)^{e_1^i} \cdots \left(\sigma_{x,n_2}^i\right)^{e_{n_2}^i} \left(\sigma_{z,1}^i\right)^{f_1^i} \cdots \left(\sigma_{z,n_2}^i\right)^{f_{n_2}^i} \,, \qquad (3.85)$$

where $\sigma_{x,j}^i$ ($\sigma_{z,j}^i$) is the Pauli operator σ_x (σ_z) acting on qubit j in block $B(i)$; $c_j^i, d_j^i, e_j^i, f_j^i = 0, 1$; and there are n_1 blocks $B(i)$ each containing n_2 qubits. For purposes of this problem it is safe to discard factors of i^λ in equations containing operators.

(a) Show that $(a_i, b_i = 0, 1)$

$$X_i^{a_i} Z_i^{b_i} = \prod_{l=1}^{n_2} \left(\sigma_{x,l}^i\right)^{a_i c_l^i + b_i e_l^i} \prod_{m=1}^{n_2} \left(\sigma_{z,m}^i\right)^{a_i d_m^i + b_i f_m^i} \,.$$

(b) Show that

$$wt(X_i^{a_i} Z_i^{b_i}) = \sum_{j=1}^{n_2} \left[\begin{array}{c} (a_i c_j^i + b_i e_j^i) + (a_i d_j^i + b_i f_j^i) \\ - (a_i c_j^i + b_i e_j^i)(a_i d_j^i + b_i f_j^i) \\ - 2 a_i b_i \left\{ c_j^i e_j^i + d_j^i f_j^i - c_j^i f_j^i - d_j^i e_j^i \right\} \end{array} \right] \,.$$

(c) Following Section 3.3, let $\mathbf{X}_i = (c^i|d^i)$ and $\mathbf{Z}_i = (e^i|f^i)$ be the $2n_2$-dimensional binary vectors corresponding to the operators X_i and Z_i, and where the n_2-dimensional binary vectors c^i, d^i, e^i, f^i are read off from eq. (3.85). Recalling the intersection operation $*$ defined in Problem 1.1, show that

$$wt(X_i) = wt(c^i) + wt(d^i) - wt(c^i * d^i)$$
$$wt(Z_i) = wt(e^i) + wt(f^i) - wt(e^i * f^i) \ .$$

(d) Introducing $\mathbf{Z}_i^\dagger \equiv (f^i|e^i)$, show that

$$wt(X_i^{a_i} Z_i^{b_i}) = a_i \, wt(X_i) + b_i \, wt(Z_i) - a_i b_i \left[2wt(\mathbf{X}_i * \mathbf{Z}_i) - wt(\mathbf{X}_i * \mathbf{Z}_i^\dagger) \right] \ .$$

(e) Noting that $Y_i = iZ_i X_i$, show that

$$wt(Y_i) = wt(X_i) + wt(Z_i) - \left[2wt(\mathbf{X}_i * \mathbf{Z}_i) - wt(\mathbf{X}_i * \mathbf{Z}_i^\dagger) \right] \ .$$

(f) Show that

$$wt(X_i^{a_i} Z_i^{b_i}) = a_i \, wt(X_i) + b_i \, wt(Z_i) + a_i b_i \left[wt(Y_i) - wt(X_i) - wt(Z_i) \right] \ .$$

3.5 Let e_* and E_* be the operators defined in eqs. (3.59) and (3.60), respectively.

(a) Show that

$$wt(E_*) = \sum_{i=1}^{n_1} wt(X_i^{a_i} Z_i^{b_i})$$
$$= \sum_{i=1}^{n_1} \left[a_i (1 - b_i) \, wt(X_i) + b_i (1 - a_i) \, wt(Z_i) + a_i b_i \, wt(Y_i) \right] \ ,$$

and

$$wt(e_*) = \sum_{i=1}^{n_1} (a_i + b_i - a_i b_i) \ .$$

(b) Recall from the proof of Theorem 3.4 that $a_i, b_i = 0, 1$ and $X_i, Y_i,$ and Z_i belong to $\mathcal{C}(S_2) - S_2$. Show that

$$wt(E_*) \geq d_2 \, wt(e_*) \ .$$

3.6 Let $\mathcal{C}$ be the concatenated code constructed using the $[5, 1, 3]$ code as $\mathcal{C}_1$ and the $[4, 2, 2]$ code as $\mathcal{C}_2$. Since k_2 does not divide n_1, apply the second construction procedure described in Section 3.4.2.

Quantum Stabilizer Codes

(a) Show that the 18 generators of C are

$$g_1 = \sigma_x^1 \sigma_z^2 \sigma_z^3 \sigma_x^4$$
$$g_2 = \sigma_y^1 \sigma_x^2 \sigma_x^3 \sigma_y^4$$
$$g_3 = \sigma_x^5 \sigma_z^6 \sigma_z^7 \sigma_x^8$$
$$g_4 = \sigma_y^5 \sigma_x^6 \sigma_x^7 \sigma_y^8$$
$$g_5 = \sigma_x^9 \sigma_z^{10} \sigma_z^{11} \sigma_x^{12}$$
$$g_6 = \sigma_y^9 \sigma_x^{10} \sigma_x^{11} \sigma_y^{12}$$
$$g_7 = \sigma_x^{13} \sigma_z^{14} \sigma_z^{15} \sigma_x^{16}$$
$$g_8 = \sigma_y^{13} \sigma_x^{14} \sigma_x^{15} \sigma_y^{16}$$
$$g_9 = \sigma_x^{17} \sigma_z^{18} \sigma_z^{19} \sigma_x^{20}$$
$$g_{10} = \sigma_y^{17} \sigma_x^{18} \sigma_x^{19} \sigma_y^{20}$$
$$g_{11} = \left(\sigma_x^1 I^2 \sigma_y^3 \sigma_y^4\right) \left(\sigma_y^5 \sigma_z^6 \sigma_y^7 I^8\right) \left(\sigma_y^9 \sigma_z^{10} \sigma_y^{11} I^{12}\right) \left(\sigma_x^{13} I^{14} \sigma_y^{15} \sigma_y^{16}\right)$$
$$g_{12} = \left(\sigma_x^5 I^6 \sigma_y^7 \sigma_y^8\right) \left(\sigma_y^9 \sigma_z^{10} \sigma_y^{11} I^{12}\right) \left(\sigma_y^{13} \sigma_z^{14} \sigma_y^{15} I^{16}\right) \left(\sigma_x^{17} I^{18} \sigma_y^{19} \sigma_y^{20}\right)$$
$$g_{13} = \left(\sigma_x^1 I^2 \sigma_y^3 \sigma_y^4\right) \left(\sigma_x^9 I^{10} \sigma_y^{11} \sigma_y^{12}\right) \left(\sigma_y^{13} \sigma_z^{14} \sigma_y^{15} I^{16}\right) \left(\sigma_y^{17} \sigma_z^{18} \sigma_y^{19} I^{20}\right)$$
$$g_{14} = \left(\sigma_y^1 \sigma_z^2 \sigma_y^3 I^4\right) \left(\sigma_x^5 I^6 \sigma_y^7 \sigma_y^8\right) \left(\sigma_x^{13} I^{14} \sigma_y^{15} \sigma_y^{16}\right) \left(\sigma_x^{17} I^{18} \sigma_y^{19} I^{20}\right)$$
$$g_{15} = \left(\sigma_x^1 I^2 \sigma_z^3 \sigma_z^4\right) \left(I^5 \sigma_x^6 \sigma_z^7 \sigma_z^8\right) \left(I^9 \sigma_x^{10} \sigma_z^{11} \sigma_z^{12}\right) \left(\sigma_x^{13} I^{14} \sigma_x^{15} \sigma_z^{16}\right)$$
$$g_{16} = \left(\sigma_x^5 I^6 \sigma_z^7 \sigma_z^8\right) \left(I^9 \sigma_x^{10} \sigma_z^{11} \sigma_z^{12}\right) \left(I^{13} \sigma_x^{14} \sigma_z^{15} \sigma_z^{16}\right) \left(\sigma_x^{17} I^{18} \sigma_z^{19} \sigma_z^{20}\right)$$
$$g_{17} = \left(\sigma_x^1 I^2 \sigma_z^3 \sigma_z^4\right) \left(\sigma_x^9 I^{10} \sigma_x^{11} \sigma_z^{12}\right) \left(I^{13} \sigma_x^{14} \sigma_z^{15} \sigma_z^{16}\right) \left(I^{17} \sigma_x^{18} \sigma_z^{19} \sigma_z^{20}\right)$$
$$g_{18} = \left(I^1 \sigma_x^2 \sigma_z^3 \sigma_z^4\right) \left(\sigma_x^5 I^6 \sigma_x^7 \sigma_z^8\right) \left(\sigma_x^{13} I^{14} \sigma_x^{15} \sigma_z^{16}\right) \left(I^{17} \sigma_x^{18} \sigma_z^{19} \sigma_z^{20}\right).$$

(b) Show that C, respectively, encodes σ_x^1, σ_x^2, σ_z^1, and σ_z^2 to

$$X_1 = \left(\sigma_x^1 I^2 \sigma_y^3 \sigma_y^4\right) \left(\sigma_x^5 I^6 \sigma_y^7 \sigma_y^8\right) \left(\sigma_x^9 I^{10} \sigma_y^{11} \sigma_y^{12}\right) \left(\sigma_x^{13} I^{14} \sigma_y^{15} \sigma_y^{16}\right) \left(\sigma_x^{17} I^{18} \sigma_y^{19} \sigma_y^{20}\right)$$
$$X_2 = \left(\sigma_x^1 I^2 \sigma_x^3 \sigma_z^4\right) \left(\sigma_x^5 I^6 \sigma_x^7 \sigma_z^8\right) \left(\sigma_x^9 I^{10} \sigma_x^{11} \sigma_z^{12}\right) \left(\sigma_x^{13} I^{14} \sigma_x^{15} \sigma_z^{16}\right) \left(\sigma_x^{17} I^{18} \sigma_x^{19} \sigma_z^{20}\right)$$
$$Z_1 = \left(\sigma_y^1 \sigma_z^2 \sigma_y^3 I^4\right) \left(\sigma_y^5 \sigma_z^6 \sigma_y^7 I^8\right) \left(\sigma_y^9 \sigma_z^{10} \sigma_y^{11} I^{12}\right) \left(\sigma_y^{13} \sigma_z^{14} \sigma_y^{15} I^{16}\right) \left(\sigma_y^{17} \sigma_z^{18} \sigma_y^{19} I^{20}\right)$$
$$Z_2 = \left(I^1 \sigma_x^2 \sigma_z^3 \sigma_z^3\right) \left(I^5 \sigma_x^6 \sigma_z^7 \sigma_z^8\right) \left(I^9 \sigma_x^{10} \sigma_z^{11} \sigma_z^{12}\right) \left(I^{13} \sigma_x^{14} \sigma_z^{15} \sigma_z^{16}\right) \left(I^{17} \sigma_x^{18} \sigma_z^{19} \sigma_z^{20}\right).$$

References

[1] Gottesman, D., Class of quantum error correcting codes saturating the quantum Hamming bound, *Phys. Rev. A* **54**, 1862, 1996.

[2] Calderbank, A. R. et al., Quantum error correction and orthogonal geometry, *Phys. Rev. Lett.* **78**, 405, 1997.

[3] Calderbank, A. R. et al. Quantum error correction via codes over GF(4), *IEEE Trans. Inf. Theor.* **44**, 1369, 1998.

[4] Steane, A. M., Simple quantum error-correcting codes, *Phys. Rev. A* **54**, 4741, 1996.

[5] Bennett, C. H., DiVincenzo, D. P., Smolin, J. A., and Wooters, W. K., Mixed-state entanglement and quantum error correction, *Phys. Rev. A* **54**, 3824, 1996.

[6] LaFlamme, R., Miguel, C., Paz, J. P., and Zurek, W. H., Perfect quantum error correcting code, *Phys. Rev. Lett.* **77**, 198, 1996.

[7] Gottesman, D., Pasting quantum codes, download at http://arXiv.org/abs/quant-ph/9607027, 1996.

[8] Steane, A. M., Quantum Reed-Muller codes, *IEEE Trans. Inf. Theory* **45**, 1701, 1999.

[9] MacWilliams, F. J. and Sloane, N. J. A., *The Theory of Error Correcting Codes*, North-Holland, New York, 1977.

[10] Knill, E. and Laflamme, R., Concatenated quantum codes, download at http://arXiv.org/abs/quant-ph/9608012, 1996.

[11] Aharonov, D. and Ben-Or, M., Fault-tolerant quantum computation with constant error, in *Proceedings of the Twenty-Ninth ACM Symposium on the Theory of Computing*, 1997, pp. 176–188.

[12] Knill, E., Laflamme, R., and Zurek, W. H., Resilient quantum computation: error models and thresholds, *Proc. R. Soc. Lond. A* **454**, 365, 1998.

[13] Knill, E., Laflamme, R., and Zurek, W. H., Resilient quantum computation, *Science* **279**, 342, 1998.

[14] Zalka, C., Threshold estimate for fault tolerant quantum computation, download at http://arXiv.org/abs/quant-ph/9612028, 1996.

[15] Gottesman, D., Stabilizer codes and quantum error correction, Ph. D. thesis, California Institute of Technology, Pasadena, CA, 1997.

[16] Preskill, J., Reliable quantum computers, *Proc. R. Soc. Lond. A* **454**, 385, 1998.

[17] Kitaev, A. Y., Quantum computation: algorithms and error correction, *Russ. Math. Surv.* **52**, 1191, 1997.

[18] Kitaev, A. Y., Quantum error correction with imperfect gates, in *Quantum Communication, Computing, and Measurement*, Plenum Press, New York, 1997, pp. 181–188.

4

Quantum Stabilizer Codes: Efficient Encoding and Decoding

The previous two chapters laid out the theoretical framework for quantum error correcting codes, and especially for quantum stabilizer codes. Following Cleve and Gottesman [1], the present chapter will explain how to construct quantum circuits that efficiently encode and decode a quantum stabilizer code. Before entering into that discussion, it proves advantageous to introduce a standard form for quantum stabilizer codes. This is done in Section 4.1. Assuming that a quantum stabilizer code in standard form has been given, Sections 4.2 and 4.3 explain how to, respectively, encode and decode its codewords. A brief summary of quantum circuits is given in Appendix C.

4.1 Standard Form

This chapter makes heavy use of the alternate formulation of quantum stabilizer codes introduced in Section 3.3. The reader is asked to review that discussion before proceeding.

We begin by noting that the formulation of Section 3.3 rests on a 1-1 correspondence between the quotient group $\mathcal{G}_n/\mathcal{C}$ and the $2n$-dimensional binary vector space F_2^{2n}, where $\mathcal{G}_n$ is the Pauli group and $\mathcal{C} = \{\pm I, \pm iI\}$ is the center of $\mathcal{G}_n$. Specifically, if $e = i^{\lambda_e}\sigma_x(a_e)\sigma_z(b_e)$ is an element of $\mathcal{G}_n$, it belongs to the coset $e\mathcal{C}$ in $\mathcal{G}_n/\mathcal{C}$, and its image under the 1-1 correspondence is the binary vector v_e:

$$v_e = \begin{pmatrix} a_e \\ b_e \end{pmatrix} = \begin{pmatrix} a_{e,1} \\ \vdots \\ a_{e,n} \\ b_{e,1} \\ \vdots \\ b_{e,n} \end{pmatrix}. \qquad (4.1)$$

Consistent with eq. (4.1), the transpose $v_e^T = (a_e \mid b_e)$ is a $2n$-component row vector that belongs to the (transposed) space $(F_2^{2n})^T$. A symplectic inner product $\langle v, w \rangle$ can be defined that maps pairs of vectors in F_2^{2n} onto $F_2 = \{0, 1\}$. Recall that two operators e and f in $\mathcal{G}_n$ commute (anticommute) if and only if the inner product of their images $\langle v_e, v_f \rangle$ is 0 (1). The generators $\{g_j = i^{\lambda_j} \sigma_x(a_j) \sigma_z(b_j) : j = 1, \ldots, n-k\}$ map to the set of linearly independent vectors

$$v_j = \begin{pmatrix} a_j \\ b_j \end{pmatrix} \quad (j = 1, \ldots, n - k) \; . \tag{4.2}$$

Since each element s of the stabilizer $\mathcal{S}$ is a product of the generators $\{g_j\}$, $s = g_1^{c_1} \cdots g_{n-k}^{c_{n-k}}$, its image v_s under the 1-1 correspondence is a linear combination of the v_j:

$$v_s = \sum_{j=1}^{n-k} c_j v_j \; . \tag{4.3}$$

Thus the $\{v_j : j = 1, \ldots, n-k\}$ span a linear subspace $S \subset F_2^{2n}$, which is the image of the stabilizer $\mathcal{S}$ under the 1-1 correspondence. Similarly, the row vectors $\{v_j^T : j = 1, \ldots, n-k\}$ span the subspace $S^T \subset (F_2^{2n})^T$, which is isomorphic to S.

It proves convenient to introduce a matrix $\mathcal{H}$ whose j^{th} row is v_j^T:

$$\mathcal{H} = \begin{pmatrix} \text{---} & v_1^T & \text{---} \\ & \vdots & \\ \text{---} & v_{n-k}^T & \text{---} \end{pmatrix}$$

$$= \left(\begin{array}{ccc|ccc} a_{1,1} & \cdots & a_{1,n} & b_{1,1} & \cdots & b_{1,n} \\ & \vdots & & & \vdots & \\ a_{n-k,1} & \cdots & a_{n-k,n} & b_{n-k,1} & \cdots & b_{n-k,n} \end{array} \right) \; . \tag{4.4}$$

By construction, S^T is the row space of $\mathcal{H}$, and $\mathcal{H}$ naturally breaks up into two $(n-k) \times n$ matrices $\mathcal{A}$ and $\mathcal{B}$:

$$\mathcal{H} = (\mathcal{A}|\mathcal{B}) \; , \tag{4.5}$$

where

$$\mathcal{A} = \begin{pmatrix} a_{1,1} & \cdots & a_{1,n} \\ & \vdots & \\ a_{n-k,1} & \cdots & a_{n-k,n} \end{pmatrix} \; ; \quad \mathcal{B} = \begin{pmatrix} b_{1,1} & \cdots & b_{1,n} \\ & \vdots & \\ b_{n-k,1} & \cdots & b_{n-k,n} \end{pmatrix} \; . \tag{4.6}$$

Because of the 1-1 correspondence, the ith columns of $\mathcal{A}$ and $\mathcal{B}$ are associated with the ith qubit of the n-qubit quantum register. Interchanging qubits i and j in the register causes an interchange of the ith and j^{th} columns within each of the matrices $\mathcal{A}$ and $\mathcal{B}$. Since such a qubit interchange effects a unitary transformation on the state of the quantum register, it produces a unitarily equivalent quantum stabilizer code. We can thus make such column interchanges freely, knowing that this will not alter the stabilizer code. In a similar vein, note that adding row j of $\mathcal{H}$ to row i ($j \neq i$) causes $v_i \to v_i + v_j$ and generator $g_i \to g_i g_j$. Such a transformation leaves the codewords and stabilizer invariant. Thus the rows of $\mathcal{H}$ can be added together freely without altering the stabilizer code. Together, these two operations allow us to do Gauss-Jordan elimination on $\mathcal{H}$. The resulting matrix will give the standard form of the generators.

Gauss-Jordan elimination of $\mathcal{H}$ begins in the usual way. One interchanges columns as needed to put suitable values in the pivot positions and then adds rows together as needed so that $\mathcal{H}$ takes the form

$$\mathcal{H} = \left(\begin{array}{cc|cc} \overbrace{I}^{r} & \overbrace{A}^{n-r} & \overbrace{B}^{r} & \overbrace{C}^{n-r} \\ 0 & 0 & D & E \end{array} \right) \begin{array}{l} \} r \\ \} n-k-r \end{array} \quad .$$

Here r is the rank of A and the r generators corresponding to the first r rows of $\mathcal{H}$ are type 1 generators, as defined by Gottesman [2]. As he showed, each basis codeword will be a superposition of 2^r (n-qubit) computational basis states. The remaining $n - k - r$ generators are either type 2 generators, or type 3 generators that have been converted to type 2. The final step in putting $\mathcal{H}$ in standard form is to carry out Gauss-Jordan elimination on E. Note that because the generators are mutually commuting, E will always have maximal rank $n-k-r$. To see this, assume the contrary so that E has rank $n-k-r-s$ and $s > 0$. Then $\mathcal{H}$ can be reduced to

$$\mathcal{H} = \left(\begin{array}{ccc|ccc} \overbrace{I}^{r} & \overbrace{A_1}^{n-k-r-s} & \overbrace{A_2}^{k+s} & \overbrace{B}^{r} & \overbrace{C_1}^{n-k-r-s} & \overbrace{C_2}^{k+s} \\ 0 & 0 & 0 & D_1 & I & E_2 \\ 0 & 0 & 0 & D_2 & 0 & 0 \end{array} \right) \begin{array}{l} \} r \\ \} n-k-r-s \\ \} s \end{array} \quad .$$

But this form for $\mathcal{H}$ indicates that the first r generators will *not* commute with the last s generators. We have our contradiction since the generators are a mutually commuting set of operators. Thus $s = 0$, E has maximal rank, and the standard form of $\mathcal{H}$ is

$$\mathcal{H} = \left(\begin{array}{ccc|ccc} \overbrace{I}^{r} & \overbrace{A_1}^{n-k-r} & \overbrace{A_2}^{k} & \overbrace{B}^{r} & \overbrace{C_1}^{n-k-r} & \overbrace{C_2}^{k} \\ 0 & 0 & 0 & D & I & E \end{array} \right) \begin{array}{l} \} r \\ \} n-k-r \end{array} \quad . \quad (4.7)$$

Pulling back the rows of $\mathcal{H}$ under the 1-1 correspondence gives the generators of $\mathcal{S}$ to within a factor of i^{λ_j}. Note that, in this chapter, whenever pulling back a (transpose) vector $v^T = (a|b)$ to an operator $i^\lambda \sigma_x(a) \sigma_z(b)$, we will always choose $i^\lambda \equiv 1$.

Next on the agenda is obtaining the standard form for the encoded Pauli operators X_i and Z_i ($i = 1, \ldots, k$). We begin with X_i, which maps to the vector $v(X_i)$ whose transpose is

$$v^T(X_i) = (\,u_1^T(i)\,u_2^T(i)\,u_3^T(i)\,|\,v_1^T(i)\,v_2^T(i)\,v_3^T(i)\,) \quad (i=1,\ldots,k) \; . \quad (4.8)$$

The components of $v^T(X_i)$ have been partitioned to match the partitioning of the columns of $\mathcal{H}$. Thus $u_1^T(i)$ and $v_1^T(i)$ have r components, $u_2^T(i)$ and $v_2^T(i)$ have $n-k-r$ components, and $u_3^T(i)$ and $v_3^T(i)$ have k components. Section 3.3 showed that the $v(X_i)$ and $v(Z_i)$ have $n-k$ degrees of freedom. They are fixed by setting $u_1(i) = 0$ and $v_2(i) = 0$. Thus $v^T(X_i) = (0\,u_2^T(i)\,u_3^T(i)\,|\,v_1^T(i)\,0\,v_3^T(i))$. Since the $X_i \in \mathcal{C}(\mathcal{S})$, they must commute with the generators of $\mathcal{S}$ and so the $v(X_i)$ must be orthogonal to the generator image vectors v_j. Recalling eq. (3.46), this requires

$$\mathcal{H}\,J\,v(X_i) = 0 \; ,$$

which expands to

$$\begin{pmatrix} 0 \\ 0 \end{pmatrix} = \begin{pmatrix} I & A_1 & A_2 & | & B & C_1 & C_2 \\ 0 & 0 & 0 & | & D & I & E \end{pmatrix} \begin{pmatrix} v_1(i) \\ 0 \\ v_3(i) \\ 0 \\ u_2(i) \\ u_3(i) \end{pmatrix} ,$$

or finally to

$$v_1(i) + A_2 v_3(i) + C_1 u_2(i) + C_2 u_3(i) = 0$$
$$u_2(i) + E u_3(i) = 0 \; . \quad (4.9)$$

Now introduce a matrix $\mathcal{X}$ whose i^{th} row is defined to be $v^T(X_i)$:

$$\mathcal{X} = \begin{pmatrix} \text{------} v^T(X_1) \text{------} \\ \vdots \\ \text{------} v^T(X_k) \text{------} \end{pmatrix} = \begin{pmatrix} 0 & u_2^T & u_3^T & | & v_1^T & 0 & v_3^T \end{pmatrix} ,$$

where

$$u_2^T = \begin{pmatrix} u_{2,1}(1) & \cdots & u_{2,n-k-r}(1) \\ \vdots & & \vdots \\ u_{2,1}(k) & \cdots & u_{2,n-k-r}(k) \end{pmatrix} ; \; u_3^T = \begin{pmatrix} u_{3,1}(1) & \cdots & u_{3,k}(1) \\ \vdots & & \vdots \\ u_{3,1}(k) & \cdots & u_{3,k}(k) \end{pmatrix} \quad (4.10)$$

$$v_1^T = \begin{pmatrix} v_{1,1}(1) & \cdots & v_{1,r}(1) \\ \vdots & & \vdots \\ v_{1,1}(k) & \cdots & v_{1,r}(k) \end{pmatrix} ; \; v_3^T = \begin{pmatrix} v_{3,1}(1) & \cdots & v_{3,k}(1) \\ \vdots & & \vdots \\ v_{3,1}(k) & \cdots & v_{3,k}(k) \end{pmatrix} . \quad (4.11)$$

The matrix $\mathcal{X}$ makes it easy to express the last remaining constraint on the X_i – they must commute among themselves. This requires

$$0 = \mathcal{X} J \mathcal{X} \; ,$$

or

$$0 = \begin{pmatrix} 0 & u_2^T & u_3^T & | & v_1^T & 0 & v_3^T \end{pmatrix} \begin{pmatrix} v_1 \\ 0 \\ v_3 \\ 0 \\ u_2 \\ u_3 \end{pmatrix}$$

$$= u_3^T v_3 + v_3^T u_3 \quad .$$

A particular solution of this equation is $u_3 = I$ and $v_3 = 0$. Plugging this back into eq. (4.9) gives $u_2 = E$ and $v_1 = C_1 E + C_2$, where the reader is reminded that addition is modulo 2. Given $\mathcal{H}$ in standard form we can identify C_1, C_2, and E, and from them, obtain $\mathcal{X}$ in standard form:

$$\mathcal{X} = \begin{pmatrix} 0 & E^T & I & | & (E^T C_1^T + C_2^T) & 0 & 0 \end{pmatrix} \quad . \tag{4.12}$$

Its rows then give the $v^T(X_i)$ in standard form.

The Z_i are handled in a similar manner. As with $v(X_i)$, the $n - k$ degrees of freedom in $v(Z_i)$ are fixed by requiring $\bar{u}_1(i) = 0$ and $\bar{v}_2(i) = 0$ so that $v^T(Z_i) = (\, 0\, \bar{u}_2^T(i)\, \bar{u}_3^T(i)\, |\, \bar{v}_1^T\, 0\, \bar{v}_3^T(i)\,)$. We introduce a matrix $\mathcal{Z}$ that is the analog of $\mathcal{X}$. Its i th row is defined to be $v^T(Z_i)$ and reduces to the form

$$\mathcal{Z} = \begin{pmatrix} 0 & \bar{u}_2^T & \bar{u}_3^T & | & \bar{v}_1^T & 0 & \bar{v}_3^T \end{pmatrix} \quad . \tag{4.13}$$

The form of $\bar{u}_2^T, \ldots, \bar{v}_3^T$ matches, respectively, that of $u_2^T, \ldots, v_3^T$ in eqs. (4.10) and (4.11), and so we do not write them out explicitly. Since the Z_i commute with the generators g_j, we must have

$$0 = \mathcal{H} J \mathcal{Z} \quad ,$$

or (after a little algebra),

$$\bar{v}_1 + A_2 \bar{v}_3 + C_1 \bar{u}_3 + C_2 \bar{u}_3 = 0$$
$$\bar{u}_2 + E \bar{u}_3 = 0 \quad . \tag{4.14}$$

The X_j must commute with Z_i for $j \neq i$ and X_i must anticommute with Z_i. This requires

$$\mathcal{X} J \mathcal{Z} = I \quad ,$$

or

$$I = \begin{pmatrix} 0 & u_2^T & I & | & v_1^T & 0 & 0 \end{pmatrix} \begin{pmatrix} \bar{v}_1 \\ 0 \\ \bar{v}_3 \\ 0 \\ \bar{u}_2 \\ \bar{u}_3 \end{pmatrix}$$

$$= \bar{v}_3 \quad ,$$

and so $\bar{v}_3 = I$. Finally, the Z_i must commute among themselves so that $\mathcal{Z}J\mathcal{Z} = 0$. Using $\bar{v}_3 = I$ gives

$$0 = \mathcal{Z}J\mathcal{Z}$$
$$= \bar{u}_3^T + \bar{u}_3 \;,$$

for which $\bar{u}_3 = 0$ is a particular solution. Plugging all these results into eq. (4.14) gives $\bar{v}_1 = A_2$ and $\bar{u}_2 = 0$. Eq. (4.13) then gives the standard form for $\mathcal{Z}$:

$$\mathcal{Z} = \begin{pmatrix} 0\;0\;0 \,|\, A_2^T\;0\;I \end{pmatrix} \;. \tag{4.15}$$

Its rows then give the $v^T(Z_i)$ in standard form. The following example determines the standard form for the [5,1,3] code.

Example 4.1
We start with eq. (3.32), which gives the generators for the [5,1,3] code. From it we obtain our initial $\mathcal{H}$:

$$\mathcal{H} = \begin{pmatrix} 1\;0\;0\;1\;0 & 0\;1\;1\;0\;0 \\ 0\;1\;0\;0\;1 & 0\;0\;1\;1\;0 \\ 1\;0\;1\;0\;0 & 0\;0\;0\;1\;1 \\ 0\;1\;0\;1\;0 & 1\;0\;0\;0\;1 \end{pmatrix} \;.$$

By appropriately adding rows together, $\mathcal{H}$ can be reduced to

$$\mathcal{H} = \begin{pmatrix} 1\;0\;0\;0\;1 & 1\;1\;0\;1\;1 \\ 0\;1\;0\;0\;1 & 0\;0\;1\;1\;0 \\ 0\;0\;1\;0\;1 & 1\;1\;0\;0\;0 \\ 0\;0\;0\;1\;1 & 1\;0\;1\;1\;1 \end{pmatrix} \;.$$

We see that $r = 4$, $A_1 = 0$, $C_1 = 0$, $E = 0$, and

$$B = \begin{pmatrix} 1\;1\;0\;1 \\ 0\;0\;1\;1 \\ 1\;1\;0\;0 \\ 1\;0\;1\;1 \end{pmatrix} \;;\quad A_2 = \begin{pmatrix} 1 \\ 1 \\ 1 \\ 1 \end{pmatrix} \;;\quad C_2 = \begin{pmatrix} 1 \\ 0 \\ 0 \\ 1 \end{pmatrix} \;.$$

Eqs. (4.12) and (4.15) then give

$$\mathcal{X} = \begin{pmatrix} 0\;0\;0\;0\;1 \,|\, 1\;0\;0\;1\;0 \end{pmatrix}$$
$$\mathcal{Z} = \begin{pmatrix} 0\;0\;0\;0\;0 \,|\, 1\;1\;1\;1\;1 \end{pmatrix} \;.$$

With our convention of setting $i^\lambda \equiv 1$ when pulling back (transpose) vectors, and recalling that $\sigma_x \sigma_z = -i\sigma_y$, it follows that the generators are

$$g_1 = (-i\sigma_y^1)\sigma_z^2\sigma_z^4(-i\sigma_y^5)$$
$$g_2 = \sigma_x^2\sigma_z^3\sigma_z^4\sigma_x^5$$
$$g_3 = \sigma_z^1\sigma_z^2\sigma_x^3\sigma_x^5$$
$$g_4 = \sigma_z^1\sigma_z^3(-i\sigma_y^4)(-i\sigma_y^5) \;,$$

Quantum Stabilizer Codes: Encoding and Decoding 121

and
$$X = \sigma_z^1 \sigma_z^4 \sigma_x^5$$
$$Z = \sigma_z^1 \sigma_z^2 \sigma_z^3 \sigma_z^4 \sigma_z^5 \ .$$

This gives the standard form for the [5,1,3] code. □

4.2 Encoding

This section explains how to construct a quantum circuit that efficiently encodes an $[n,k,d]$ quantum stabilizer code [1]. The reader may wish to review Sections 1.4.1 and 3.1.4 before proceeding.

We begin by choosing the unencoded k-qubit computational basis (CB) states $|\delta_1 \cdots \delta_k\rangle$ to be simultaneous eigenstates of the Pauli operators σ_z^i:

$$\sigma_z^i |\delta_1 \cdots \delta_k\rangle = (-1)^{\delta_i} |\delta_1 \cdots \delta_k\rangle \ , \qquad (4.16)$$

where $i = 1, \ldots, k$ and $\delta_i = 0, 1$. The state $|\delta_1 \cdots \delta_k\rangle$ can be constructed from the Pauli operators $\{\sigma_x^j : j = 1, \ldots, k\}$ and the k-qubit CB state $|0 \cdots 0\rangle_k$, which has $\delta_i = 0$ for $i = 1, \ldots, k$:

$$|\delta_1 \cdots \delta_k\rangle = \left(\sigma_x^1\right)^{\delta_1} \cdots \left(\sigma_x^k\right)^{\delta_k} |0 \cdots 0\rangle_k \ . \qquad (4.17)$$

The subscript k on $|0 \cdots 0\rangle_k$ is a reminder that it is a k-qubit state. The n-qubit CB states $|\delta_1 \cdots \delta_n\rangle$ are also defined to be eigenstates of the σ_z^i with $i = 1, \ldots, n$. They satisfy equations analogous to eqs. (4.16) and (4.17), though we will not write them out explicitly.

The encoding operator ξ maps (i) the unencoded (k-qubit) CB state $|\delta_1 \cdots \delta_k\rangle$ to the (n-qubit) basis codeword $|\overline{\delta_1 \cdots \delta_k}\rangle = \xi |\delta_1 \cdots \delta_k\rangle$, and (ii) the single-qubit Pauli operator σ_x^i to the n-qubit encoded operator $X_i = \xi \sigma_x^i \xi^\dagger$. It follows from this and eq. (4.17) that

$$\begin{aligned}
|\overline{\delta_1 \cdots \delta_k}\rangle &= \xi |\delta_1 \cdots \delta_k\rangle \\
&= \xi \left[\left(\sigma_x^1\right)^{\delta_1} \cdots \left(\sigma_x^k\right)^{\delta_k} \right] (\xi^\dagger \xi) |0 \cdots 0\rangle_k \\
&= (X_1)^{\delta_1} \cdots (X_k)^{\delta_k} |\overline{0 \cdots 0}\rangle \ ,
\end{aligned} \qquad (4.18)$$

where $|\overline{0 \cdots 0}\rangle = \xi |0 \cdots 0\rangle_k$ and the identity operator $\xi^\dagger \xi$ has been inserted into the second line. As with the examples in Section 3.2, it proves convenient to define the basis codeword $|\overline{0 \cdots 0}\rangle$ to be

$$|\overline{0 \cdots 0}\rangle = \sum_{s \in \mathcal{S}} s |0 \cdots 0\rangle_n \ . \qquad (4.19)$$

The result of the following Exercise will allow us to rewrite the RHS of this equation in a more useful form.

Exercise 4.1 *Let $\{g_j : j = 1, \ldots, n-k\}$ be the generators of S and I_n the n-qubit identity operator. Show that*

$$\sum_{s \in S} s = \prod_{j=1}^{n-k} (I_n + g_j) \ . \tag{4.20}$$

(Hint: Recall eq. (1.79).)

Using eq. (4.20) in eq. (4.19) gives

$$\overline{|0 \cdots 0\rangle} = \prod_{j=1}^{n-k} (I_n + g_j) |0 \cdots 0\rangle_n \ . \tag{4.21}$$

Plugging eq. (4.21) into eq. (4.18) and recalling that $X_i \in C(S)$ for $i = 1, \ldots, k$ gives

$$\overline{|\delta_1 \cdots \delta_k\rangle} = \prod_{j=1}^{n-k} (I + g_j) \, X_1^{\delta_1} \cdots X_k^{\delta_k} |0 \cdots 0\rangle_n \ . \tag{4.22}$$

(The subscript n on I_n has been suppressed. It will continue to be suppressed for the remainder of this Section.) The encoding construction that we are about to describe will implement the action of eq. (4.22) using a quantum circuit that applies an appropriate sequence of single-qubit and controlled multi-qubit operations to the n-qubit input state $|0 \cdots 0 \, \delta_1 \cdots \delta_k\rangle$ to yield the basis codeword $\overline{|\delta_1 \cdots \delta_k\rangle}$. Here the state $|0 \cdots 0 \, \delta_1 \cdots \delta_k\rangle$ is the n-qubit CB state

$$|0 \cdots 0 \, \delta_1 \cdots \delta_k\rangle \equiv |0 \cdots 0\rangle_{n-k} \otimes |\delta_1 \cdots \delta_k\rangle \ , \tag{4.23}$$

where the state $|0 \cdots 0\rangle_{n-k}$ is for qubits 1 through $n-k$, and $|\delta_1 \cdots \delta_k\rangle$ is for qubits $n-k+1$ through n.

(1) The construction begins by working out the action of the X_j in eq. (4.22). Recall from Section 4.1 that for X_j in standard form,

$$v^T(X_j) = (0 \ u_2^T(j) \ u_3^T(j) \mid v_1^T(j) \ 0 \ 0) \ , \tag{4.24}$$

where $u_3^T(j) = (0 \cdots 1_j \cdots 0)$ and the subscript j on 1_j indicates that the 1 sits in column j. The operator X_j is then (recall we are setting $i^\lambda \equiv 1$)

$$X_j = S_x[u_2(j)] \, S_z[v_1(j)] \, \sigma_x^{n-k+j} \ , \tag{4.25}$$

where

$$S_x[u_2(j)] \equiv (\sigma_x^{r+1})^{u_{2,1}(j)} \cdots (\sigma_x^{n-k})^{u_{2,n-k-r}(j)} \tag{4.26}$$

$$S_z[v_1(j)] \equiv (\sigma_z^1)^{v_{1,1}(j)} \cdots (\sigma_z^r)^{v_{1,r}(j)} \ . \tag{4.27}$$

Eqs. (4.26) and (4.27) indicate that $S_x[u_2(j)]$ acts on qubits $r+1$ to $n-k$, and $S_z[v_1(j)]$ acts on qubits 1 to r. If we set $j = k$ in eq. (4.25), then

$$X_k^{\delta_k}|0\cdots00\rangle_n = \begin{cases} |0\cdots00\rangle_n & \text{for } \delta_k = 0 \\ S_x[u_2(k)]\,|0\cdots01\rangle_n & \text{for } \delta_k = 1 \end{cases}, \quad (4.28)$$

where we have used that $S_z[v_1(k)]|0\cdots01\rangle_n = |0\cdots01\rangle_n$ to arrive at the result on the the second line of this equation. Eq. (4.28) can be written more compactly as

$$X_k^{\delta_k}|0\cdots0\rangle_n = \{\,S_x[u_2(k)]\,\}^{\delta_k}\,|0\cdots\delta_k\rangle_n\;, \quad (4.29)$$

with $\delta_k = 0, 1$. Iterating this argument gives

$$X_1^{\delta_1}\cdots X_k^{\delta_k}|0\cdots0\rangle_n = \prod_{j=1}^{k}\{\,S_x[u_2(j)]\,\}^{\delta_j}\,|0\cdots0\,\delta_1\cdots\delta_k\rangle\;. \quad (4.30)$$

Notice that the operator $\check{U}_j \equiv \{\,S_x[u_2(j)]\,\}^{\delta_j}$ is a controlled–$S_x[u_2(j)]$ operation. Its action on the state $|0\cdots0\,\delta_1\cdots\delta_k\rangle$ is to leave it alone when $\delta_j = 0$ and to apply $S_x[u_2(j)]$ when $\delta_j = 1$. Since δ_j is associated with qubit $n-k+j$ in the state $|0\cdots0\,\delta_1\cdots\delta_k\rangle$, this qubit is the control qubit, and qubits $r+1$ to $n-k$ that are acted on by $S_x[u_2(j)]$ (see eq. (4.26)) are the target qubits. In terms of the $\check{U}_j$, eq. (4.30) can be rewritten as

$$X_1^{\delta_1}\cdots X_k^{\delta_k}|0\cdots0\rangle_n = \check{U}_1\cdots\check{U}_k\,|0\cdots0\,\delta_1\cdots\delta_k\rangle\;. \quad (4.31)$$

The encoding procedure will implement the action of the $X_j^{\delta_j}$ on $|0\cdots0\rangle_n$ by implementing the sequence of controlled operations on the RHS of eq. (4.31) on $|0\cdots0\,\delta_1\cdots\delta_k\rangle$. Each controlled operation $\check{U}_j$ has $n-k-r$ target qubits. Thus the maximum number of 2-qubit gates needed to implement it is $n-k-r$, and so implementing the RHS of eq. (4.31) requires at most $k(n-k-r)$ 2-qubit gates.

(2) At this stage in the encoding procedure, the state of the quantum register is

$$|\psi\rangle = X_1^{\delta_1}\cdots X_k^{\delta_k}\,|0\cdots0\rangle_n\;. \quad (4.32)$$

To obtain the basis codewords $|\overline{\delta_1\cdots\delta_k}\rangle$ (see eq. (4.22)), we must still apply the operator

$$\mathcal{G} = \prod_{j=1}^{n-k}(I + g_j) \quad (4.33)$$

to $|\psi\rangle$. Since the factors in $\mathcal{G}$ commute among themselves, we can write $\mathcal{G} = \mathcal{G}_1 \mathcal{G}_2$, where

$$\mathcal{G}_1 = \prod_{j=1}^{r} (I + g_j) \qquad (4.34)$$

$$\mathcal{G}_2 = \prod_{j=r+1}^{n-k} (I + g_j) \ . \qquad (4.35)$$

Recall from Section 4.1 that the generators g_j with $j = 1, \ldots, r$ are type 1 generators and those with $j = r+1, \ldots, n-k$ are type 2 generators. The type 1 generators correspond to the first r rows of $\mathcal{H}$ in eq. (4.7), and the type 2 generators correspond to rows $r+1$ to $n-k$. Applying $\mathcal{G}$ to $|\psi\rangle$ and noting that $\mathcal{G}_2$ can be commuted past the $X_i^{\delta_i}$ gives

$$|\overline{\delta_1 \cdots \delta_k}\rangle = \mathcal{G}_1 X_1^{\delta_1} \cdots X_k^{\delta_k} \mathcal{G}_2 |0 \cdots 0\rangle_n \ . \qquad (4.36)$$

Notice from eq. (4.7) that type 2 generators are constituted solely from σ_z^i operators and that $\sigma_z^i |0\cdots 0\rangle_n = |0\cdots 0\rangle_n$ so that $\mathcal{G}_2$ fixes the state $|0\cdots 0\rangle_n$. Thus,

$$\begin{aligned} |\overline{\delta_1 \cdots \delta_k}\rangle &= \mathcal{G}_1 X_1^{\delta_1} \cdots X_k^{\delta_k} |0\cdots 0\rangle_n \\ &= \mathcal{G}_1 \check{U}_T |0\cdots 0\, \delta_1 \cdots \delta_k\rangle \ , \end{aligned} \qquad (4.37)$$

where eq. (4.31) was used to go from the first line to the second and we have defined

$$\check{U}_T \equiv \check{U}_1 \cdots \check{U}_k \ . \qquad (4.38)$$

Now let us focus on the factor $(I + g_j)$ in $\mathcal{G}_1$, where $j = 1, \ldots, r$. Associated with g_j is the binary vector v_j^T:

$$v_j^T = (\, 0 \cdots 1_j \cdots 0 \; A_1(j) \; A_2(j) \;|\; B(j) \; C_1(j) \; C_2(j) \,) \ , \qquad (4.39)$$

which is the jth row of $\mathcal{H}$ (see eqs. (4.4) and (4.7)). Pulling back v_j^T gives (recall $i^{\lambda_j} \equiv 1$),

$$g_j = T_j \, \sigma_x^j \left(\sigma_z^j\right)^{B_j(j)} \ . \qquad (4.40)$$

General formulas for T_j ($j = 1, \ldots, r$) are worked out in Problem 4.3, although given a specific generator g_j in standard form, eq. (4.40) identifies T_j as the operator that remains when we factor out all operators associated with qubit j. For example, for the [5,1,3] code, g_1 in standard form is $(-i\sigma_y^1)\sigma_z^2\sigma_z^4(-i\sigma_y^5)$ (see Example 4.1). Factoring out the operator $-i\sigma_y^1$ associated with qubit $j = 1$ gives $T_1 = \sigma_z^2 \sigma_z^4 (-i\sigma_y^5)$. This is usually the simplest way to determine the T_j operators.

We are ready to apply $(I + g_j)$ to $|\psi\rangle$. Using eqs. (4.31), (4.38), and (4.40) in eq. (4.32) gives

$$(I + g_j)|\psi\rangle = \check{U}_T |0 \cdots 0 \delta_1 \cdots \delta_k\rangle + T_j \, \sigma_x^j \left(\sigma_z^j\right)^{B_j(j)} \check{U}_T |0 \cdots 0 \delta_1 \cdots \delta_k\rangle. \qquad (4.41)$$

Since $\sigma_x^j \left(\sigma_z^j\right)^{B_j(j)}$ acts on qubit j with $j = 1, \ldots, r$ and $\check{U}_T$ acts on qubits $r+1$ to $n-k$, these operators commute. Thus $\sigma_x^j \left(\sigma_z^j\right)^{B_j(j)}$ can be moved to the right past $\check{U}_T$ in the second term on the RHS of eq. (4.41). Notice that

$$\left(\sigma_z^j\right)^{B_j(j)} |0 \cdots 0_j \cdots 0\, \delta_1 \cdots \delta_k\rangle = |0 \cdots 0_j \cdots 0\, \delta_1 \cdots \delta_k\rangle \;,$$

and that σ_x^j acts to flip the 0 in column j to a 1 in the state $|0 \cdots 0\, \delta_1 \cdots \delta_k\rangle$:

$$\sigma_x^j \left(\sigma_z^j\right)^{B_j(j)} |0 \cdots 0_j \cdots 0\, \delta_1 \cdots \delta_k\rangle = |0 \cdots 1_j \cdots 0\, \delta_1 \cdots \delta_k\rangle \;.$$

Thus eq. (4.41) can be rewritten as

$$(I + g_j)|\psi\rangle = \check{U}_T |0 \cdots 0_j \cdots 0\, \delta_1 \cdots \delta_k\rangle \\ + T_j \check{U}_T |0 \cdots 1_j \cdots 0\, \delta_1 \cdots \delta_k\rangle \;. \quad (4.42)$$

We now show how to produce the state $(I + g_j)|\psi\rangle$ using well-known operations from the theory of quantum circuits. To that end, we first apply the operator H_j to $|\psi\rangle$, where $j = 1, \cdots, r$, and H_j is the one-qubit Hadamard operator (see Appendix C) whose action on the eigenstates $|\delta_j\rangle$ of σ_z^j is

$$H_j |\delta_j\rangle = \frac{1}{\sqrt{2}} \left[|0\rangle + (-1)^{\delta_j} |1\rangle\right] \quad (\delta_j = 0, 1) \;.$$

Then, using eqs. (4.31), (4.32), and (4.38) gives

$$H_j |\psi\rangle = \check{U}_T H_j |0 \cdots 0_j \cdots 0\, \delta_1 \cdots \delta_k\rangle \\ = \check{U}_T \{|0 \cdots 0_j \cdots 0\, \delta_1 \cdots \delta_k\rangle \\ + |0 \cdots 1_j \cdots 0\, \delta_1 \cdots \delta_k\rangle\} \;, \quad (4.43)$$

where we used that $\check{U}_T$ commutes with H_j since the two operators act on different qubits, and we have suppressed the factor of $1/\sqrt{2}$ arising from the action of H_j. Finally we apply a controlled-T_j gate, $\check{W}_j \equiv (T_j)^{\alpha_j}$, with $j = 1, \cdots, r$ and $\alpha_j = 0, 1$ to eq. (4.43):

$$\check{W}_j H_j |\psi\rangle = \check{U}_T |0 \cdots 0_j \cdots 0\, \delta_1 \cdots \delta_k\rangle + T_j \check{U}_T |0 \cdots 1_j \cdots 0\, \delta_1 \cdots \delta_k\rangle \;, \quad (4.44)$$

where we recall that $\check{U}_T$ acts on qubits $r+1$ to $n-k$ and so does not alter the value of α_j, which remains equal to 0_j in the first term on the RHS and to 1_j in the second. Thus $\check{W} = I$ in the first term and $\check{W}_j = T_j$ in the second. Comparing the RHS's of eqs. (4.42) and (4.44) we see that

$$(I + g_j)|\psi\rangle = \check{W}_j H_j |\psi\rangle \;. \quad (4.45)$$

Thus, to apply $(I + g_j)$ to $|\psi\rangle$, the encoding circuit first applies the Hadamard gate H_j to $|\psi\rangle$ $(j = 1, \ldots, r)$, followed by the controlled-T_j gate $\check{W}_j$. Note from eq. (4.40) that T_j acts on at most $n-1$ qubits and so $\check{W}_j$ applies at most $n-1$ two-qubit gates. The one-qubit gate H_j is also applied. Thus applying $(I+g_j)$ to $|\psi\rangle$ requires that at most $(n-1)+1 = n$ quantum gates be applied. Iterating this procedure gives

$$|\overline{\delta_1 \cdots \delta_k}\rangle = \prod_{j=1}^{n-k}(I+g_j)\, X_1^{\delta_1}\cdots X_k^{\delta_k}|0\cdots 0\rangle_n$$

$$= \left\{\prod_{j=1}^{r}\check{W}_j H_j\right\}\left(\prod_{m=1}^{k}\check{U}_m\right)|0\cdots 0\,\delta_1\cdots\delta_k\rangle, \qquad (4.46)$$

where eqs. (4.31), (4.32), and (4.38) were used in going from the first to the second line. Implementing the RHS of this equation requires r one-qubit gates to be used; and at most $k(n-k-r)$ two-qubit gates to be used for the product of the $\check{U}_m$, and at most $r(n-1)$ two-qubit gates for the product of the $\check{W}_l$. The complete encoding circuit thus uses at most

$$r + r(n-1) + k(n-k-r) = (k+r)(n-k) \leq n(n-k) \qquad (4.47)$$

quantum gates. Notice that $n(n-k)$ is also an upper bound on the number of one-qubit operators needed to form the $n-k$ generators $\{g_j : j = 1,\ldots,n-k\}$, which is one way to quantify the resources needed to specify a quantum stabilizer code. We see that the above encoding procedure [1] is linear in this resource measure and thus is an efficient encoding scheme.

The following example works out the encoding circuit for the [5,1,3] code.

Example 4.2
From Example 4.1 the standard form for the generators of the [5,1,3] code is

$$g_1 = (-i\sigma_y^1)\sigma_z^2\sigma_z^4(-i\sigma_y^5)$$
$$g_2 = \sigma_x^2\sigma_z^3\sigma_z^4\sigma_x^5$$
$$g_3 = \sigma_z^1\sigma_z^2\sigma_x^3\sigma_x^5$$
$$g_4 = \sigma_z^1\sigma_z^3(-i\sigma_y^4)(-i\sigma_y^5)\ .$$

From this and eq. (4.40) we read off

$$T_1 = \sigma_z^2\sigma_z^4(-i\sigma_y^5)$$
$$T_2 = \sigma_z^3\sigma_z^4\sigma_x^5$$
$$T_3 = \sigma_z^1\sigma_z^2\sigma_x^5$$
$$T_4 = \sigma_z^1\sigma_z^3(-i\sigma_y^5)\ .$$

Finally, $v^T(X)$ was shown to be $(00001\,|\,10010)$, which identifies $u_2^T = 0$ and so

$$S_x[u_2] = I\ ,$$

Quantum Stabilizer Codes: Encoding and Decoding

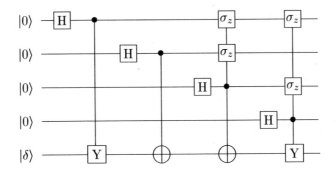

FIGURE 4.1
The encoding circuit for the [5,1,3] quantum stabilizer code. Here $Y = -i\sigma_y^5$. Adapted from (i) Figure 4.2, Ref. [3]; and (ii) Ref. [4], ©Daniel Gottesman (2000), used with permission.

and
$$\check{U}_T = I \ .$$
Eq. (4.46) then gives the basis codewords $|\bar{\delta}\rangle$ ($\delta = 0, 1$):

$$|\bar{\delta}\rangle = \left[\prod_{j=1}^{4} (T_j)^\delta H_j \right] |0000\,\delta\rangle \ . \tag{4.48}$$

The encoding circuit corresponding to eq. (4.48) appears in Figure 4.1, where qubit 1 (5) is associated with the top (bottom) line in the figure. Note that the σ_z factors in T_1^δ and T_2^δ act trivially on $|0000\delta\rangle$ and so do not appear in the encoding circuit. □

4.3 Decoding

As noted in Ref. [1], the most straightforward approach to decoding is simply to run the encoding circuit of Section 4.2 in reverse. However, this approach is not always the most efficient way to decode. Gottesman [3] provides an alternative scheme for decoding an [n,k,d] quantum stabilizer code, which under appropriate circumstances can be more efficient than reverse encoding. This scheme introduces k ancilla qubits that are placed in the state $|0\cdots0\rangle_k$. Decoding takes the initial state $|\psi_{in}\rangle = |\overline{\delta_1 \cdots \delta_k}\rangle \otimes |0\cdots0\rangle_k$ to the final state

$$\begin{aligned}|\psi_f\rangle &= U_{decode} |\overline{\delta_1 \cdots \delta_k}\rangle \otimes |0\cdots0\rangle_k \\ &= |\overline{0\cdots0}\rangle \otimes |\delta_1 \cdots \delta_k\rangle \ .\end{aligned} \tag{4.49}$$

A few remarks are in order before discussing how U_{decode} is implemented. First, since U_{decode} is a linear operator its action on general codewords is determined once its action on basis codewords is known. Thus we can restrict our discussion of decoding to basis codewords without loss of generality. Second, the decoding procedure outputs the decoded state $|\delta_1 \cdots \delta_k\rangle = \xi^\dagger |\overline{\delta_1 \cdots \delta_k}\rangle$ as the final state of the ancilla, and also returns n qubits in the state $|0 \cdots 0\rangle$. Having n qubits in this state is an important bonus as it allows the encoding circuit of Section 4.2 to be simplified. Recall that the operator $\mathcal{G}$ (see eqs. (4.21) and (4.33)) was used to construct $|0 \cdots 0\rangle$ from the state $|0 \cdots 0\rangle_n$. Since $|0 \cdots 0\rangle$ can now be input directly, $\mathcal{G}$ no longer needs to be applied, reducing by rn the number of gates needed for encoding (see Section 4.2).

To begin our discussion of U_{decode}, recall that the standard form for Z_i is (see eq. (4.15))

$$Z_i = (\sigma_z^1)^{A_{2,1}(i)} \cdots (\sigma_z^r)^{A_{2,r}(i)} \sigma_z^{n-k+i}$$
$$\equiv \sigma_z(V(Z_i)) \ , \tag{4.50}$$

where the components of the binary vector $V(Z_i)$ are

$$V_j(Z_i) = \begin{cases} A_{2,j}(i) & (j = 1, \ldots, r) \\ \delta_{j,n-k+i} & (j = r+1, \ldots, n) \end{cases} . \tag{4.51}$$

Let the basis codeword $|\overline{\delta}\rangle = |\overline{\delta_1 \cdots \delta_k}\rangle$ have the following expansion in the n-qubit CB states $\{|d_1 \cdots d_n\rangle : d_i = 0, 1; i = 1, \ldots, n\}$:

$$|\overline{\delta_1 \cdots \delta_k}\rangle = \sum_{d \in F_2^n} C_{\overline{\delta}}(d) |d_1 \cdots d_n\rangle \ , \tag{4.52}$$

where $d \equiv d_1 \cdots d_n$. Recalling that the basis codeword $|\overline{\delta_1 \cdots \delta_k}\rangle$ is an eigenket of Z_i (see eq. (1.77)), we also have that

$$Z_i |\overline{\delta_1 \cdots \delta_k}\rangle = (-1)^{\delta_i} |\overline{\delta_1 \cdots \delta_k}\rangle \ . \tag{4.53}$$

Combining eqs. (4.50) and (4.52) gives

$$Z_i |\overline{\delta_1 \cdots \delta_k}\rangle = \sum_{d \in F_2^n} C_{\overline{\delta}}(d) \, \sigma_z(V(Z_i)) |d_1 \cdots d_n\rangle$$
$$= \sum_{d \in F_2^n} C_{\overline{\delta}}(d) \, (-1)^{V(Z_i) \cdot d} |d_1 \cdots d_n\rangle \ , \tag{4.54}$$

where addition in the scalar product is modulo 2. Since eqs. (4.53) and (4.54) must agree, we must have

$$\delta_i = V(Z_i) \cdot d \tag{4.55}$$

for all d for which $C_{\overline{\delta}}(d) \neq 0$.

Quantum Stabilizer Codes: Encoding and Decoding

We are ready to describe the first stage of the decoding procedure. At this point the state of the n qubits plus k ancilla qubits is $|\overline{\delta_1 \cdots \delta_k}\rangle \otimes |0 \cdots 0\rangle_k$. Now imagine that we apply a CNOT operation to ancilla 1 conditioned on the first encoded qubit δ_1:

$$(\sigma_x^{a_1})^{\delta_1} |\overline{\delta_1 \cdots \delta_k}\rangle \otimes |0 \cdots 0\rangle_k = \sum_{d \in F_2^n} C_{\overline{\delta}}(d) |d_1 \cdots d_n\rangle \otimes \left[(\sigma_x^{a_1})^{\delta_1} |0 \cdots 0\rangle_n \right] . \tag{4.56}$$

From eq. (4.55) we can write

$$(\sigma_x^{a_1})^{\delta_1} |\overline{\delta_1 \cdots \delta_k}\rangle \otimes |0 \cdots 0\rangle_k$$
$$= \sum_{d \in F_2^n} C_{\overline{\delta}}(d) |d_1 \cdots d_n\rangle \otimes (\sigma_x^{a_1})^{V(Z_1) \cdot d} |0 \cdots 0\rangle_k . \tag{4.57}$$

Notice that a $\sigma_x^{a_1}$ is applied to the first ancilla for each qubit j for which $V_j(Z_1) \neq 0$ and for which $d_j = 1$. Thus eq. (4.57) can be implemented by applying a CNOT gate from each of these qubits to the first ancilla. Direct calculation, together with eq. (4.55), gives

$$(\sigma_x^{a_1})^{V(Z_1) \cdot d} |0\rangle = (\sigma_x^{a_1})^{V_1(Z_1) d_1} \cdots (\sigma_x^{a_1})^{V_n(Z_1) d_n} |0\rangle$$
$$= |V(Z_1) \cdot d\rangle$$
$$= |\delta_1\rangle , \tag{4.58}$$

and substituting this back into eq. (4.57) gives

$$(\sigma_x^{a_1})^{\delta_1} |\overline{\delta_1 \cdots \delta_k}\rangle \otimes |0 \cdots 0\rangle_k = \sum_{d \in F_2^n} C_{\overline{\delta}}(d) |d_1 \cdots d_n\rangle \otimes |\delta_1 0 \cdots 0\rangle_k$$
$$= |\overline{\delta_1 \cdots \delta_k}\rangle \otimes |\delta_1 0 \cdots 0\rangle_k . \tag{4.59}$$

Iterating this procedure for the remaining $k-1$ ancilla qubits gives

$$\prod_{j=1}^{k} (\sigma_x^{a_j})^{\delta_j} |\overline{\delta_1 \cdots \delta_k}\rangle \otimes |0 \cdots 0\rangle_k = |\overline{\delta_1 \cdots \delta_k}\rangle \otimes |\delta_1 \cdots \delta_k\rangle . \tag{4.60}$$

By applying CNOT gates to the ancilla in this way we have managed to put the ancilla into the decoded state $|\delta_1 \cdots \delta_k\rangle$. Eq. (4.60) represents completion of the first stage of the decoding procedure implemented by U_{decode}. We will see in Chapter 5 that this first stage procedure is not fault-tolerant, although a straightforward modification of it allows this serious deficiency to be removed.

We now focus on putting the other n qubits into the state $|\overline{0 \cdots 0}\rangle$. This can be done by applying a controlled-X_i operation to each encoded qubit i conditioned on the ith ancilla qubit. For example, for $i = 1$,

$$(X_1)^{\delta_1} |\overline{\delta_1 \cdots \delta_k}\rangle \otimes |\delta_1 \cdots \delta_k\rangle = |\overline{(\delta_1 \oplus \delta_1) \delta_2 \cdots \delta_k}\rangle \otimes |\delta_1 \cdots \delta_k\rangle$$
$$= |\overline{0 \delta_2 \cdots \delta_k}\rangle \otimes |\delta_1 \cdots \delta_k\rangle . \tag{4.61}$$

Iterating over the remaining $k-1$ encoded qubit/ancilla pairs completes the decoding operation U_{decode}:

$$U_{decode}|\overline{\delta_1 \cdots \delta_k}\rangle \otimes |0\cdots 0\rangle_k = \prod_{i=1}^{k}(X_i)^{\delta_i} \prod_{j=1}^{k}(\sigma_x^{a_j})^{\delta_j}|\overline{\delta_1 \cdots \delta_k}\rangle \otimes |0\cdots 0\rangle_k \quad (4.62)$$

$$= \prod_{i=1}^{k}(X_i)^{\delta_i}|\overline{\delta_1 \cdots \delta_k}\rangle \otimes |\delta_1 \cdots \delta_k\rangle \quad (4.63)$$

$$= |0\cdots 0\rangle \otimes |\delta_1 \cdots \delta_k\rangle \ . \quad (4.64)$$

Applying $(\sigma_x^{a_i})^{V(Z_i)\cdot d}$ requires at most $r+1$ operations since Z_i acts on at most $r+1$ qubits. Thus the first phase of decoding (eq. (4.60)) requires at most $k(r+1)$ quantum gates. The final phase of decoding (eq. (4.63)) requires k controlled-X_i operations, each requiring at most $n-k+1$ two-qubit gates (see Section 4.2). The total number of quantum gates used in this decoding scheme is then at most

$$k(r+1) + k(n-k+1) = k(n-k+r+2) \ .$$

Since reverse encoding uses no more than $(k+r)(n-k)$ gates, one anticipates that Gottesman decoding should start to become more efficient than reverse encoding when

$$k(n-k+r+2) < (k+r)(n-k) \quad (4.65)$$

or

$$2k(r+1) < nr \ . \quad (4.66)$$

Since eq. (4.65) relates upper bounds, codes may exist where eq. (4.66) is satisfied, but reverse encoding uses fewer gates than Gottesman decoding.

The following example applies this decoding scheme to the [4,2,2] code.

Example 4.3

The needed ingredients for working out the decoding circuit for the [4,2,2] code appear in Problem 4.1. There we find the standard forms for Z_1, Z_2, X_1, and X_2:

$$Z_1 = \sigma_z^2 \sigma_z^3 \quad ; \quad X_1 = \sigma_z^1 \sigma_z^2 \sigma_x^3$$
$$Z_2 = \sigma_z^1 \sigma_z^4 \quad ; \quad X_2 = \sigma_z^2 \sigma_x^4 \ ,$$

from which we read off

$$V(Z_1) = (0110)$$

and

$$V(Z_2) = (1001) \ .$$

From the $V(Z_i)$ we see that the first stage of decoding will apply CNOT gates to the first ancilla qubit using qubits 2 and 3 as controls and CNOT gates to

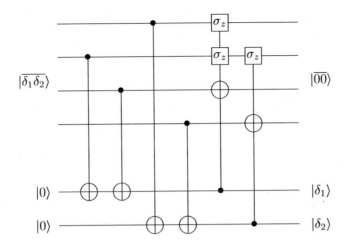

FIGURE 4.2
Decoding circuit for the [4,2,2] quantum stabilizer code.

the second ancilla qubit using qubits 1 and 4 as controls. The second stage of decoding then applies a controlled-X_1 gate using the first ancilla qubit as control and a controlled-X_2 gate using the second ancilla qubit as control. The decoding circuit that applies this sequence of gates is given in Figure 4.2.
□

Problems

4.1 The generators for the [4,2,2] code are given in eq. (3.37).

(a) Show that the standard form for $\mathcal{H}$ is

$$\mathcal{H} = \begin{pmatrix} 1\,0\,0\,1 & 0\,1\,1\,0 \\ 0\,1\,1\,0 & 1\,1\,1\,1 \end{pmatrix} .$$

Pulling back the rows of $\mathcal{H}$ gives the generators $\{g_j : j = 1, 2\}$ in standard form (recall we set $i^{\lambda_j} \equiv 1$ and $\sigma_x \sigma_z = -i\sigma_y$):

$$g_1 = \sigma_x^1 \sigma_z^2 \sigma_z^3 \sigma_x^4$$
$$g_2 = \sigma_z^1 (-i\sigma_y^2)(-i\sigma_y^3) \sigma_z^4$$

(b) Show that $\mathcal{X}$ and $\mathcal{Z}$ in standard form are

$$\mathcal{X} = \begin{pmatrix} 0\ 0\ 1\ 0 & 1\ 1\ 0\ 0 \\ 0\ 0\ 0\ 1 & 0\ 1\ 0\ 0 \end{pmatrix} \; ;$$

$$\mathcal{Z} = \begin{pmatrix} 0\ 0\ 0\ 0 & 0\ 1\ 1\ 0 \\ 0\ 0\ 0\ 0 & 1\ 0\ 0\ 1 \end{pmatrix} \; .$$

Pulling back the rows of $\mathcal{X}$ and $\mathcal{Z}$ gives, respectively, the operators X_i and Z_i ($i=1,2$) in standard form (recall we set $i^\lambda \equiv 1$):

$$\begin{array}{ll} X_1 = \sigma_z^1 \sigma_z^2 \sigma_x^3 & Z_1 = \sigma_z^2 \sigma_z^3 \\ X_2 = \sigma_z^2 \sigma_x^4 & Z_2 = \sigma_z^1 \sigma_z^4 \end{array} \; .$$

4.2 The generators for the [8,3,3] code are given in eq. (3.40).

(a) Show that the standard form for $\mathcal{H}$ is

$$\mathcal{H} = \begin{pmatrix} 1\ 0\ 0\ 0\ 1\ 1\ 1\ 0 & 0\ 1\ 0\ 0\ 1\ 1\ 0\ 1 \\ 0\ 1\ 0\ 0\ 1\ 1\ 0\ 1 & 0\ 0\ 1\ 0\ 1\ 0\ 1\ 1 \\ 0\ 0\ 1\ 0\ 1\ 0\ 1\ 1 & 0\ 1\ 0\ 1\ 1\ 0\ 1\ 0 \\ 0\ 0\ 0\ 1\ 0\ 1\ 1\ 1 & 0\ 0\ 1\ 1\ 1\ 1\ 0\ 0 \\ 0\ 0\ 0\ 0\ 0\ 0\ 0\ 0 & 1\ 1\ 1\ 1\ 1\ 1\ 1\ 1 \end{pmatrix} \; .$$

Pulling back the rows of $\mathcal{H}$ gives the generators $\{g_j : j=1,\ldots,5\}$ in standard form (recall we set $i^{\lambda_j} \equiv 1$ and $\sigma_x \sigma_z = -i\sigma_y$):

$$g_1 = \sigma_x^1 \sigma_z^2 (-i\sigma_y^5)(-i\sigma_y^6) \sigma_x^7 \sigma_z^8$$
$$g_2 = \sigma_x^2 \sigma_z^3 (-i\sigma_y^5) \sigma_x^6 \sigma_z^7 (-i\sigma_y^8)$$
$$g_3 = \sigma_z^2 \sigma_z^3 \sigma_x^4 (-i\sigma_y^5)(-i\sigma_z^7) \sigma_x^8$$
$$g_4 = \sigma_z^3 (-i\sigma_y^4) \sigma_z^5 (-i\sigma_y^6) \sigma_x^7 \sigma_x^8$$
$$g_5 = \sigma_z^1 \sigma_z^2 \sigma_z^3 \sigma_z^4 \sigma_z^5 \sigma_z^6 \sigma_z^7 \sigma_z^8 \; .$$

(b) Show that $\mathcal{X}$ and $\mathcal{Z}$ in standard form are

$$\mathcal{X} = \begin{pmatrix} 0\ 0\ 0\ 0\ 1\ 1\ 0\ 0 & 0\ 1\ 1\ 0\ 0\ 0\ 0\ 0 \\ 0\ 0\ 0\ 0\ 1\ 0\ 1\ 0 & 1\ 0\ 0\ 1\ 0\ 0\ 0\ 0 \\ 0\ 0\ 0\ 0\ 1\ 0\ 0\ 1 & 0\ 0\ 1\ 1\ 0\ 0\ 0\ 0 \end{pmatrix} \; ;$$

$$\mathcal{Z} = \begin{pmatrix} 0\ 0\ 0\ 0\ 0\ 0\ 0\ 0 & 1\ 1\ 0\ 1\ 0\ 1\ 0\ 0 \\ 0\ 0\ 0\ 0\ 0\ 0\ 0\ 0 & 1\ 0\ 1\ 1\ 0\ 0\ 1\ 0 \\ 0\ 0\ 0\ 0\ 0\ 0\ 0\ 0 & 0\ 1\ 1\ 1\ 0\ 0\ 0\ 1 \end{pmatrix} \; .$$

Pulling back the rows of $\mathcal{X}$ and $\mathcal{Z}$ gives, respectively, the operators X_i and Z_i ($i=1,2,3$) in standard form (recall we set $i^\lambda \equiv 1$):

$$\begin{array}{ll} X_1 = \sigma_z^2 \sigma_z^3 \sigma_x^5 \sigma_x^6 & Z_1 = \sigma_z^1 \sigma_z^2 \sigma_z^4 \sigma_z^6 \\ X_2 = \sigma_z^1 \sigma_z^4 \sigma_x^5 \sigma_x^7 & Z_2 = \sigma_z^1 \sigma_z^3 \sigma_z^4 \sigma_z^7 \\ X_3 = \sigma_z^3 \sigma_z^4 \sigma_x^5 \sigma_x^8 & Z_3 = \sigma_z^2 \sigma_z^3 \sigma_z^4 \sigma_z^8 \end{array} \; .$$

4.3 Given eq. (4.39) for v_j^T, and setting all factors $i^{\lambda_j} \equiv 1$, show that the generator g_j is given by eq. (4.40), with the operator T_j given by

$$T_j = S_x\left[A_1(j)\right] S_x'\left[A_2(j)\right] R_z^j R_z'\left[C_1(j)\right] R_z''\left[C_2(j)\right] , \quad (4.67)$$

with

$$S_x\left[A_1(j)\right] = \left(\sigma_x^{r+1}\right)^{A_{1,1}(j)} \cdots \left(\sigma_x^{n-k}\right)^{A_{1,n-k-r}(j)} \quad (4.68)$$

$$S_x'\left[A_2(j)\right] = \left(\sigma_x^{n-k+1}\right)^{A_{2,1}(j)} \cdots \left(\sigma_x^{n}\right)^{A_{2,k}(j)} \quad (4.69)$$

$$R_z'\left[C_1(j)\right] = \left(\sigma_z^{r+1}\right)^{C_{1,1}(j)} \cdots \left(\sigma_z^{n-k}\right)^{C_{1,n-k-r}(j)} \quad (4.70)$$

$$R_z''\left[C_2(j)\right] = \left(\sigma_z^{n-k+1}\right)^{C_{2,1}(j)} \cdots \left(\sigma_z^{n}\right)^{C_{2,k}(j)} \quad (4.71)$$

$$R_z^j = \left(\sigma_z^j\right)^{B_j(j)} S_z\left[B(j)\right] . \quad (4.72)$$

Here $S_z[\cdots]$ is given by eq. (4.27) and the binary vectors $A_1(j)$, $A_2(j)$, $C_1(j)$, $C_2(j)$, and $B(j) = B_1(j) \cdots B_j(j) \cdots B_r(j)$ are read off from eq. (4.39).

4.4 Given the results of Problem 4.1, show that the encoding circuit for the [4,2,2] quantum stabilizer code is that given in Figure 4.3.

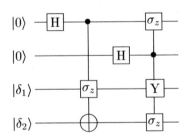

FIGURE 4.3
The encoding circuit for the [4,2,2] quantum stabilizer code. Here $Y = -i\sigma_y^3$.

4.5 Given the results of Problem 4.2, show that the encoding circuit for the [8,3,3] quantum stabilizer code is that given in Figure 4.4.

4.6 Show that the decoding scheme of Section 4.3 yields the decoding circuit shown in Figure 4.5 for the [5,1,3] quantum stabilizer code.

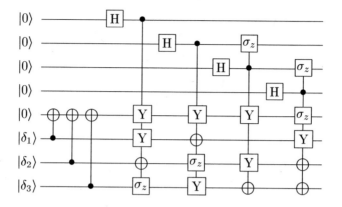

FIGURE 4.4
The encoding circuit for the [8,3,3] quantum stabilizer code. Here $Y = -i\sigma_y$.

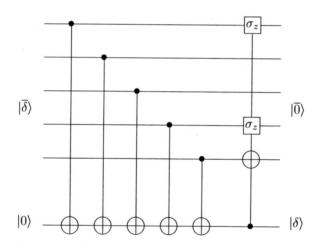

FIGURE 4.5
The decoding circuit for the [5,1,3] quantum stabilizer code.

References

[1] Cleve, R. and Gottesman, D., Efficient computations of encodings for quantum error correction, *Phys. Rev. A* **56**, 76, 1997.

[2] Gottesman, D., Class of quantum error correcting codes saturating the quantum Hamming bound, *Phys. Rev. A* **54**, 1862, 1996.

[3] Gottesman, D., Stabilizer codes and quantum error correction, Ph. D. thesis, California Institute of Technology, Pasadena, CA, 1997.

[4] See Errata for Ref. [3] at http://www.perimeterinstitute.ca/personal/dgottesman/thesis-errata.html .

5

Fault-Tolerant Quantum Computing

The first theme of this book—quantum error correction—has now been presented. We carry on in this chapter with a discussion of fault-tolerant quantum computing (FTQC), which is our second theme. The discussion will take place in the context of quantum stabilizer codes and is based on Refs. [1–6]. See also Refs. [7, 8]. Section 5.1 introduces the idea of fault-tolerance and makes basic definitions, while Section 5.2 presents a procedure for doing fault-tolerant quantum error correction. The theory of FTQC is then taken up in the following five sections. Section 5.3 examines the action of unitary operators on a quantum stabilizer code. It introduces two important groups: (i) the normalizer of the stabilizer S whose elements are encoded unitary operations; and (ii) the Clifford group which is generated by the Hadamard, phase, and CNOT gates. The discussion focuses on unitary operations that belong to the intersection of these two groups. By combining measurements on ancilla qubits (Section 5.4) with a four-qubit unitary operation introduced in Section 5.5, it will be possible to fault-tolerantly implement these gates on encoded data for stabilizer codes that encode a single qubit. Section 5.6 presents an extension of this approach that works for stabilizer codes encoding multiple qubits, and Section 5.7 completes the theoretical development by showing how a fault-tolerant encoded Toffoli gate can be constructed for any quantum stabilizer code. By this point a theoretical framework is in place that can construct a set of fault-tolerant encoded quantum gates that is capable of universal quantum computation using any quantum stabilizer code. The final two sections illustrate the theory by applying it to the [5,1,3] and [4,2,2] stabilizer codes. A brief summary of universal sets of quantum gates is given in Appendix C.

5.1 Fault-Tolerance

Noise and imperfect quantum gates hamper the operation of a quantum computer in both direct and indirect ways. Directly in that they cause errors to appear in the computational data, and indirectly as computation usually requires the use of controlled–operations that are controlled by the corrupted data. The corrupted data causes wrong operations to be applied, and so new errors are generated from existing errors.

Example 5.1
(a) To illustrate how such proliferation of errors can occur, consider two qubits, the first in state $|1\rangle$ and the second in the generic state $a|0\rangle + b|1\rangle$. The two-qubit state is then $|1\rangle \otimes [a|0\rangle + b|1\rangle]$. Suppose noise induces a bit-flip error on the first qubit ($E = \sigma_x^1$) with probability p. This switches the two-qubit state to

$$|\psi_{in}\rangle = |0\rangle \otimes [a|0\rangle + b|1\rangle] \quad .$$

Now suppose a CNOT gate is applied with the first (second) qubit as the control (target). The output state is then

$$\begin{aligned}|\psi_{out}\rangle &= U_{CNOT}|\psi_{in}\rangle \\ &= |0\rangle \otimes [a|0\rangle + b|1\rangle] \\ &= \sigma_x^1 \sigma_x^2 |\psi_c\rangle \quad ,\end{aligned} \quad (5.1)$$

where $|\psi_c\rangle = |1\rangle \otimes [a|1\rangle + b|0\rangle]$ is the state that would have resulted from the CNOT gate if the original error on qubit 1 had not occurred. We see from eq. (5.1) that the original bit-flip error σ_x^1 on the control qubit has generated a second bit-flip error σ_x^2 on the target qubit even though the CNOT gate executed *correctly*. Thus two errors have appeared in $|\psi_{out}\rangle$ with probability p! The error model of Section 1.4.1 tacitly assumed that qubits were sitting passively in storage so that two errors in the data required the occurence of two independent errors that would happen with probability p^2. As we have just seen, because a quantum computer actively manipulates its data, errors can multiply and so it is possible for two or more errors to appear in the data with probability p.

(b) Now let qubit 1 be in the state $a|0\rangle + b|1\rangle$ and qubit 2 in the state $(1/\sqrt{2})[|0\rangle + |1\rangle]$. The initial state is then $[a|0\rangle + b|1\rangle] \otimes \{(1/\sqrt{2})[|0\rangle + |1\rangle]\}$. Imagine that a phase error $E = \sigma_z^2$ acts on qubit 2 with probability p. The two-qubit state becomes

$$|\psi_{in}\rangle = [a|0\rangle + b|1\rangle] \otimes \left\{\frac{1}{\sqrt{2}}[|0\rangle - |1\rangle]\right\} \quad .$$

Apply again a CNOT gate from qubit 1 to qubit 2 so that

$$\begin{aligned}|\psi_{out}\rangle &= U_{CNOT}|\psi_{in}\rangle \\ &= \frac{1}{\sqrt{2}}[a|0\rangle \otimes \{|0\rangle - |1\rangle\} + b|1\rangle \otimes \{|1\rangle - |0\rangle\}] \\ &= [a|0\rangle - b|1\rangle] \otimes \left\{\frac{1}{\sqrt{2}}[|0\rangle - |1\rangle]\right\} \\ &= \sigma_z^1 \sigma_z^2 |\psi_c\rangle \quad ,\end{aligned} \quad (5.2)$$

where $|\psi_c\rangle = [a|0\rangle + b|1\rangle] \otimes \{(1/\sqrt{2})[|0\rangle + |1\rangle]\}$ is the state that would have been output from the CNOT gate had the original error on qubit 2 not occurred. We see that two phase errors appear in the final state with probability

Fault-Tolerant Quantum Computing 139

p. Here a phase error on the target qubit has spread to a second phase error on the control qubit. Again, the initial error has multiplied even though the CNOT gate worked properly. □

5.1.1 To Encode or Not to Encode ...

One might wonder whether using a QECC would improve the reliability of the data processed by a quantum computer when noise and imperfect quantum gates are present. Surprisingly, the answer is not an absolute yes. In fact, QECCs encounter problems similar to those observed in Example 5.1:

1. As we saw in Chapter 4, encoding and decoding is done using quantum circuits that utilize controlled–operations. Thus errors present in either the input state, or that appear during operation of the circuit, can generate more errors through execution of the controlled–operations.

2. We saw in Section 1.3.5 that error correction requires syndrome extraction, which applies a unitary operator U to entangle a code block with a set of ancilla qubits. We shall see in Section 5.2 that U is implemented using a quantum circuit that contains controlled–operations. The controlled–operations can again generate new errors from existing errors.

3. Finally, if we want to maximize the degree of protection of a QECC, we will need to find a sufficiently powerful set of encoded operations that will allow us to do arbitrary computations directly on the encoded data. We will see that controlled–operations will be part of this set and so proliferation of errors will be an issue here too.

5.1.2 Fault-Tolerant Design

We see that introducing a QECC does not automatically improve the reliability of the encoded data. It is the goal of *fault-tolerant design* to provide quantum circuits that control the manner in which errors proliferate during (i) encoding/decoding, (ii) error correction and recovery, and (iii) encoded operations. As we shall see in Chapter 6, when a sufficiently layered concatenated code is used for the QECC, and the imperfect quantum gates are sufficiently accurate, these circuits will allow a quantum computation of arbitrary duration to be done with arbitrarily small error probability.

In Section 1.4.1 we introduced an error model that is appropriate for describing noise-induced storage errors. Its assumptions were (i) errors on different qubits occur independently; (ii) single-qubit errors σ_x^j, σ_y^j, and σ_z^j are equally likely; and (iii) the single-qubit error probability p_j is the same for all qubits: $p_j = p$. One more assumption will now be added to this model to describe the effect of imperfect quantum gates: (iv) gate errors only affect the qubits acted on by the gate. Clearly, other error models are possible. The one presented here has the merit of being simple to analyze while still being sufficiently realistic to be physically interesting. We are now ready to define a fault-tolerant quantum operation.

DEFINITION 5.1 *Suppose we are given a quantum register composed of several blocks of qubits and a QECC with which to encode data into each block. An operation is said to be fault-tolerant if the occurrence of a single gate error or storage error during the course of the operation produces no more than one error in each (code) block.*

For QECCs that can correct more than one error, the restriction to only one error per block might seem overly conservative. Perhaps so, though this "one error per block" restriction simplifies the application of Definition 5.1 to specific operations. As will be shown in the remainder of this chapter, even with this tougher condition, FTQC will prove possible.

For later applications, it proves convenient to make two simple restatements of Definition 5.1. Recall that for our error model, a single error occurs with probability p:

(1) An operation is fault-tolerant if, to first-order in p (viz. $\mathcal{O}(p)$), the operation generates no more than one error per code block.

(2) An operation fails to be fault-tolerant if, to $\mathcal{O}(p)$, it can generate more than one error per code block.

Example 5.2

The SWAP gate will prove to be very useful. This gate interchanges the states of two qubits:

$$U_{swp}^{(12)} |\alpha\rangle|\beta\rangle = |\beta\rangle|\alpha\rangle \quad , \tag{5.3}$$

where the superscript on $U_{swp}^{(12)}$ indicates that qubits 1 and 2 are having their states swapped. Applied to qubits within the same block, this operation is not fault-tolerant since an error in the application of the SWAP gate can produce errors on both qubits (assumption (iv) of our error model). Thus one gate error can produce two errors in a block, violating Definition 5.1. Still, it is possible to construct a fault-tolerant SWAP gate that interchanges the states of two qubits in the same block if the two states are never directly swapped. For example, let qubits 1 and 2 belong to block 1 and qubit 3 belong to block 2, and let the initial three-qubit state be $|\alpha\rangle|\beta\rangle|\gamma\rangle$. We want to swap the states $|\alpha\rangle$ and $|\beta\rangle$. This can be done fault-tolerantly by first swapping the states of qubits 1 and 3, then those of qubits 1 and 2, and finally the states of qubits 2 and 3:

$$\begin{aligned} U_{sp}^{(23)} U_{sp}^{(12)} U_{sp}^{(13)} |\alpha\rangle|\beta\rangle|\gamma\rangle &= U_{sp}^{(23)} U_{sp}^{(12)} |\gamma\rangle|\beta\rangle|\alpha\rangle \\ &= U_{sp}^{(23)} |\beta\rangle|\gamma\rangle|\alpha\rangle \\ &= |\beta\rangle|\alpha\rangle|\gamma\rangle \quad . \end{aligned}$$

As desired, the states $|\alpha\rangle$ and $|\beta\rangle$ have been swapped. To see that this operation is fault-tolerant, suppose that only the second SWAP gate executed incorrectly so that $|\gamma\rangle \to |\beta'\rangle$ and $|\beta\rangle \to |\gamma'\rangle$. The output of the second SWAP gate is then $|\beta'\rangle|\gamma'\rangle|\alpha\rangle$. Applying the final SWAP gate gives $|\beta'\rangle|\alpha\rangle|\gamma'\rangle$. Thus block 1 contains only one error and block 2 also contains just one error—one gate error has produced only one error per block. So far, so good. Now suppose a storage error occurs on qubit 1 so that $|\alpha\rangle \to |\alpha'\rangle$ before the SWAP gates are applied. Applying the three SWAP gates yields the final state $|\beta\rangle|\alpha'\rangle|\gamma\rangle$ so that only one error is present in block 1 and no error occurs in block 2. The reader can check that the remaining ways in which one gate or storage error can occur will lead to no more than one error per block. The end result is that this three-step SWAP operation satisfies Definition 5.1 and so is fault-tolerant. □

The following definition introduces a transversal operation, which is then shown to be fault-tolerant.

DEFINITION 5.2 *A transversal operation satisfies one of the following two conditions. (1) It only applies one-qubit gates to the qubits in a code block. (2) It only interacts the ith qubit in one code block with the ith qubit in a different code block or block of ancilla.*

It is straightforward to check the fault-tolerance of transversal operations. For one-qubit transversal operations, fault-tolerance is automatic since, if only one gate error or storage error occurs, only one qubit is affected. Thus only one error appears in the code block and Definition 5.1 is satisfied. For two-qubit transversal operations, suppose the operation that interacts the ith qubits in the two different blocks fails. By assumption (iv) of our error model, only these two qubits can develop errors. Since they are in separate blocks, at most one error appears in each block and so Definition 5.1 is again satisfied. Now consider the case of a storage error. Suppose one appears on qubit i in the first block before it is interacted with the ith qubit in the second block. Upon interaction, this error cannot propagate beyond these two qubits since only they take part in the interaction. Since they are in separate blocks, at most one error appears in each block and so the definition of fault-tolerance is satisfied. Finally, if the storage error occurs after the two i^{th} qubits have been interacted, only one code block picks up an error and so Definition 5.1 is satisfied in this case as well. Transversal operations are thus seen to be fault-tolerant.

5.2 Error Correction

This section presents a fault-tolerant protocol for quantum error correction [1–3]. The reader is asked to review Sections 1.3.5, 1.4.1, and 3.1.3 before proceeding. We begin by establishing some preliminary results that will be used below.

(a) Recall that for an error $E \in \mathcal{G}_n$ and an $[n,k,d]$ quantum stabilizer code, the syndrome for E is the bit string $S(E) = l_1 \cdots l_{n-k}$, where the syndrome bit l_i is determined by whether E commutes or anticommutes with the generator g_i:

$$g_i E = (-1)^{l_i} E g_i \qquad (i = 1, \ldots, n-k) \ . \qquad (5.4)$$

As shown in the proof of Theorem 1.11, the corrupted codeword $E|c\rangle$ is a simultaneous eigenstate of the generators $\{g_1, \ldots, g_{n-k}\}$:

$$g_i \{E|c\rangle\} = (-1)^{l_i} \{E|c\rangle\} \qquad (i = 1, \ldots, n-k) \ . \qquad (5.5)$$

Measuring g_i when the quantum register is in the state $E|c\rangle$ returns the eigenvalue $(-1)^{l_i}$ and so determines l_i. The syndrome $S(E) = l_1 \cdots l_{n-k}$ is then found by measuring all $n-k$ generators. Using the notation established in Section 3.1.1 (and suppressing factors of i^λ), let $E = \sigma_x(e_x)\sigma_z(e_z)$ and $g_i = \sigma_x(a_i)\sigma_z(b_i)$. Direct calculation (see eq. (3.6)) gives

$$g_i E = (-1)^{e_x \cdot b_i + e_z \cdot a_i} E g_i \ . \qquad (5.6)$$

Comparing eqs. (5.4) and (5.6) gives

$$l_i = e_x \cdot b_i + e_z \cdot a_i \qquad (i = 1, \ldots, n-k) \ , \qquad (5.7)$$

where the scalar product was defined in Section 3.1.1, and all addition operations are modulo 2.

(b) Let U be a unitary operator. Applying it to eq. (5.5) gives

$$\bar{g}_i \{\overline{E}|\bar{c}\rangle\} = (-1)^{l_i} \{\overline{E}|\bar{c}\rangle\} \qquad (i = 1, \ldots, n-k) \ , \qquad (5.8)$$

where $\bar{g}_i = U g_i U^\dagger$, $\overline{E} = U E U^\dagger$, and $|\bar{c}\rangle = U|c\rangle$. We see that the unitary image of a corrupted codeword $\overline{E}|\bar{c}\rangle$ is a simultaneous eigenstate of the generator images $\bar{g}_i$. Notice that the eigenvalue of $\bar{g}_i$ still determines the syndrome bit l_i.

(c) Let $g_i = \sigma_x(a_i)\sigma_z(b_i)$ be one of the code generators. To simplify the following analysis, assume that g_i contains no factors of σ_y. This restriction will be removed in Section 5.2.1. Writing $a_i = (a_{i,1}, \ldots, a_{i,n})$, and recalling that the Hadamard gate H_j for the jth qubit transforms $\sigma_x^j \to \sigma_z^j$, we see that applying the transversal operation

$$\mathbf{H}_i = \prod_{j=1}^n (H_j)^{a_{i,j}} \qquad (5.9)$$

Fault-Tolerant Quantum Computing

to g_i gives

$$\bar{g}_i = \mathbf{H}_i g_i \mathbf{H}_i^\dagger$$
$$= \sigma_z(a_i)\sigma_z(b_i)$$
$$= \sigma_z(a_i + b_i) \ . \qquad (5.10)$$

Since $\mathbf{H}_i$ is unitary, the $\bar{g}_i$ satisfy eq. (5.8) and can be used to determine the error syndrome. Note also that since $\mathbf{H}_i$ is transversal, it is a fault-tolerant operation.

(d) Recall from Problem 1.1 that the intersection $x * y$ of two binary vectors x and y is the binary vector

$$x * y = (x_1 y_1, \ldots, x_n y_n) \ .$$

Thus $x * y$ has non-zero components only for those j for which $x_j = y_j = 1$. Let $a = (a_1, \ldots, a_n)$ be a binary vector. We define its bit-wise complement $\bar{a}$ as:

$$\bar{a} = (a_1 + 1, \ldots, a_n + 1) \ , \qquad (5.11)$$

where all addition operations are modulo 2. If $E = \sigma_x(e_x)\sigma_z(e_z)$, it follows that

$$\bar{E} = \mathbf{H}_i E \mathbf{H}_i$$
$$= \sigma_x(e_x * \bar{a}_i)\sigma_z(e_x * a_i)\sigma_z(e_z * \bar{a}_i)\sigma_x(e_z * a_i) \ , \qquad (5.12)$$

where we have used that $\mathbf{H}_i$ is Hermitian.

Exercise 5.1 *Derive eq. (5.12).*

(e) We will need one more result. Applying $\mathbf{H}_i$ to both sides of $g_i|c\rangle = |c\rangle$ gives

$$\bar{g}_i|\bar{c}\rangle = |\bar{c}\rangle \ , \qquad (5.13)$$

where $\bar{g}_i$ and $|\bar{c}\rangle$ were defined in *(b)* above. Expanding $|\bar{c}\rangle$ in the CB states $|k\rangle = |k_1 \cdots k_n\rangle$ gives

$$|\bar{c}\rangle = \sum_k \bar{c}(k)|k\rangle \ . \qquad (5.14)$$

Recalling that the CB states are simultaneous eigenstates of the $\{\sigma_z^i : i = 1, \ldots, n\}$, eqs. (5.10) and (5.14) give

$$\bar{g}_i|\bar{c}\rangle = \sum_k \bar{c}(k)(-1)^{(a_i+b_i)\cdot k}|k\rangle \ . \qquad (5.15)$$

Inserting eqs. (5.14) and (5.15) in eq. (5.13), we see that

$$(a_i + b_i) \cdot k = 0 \pmod{2} \ . \qquad (5.16)$$

Thus $|\bar{c}\rangle$ only contains bit strings $k = k_1 \cdots k_n$ that have even parity relative to the generator $\bar{g}_i = \sigma_z(a_i + b_i)$.

5.2.1 Syndrome Extraction

In syndrome extraction ancilla qubits are interacted with a code block in such a way that the syndrome $S(E)$ is encoded into the state of the ancilla. An appropriate measurement of the ancilla state then yields $S(E)$. This subsection explains how this can be done using a quantum circuit that only applies transversal operations so that the circuit itself is fault-tolerant. Further work is needed, however, to make preparation of the ancilla state and verification of the measured syndrome value fault-tolerant. These tasks are taken up in Sections 5.2.2 and 5.2.3, respectively.

Let g_i be a generator with weight w_i. To determine its associated syndrome bit l_i we introduce w_i ancilla qubits. Assume a code block has an error E and is in the state $E|c\rangle$, where $|c\rangle$ is a codeword. Since $\overline{E|c\rangle} = \mathbf{H}_i E|c\rangle$ is an eigenstate of $\overline{g}_i = \mathbf{H}_i g_i \mathbf{H}_i$ (recall $\mathbf{H}_i$ is Hermitian) with eigenvalue $(-1)^{l_i}$, and $\overline{g}_i$ only contains factors of σ_z, it proves convenient to work with $\overline{E|c\rangle}$, rather than with $E|c\rangle$. We begin then by applying $\mathbf{H}_i$ to the code block that puts it in the state $\overline{E|c\rangle}$. Using eqs. (5.12) and (5.14) gives

$$\overline{E|c\rangle} = \sigma_x(e_x * \not{a}_i)\sigma_z(e_x * a_i)\sigma_z(e_z * \not{a}_i)\sigma_x(e_z * a_i) \sum_k \overline{c}(k)|k\rangle \ .$$

Commuting $\sigma_x(e_x * \not{a}_i)$ past the σ_z operators and then applying the σ_x operators to the CB states $|k\rangle$ gives

$$\overline{E|c\rangle} = (-1)^\phi \sigma_z(e_x * a_i)\sigma_z(e_z * \not{a}_i) \sum_k \overline{c}(k)|k + e_x * \not{a}_i + e_z * a_i\rangle \ , \quad (5.17)$$

where $\phi = (e_x * \not{a}_i) \cdot (e_z * \not{a}_i)$ and the factor containing it on the RHS arises from moving $\sigma_x(e_x * \not{a}_i)$ past $\sigma_z(e_z * \not{a}_i)$. Note that $\sigma_x(e_x * \not{a}_i)$ and $\sigma_z(e_x * a_i)$ act on different qubits and so commute.

Exercise 5.2 *Following the proof of Theorem 3.1-(3), show that*

$$\sigma_x(e_x * \not{a}_i)\sigma_z(e_x * a_i)\sigma_z(e_z * \not{a}_i)\sigma_x(e_z * a_i) =$$
$$(-1)^\phi \sigma_z(e_x * a_i)\sigma_z(e_z * \not{a}_i)\sigma_x(e_x * \not{a}_i)\sigma_x(e_z * a_i) \ ,$$

where $\phi = (e_x * \not{a}_i) \cdot (e_z * \not{a}_i)$; $\not{a}_i = (a_{i,1} + 1, \ldots, a_{i,n} + 1) = a_i + (1, \ldots, 1)$; and the addition operations are modulo 2.

At this point the w_i ancilla qubits are introduced and are assumed to have been prepared in the Shor state $|A\rangle$,

$$|A\rangle = 2^{-(w_i-1)/2} \sum_{A_e} |A_e\rangle \ , \quad (5.18)$$

where the sum is over all even-parity bit strings A_e of length w_i. Recall that we are assuming that each generator $g_i = \sigma_x(a_i)\sigma_z(b_i)$ contains no factors of σ_y. Writing $a_i = (a_{i,1}, \ldots, a_{i,n})$ and $b_i = (b_{i,1}, \ldots, b_{i,n})$, this means that $a_{i,j}b_{i,j} = 0$ for all j. Consequently, the weight of $\bar{g}_i = \sigma_z(a_i + b_i)$ is also w_i. The next step is to apply CNOT gates from the w_i qubits in the code block for which $a_{i,j} + b_{i,j} = 1$ to the ancilla qubits transversally. Writing

$$U_{CNOT}(a_i + b_i) \equiv \prod_{j=1}^{n} (U_{CNOT})^{a_{i,j}+b_{i,j}}, \qquad (5.19)$$

using eqs. (5.17) and (5.18), and suppressing normalization factors on the Shor state for the remainder of this subsection gives

$$U_{CNOT}(a_i + b_i)\overline{E}|\bar{c}\rangle \otimes |A\rangle =$$
$$(-1)^\phi \sigma_z(e_x * a_i)\sigma_z(e_z * d_i)$$
$$\times \sum_k \bar{c}(k)|k + e_x * d_i + e_z * a_i\rangle$$
$$\otimes \sum_{A_e} |A_e + (a_i + b_i) * (k + e_x * d_i + e_z * a_i)\rangle . \qquad (5.20)$$

Exercise 5.3 Derive eq. (5.20).

Recall that $(a_i + b_i) \cdot k = 0 \pmod{2}$ which was one of the preliminary results found above. It follows that the bit string $(a_i + b_i) * k$ has even parity and can be absorbed into A_e in eq. (5.20). Thus eq. (5.20) can be rewritten as

$$U_{CNOT}(a_i + b_i)\overline{E}|\bar{c}\rangle \otimes |A\rangle =$$
$$(-1)^\phi \sigma_z(e_x * a_i)\sigma_z(e_z * d_i)$$
$$\times \sum_k \bar{c}(k)|k + e_x * d_i + e_z * a_i\rangle$$
$$\otimes \sum_{A_e} |A_e + (a_i + b_i) * (e_x * d_i + e_z * a_i)\rangle . \qquad (5.21)$$

Exercise 5.4 Show that

$$(a_i + b_i) * (e_x * d_i + e_z * a_i) = e_z * a_i + e_x * b_i .$$

Using the result of Exercise 5.4 and recalling that $|i + 1\rangle = \sigma_x|i\rangle$ allows us to write eq. (5.21) as

$$U_{CNOT}(a_i + b_i)\overline{E}|\bar{c}\rangle \otimes |A\rangle =$$
$$(-1)^\phi \sigma_z(e_x * a_i)\sigma_z(e_z * d_i)\sigma_x(e_x * d_i)\sigma_x(e_z * a_i)$$
$$\times \sum_k \bar{c}(k)|k\rangle \otimes \sum_{A_e} |A_e + e_z * a_i + e_x * b_i\rangle$$
$$= \overline{E}|\bar{c}\rangle \otimes \sum_{A_e} |A_e + e_z * a_i + e_x * b_i\rangle . \qquad (5.22)$$

Finally, applying the transversal operator $\mathbf{H}_i$ to eq. (5.22) gives

$$\mathbf{H}_i U_{CNOT}(a_i + b_i)\mathbf{H}_i\, E|c\rangle \otimes |A\rangle = E|c\rangle \otimes \sum_{A_e} |A_e + e_z * a_i + e_x * b_i\rangle \ . \quad (5.23)$$

Notice that this sequence of operations has (i) left the code and ancilla blocks in a *non-entangled* state, and (ii) left the code block in the state $E|c\rangle$. Because the final state is non-entangled, measuring the ancilla block will not cause any superposition of states in the code block to collapse. If we measure the ancilla qubits so that the modified Shor state collapses to the state, say $|\overline{A}_e + e_z * a_i + e_x * b_i\rangle$, the measurement result m_i will be

$$m_i = (-1)^{e_z \cdot a_i + e_x \cdot b_i + \text{wt}(\overline{A}_e)}$$
$$= (-1)^{l_i} \ , \quad (5.24)$$

where we have used eq. (5.7) and that $\text{wt}(\overline{A}_e)$ is an even integer. Note that since $\text{wt}(A_e)$ is an even interger for all A_e, no matter which state the modified Shor state collapses to, the measurement result m_i will always be $(-1)^{l_i}$. This procedure has thus extracted the syndrome bit l_i. Repeating the procedure for each generator yields the complete syndrome $S(E) = l_1 \cdots l_{n-k}$.

If g_i contains factors of σ_y, then the above procedure must be modified. If σ_y^k is present in g_i, then instead of applying the Hadamard gate H_k to the kth qubit when constructing $\overline{g}_i$, we must apply

$$\tilde{H}_k = \frac{1}{\sqrt{2}}\begin{pmatrix} 1 & -i \\ -i & 1 \end{pmatrix} = \exp\left[-\frac{i\pi}{4}\sigma_x\right] \ , \quad (5.25)$$

which maps $\sigma_y \to \sigma_z$. If we do this for each qubit for which g_i has a factor of σ_y, we obtain the new generator $\overline{g}_i = \sigma_z(a_i + b_i + a_i * b_i)$. Problem 5.1 works out the modifications to syndrome extraction that arise when the generators contain factors of σ_y.

Before moving on to consider fault-tolerant preparation of the Shor state, we illustrate syndrome extraction using the [5,1,3] code.

Example 5.3

From Section 3.2.1 the generators for the [5,1,3] code are

$$\begin{aligned}
g_1 &= \sigma_x(10010)\sigma_z(01100) \\
g_2 &= \sigma_x(01001)\sigma_z(00110) \\
g_3 &= \sigma_x(10100)\sigma_z(00011) \\
g_4 &= \sigma_x(01010)\sigma_z(10001) \ ,
\end{aligned} \quad (5.26)$$

which allows the bit strings $a_i = (a_{i,1}, \ldots, a_{i,5})$ and $b_i = (b_{i,1}, \ldots, b_{i,5})$ to be read off for each generator g_i and determines the weights to be $w_i = 4$ for

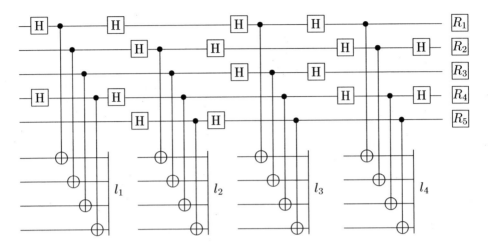

FIGURE 5.1
Quantum circuit for syndrome extraction for the [5,1,3] quantum stabilizer code. The upper five lines correspond to the code block and each set of lower four lines to an ancilla block. Qubit indicies increase in going from top to bottom in each block. The measured syndrome value $S(E) = l_1 l_2 l_3 l_4$ determines the recovery operation $\mathbf{R} = R_1 R_2 R_3 R_4 R_5$.

$i = 1, \ldots, 4$. It follows from eq. (5.9) that

$$\begin{aligned}
\mathbf{H}_1 &= H_1 H_4 \\
\mathbf{H}_2 &= H_2 H_5 \\
\mathbf{H}_3 &= H_1 H_3 \\
\mathbf{H}_4 &= H_2 H_4
\end{aligned} \tag{5.27}$$

and so

$$\begin{aligned}
\bar{g}_1 &= \sigma_z(11110) \\
\bar{g}_2 &= \sigma_z(01111) \\
\bar{g}_3 &= \sigma_z(10111) \\
\bar{g}_4 &= \sigma_z(11011) \; .
\end{aligned} \tag{5.28}$$

Eqs. (5.23) and (5.28) determine the quantum circuit for syndrome extraction that appears in Figure 5.1. This circuit encodes the syndrome bits into four ancilla blocks where they can be measured, and the result used to determine the transversal (and hence fault-tolerant) recovery operation $\mathbf{R} = R_1 R_2 R_3 R_4 R_5$. The syndrome bit l_i is transferred to the ith ancilla block in accordance with eq. (5.23). Plenio et al. [3] point out that there are

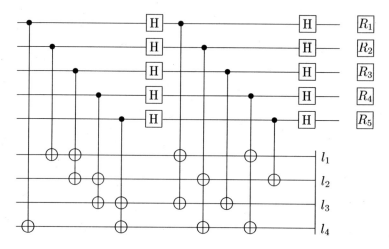

FIGURE 5.2
Quantum circuit for syndrome extraction for the [5,1,3] quantum stabilizer code following Ref. [3]. Note that each four qubit ancilla block has been compressed into a single line to make the circuit easier to examine, and the four CNOT gates applied to each ancilla block are applied transversally. The measured syndrome value $S(E) = l_1 \cdots l_4$ determines the recovery operation $\mathbf{R} = R_1 \cdots R_5$.

advantages to rearranging the sequence of CNOT gates in Figure 5.1. Instead of using the operators $\{\mathbf{H}_i U_{CNOT}(a_i + b_i)\mathbf{H}_i : i = 1, \ldots, 4\}$ to extract the syndrome, they use

$$\mathbf{H} U_{CNOT}(a_i) \, \mathbf{H} U_{CNOT}(b_i) \quad (i = 1, \ldots, 4) \ , \tag{5.29}$$

where

$$\mathbf{H} = \prod_{j=1}^{5} H_j \tag{5.30}$$

is the transversal application of Hadamard gates to all qubits in the code block. The corresponding circuit appears in Figure 5.2. Note that each ancilla block has been compressed into a single line to make the circuit easier to examine. Each ancilla block still contains four qubits. □

5.2.2 Shor State Verification

Although the syndrome extraction circuit only uses transversal operations we do not yet have a fault-tolerant procedure for quantum error correction. The

Fault-Tolerant Quantum Computing

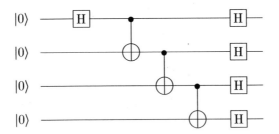

FIGURE 5.3
Quantum circuit to prepare the Shor state $|A\rangle$ for $w_i = 4$. The first Hadamard gate and three CNOT gates place the ancilla block in the cat-state. The remaining four Hadamard gates transform this state into $|A\rangle$.

remaining difficulties stem from errors that may be present in the Shor state. As we will see in Section 5.2.3, bit-flip errors in the Shor state can cause an error in the measured syndrome value. Such an error will generate new errors in the data by causing the wrong recovery operation to be applied to the code block. A fault-tolerant procedure is thus needed to verify the measured syndrome value. Phase errors in the Shor state can also ruin fault-tolerance by spreading to the code block through the CNOT gates used during syndrome extraction. This subsection will deal with the difficulties arising from phase errors in the Shor state, while the following subsection will examine the case of bit-flip errors.

The ancilla block can be placed in the Shor state $|A\rangle$ using a two-step procedure [1]. First the block is prepared in the cat-state (named for Schrodinger's paradoxical cat [9]):

$$|\psi_{cat}\rangle = \frac{1}{\sqrt{2}} \{ |0 \cdots 0\rangle + |1 \cdots 1\rangle \} , \qquad (5.31)$$

which is a uniform superposition of states in which the ancilla qubits are all in the $|0\rangle$ CB-state or all in the $|1\rangle$ CB-state. The procedure is completed by applying a Hadamard gate to each ancilla qubit. Direct calculation shows that the Hadamard gates transform $|\psi_{cat}\rangle \to |A\rangle$.

Exercise 5.5 *For an ancilla block containing w_i qubits, show that*

$$|A\rangle = \prod_{j=1}^{w_i} H_j |\psi_{cat}\rangle .$$

A quantum circuit to prepare $|A\rangle$ is given in Figure 5.3. To make the discussion concrete we have set $w_i = 4$ as would be appropriate for the [5,1,3] code examined in Example 5.3. Generalizing to arbitrary w_i is straightforward. Initially all ancilla qubits are in the $|0\rangle$ CB-state. The first Hadamard gate and three CNOT gates put the ancilla block in the cat-state and the final four Hadamard gates transform the cat-state into the Shor state $|A\rangle$. Although the final Hadamard gates are applied transversely and so operate fault-tolerantly, unfortunately the same is not true of the portion of the circuit that creates the cat-state. Failure of one of the CNOT gates can lead to multiple errors in both the ancilla and code blocks. Specifically, suppose the second CNOT gate fails, causing a bit-flip error on its target qubit. Since this qubit becomes the control for the third CNOT gate, its error spreads to the target qubit (see Example 5.1). Thus one gate error has produced two errors in the ancilla block with probability of $\mathcal{O}(p)$. But things get worse. After the final Hadamard gates are applied, the two bit-flip errors become two phase errors in the Shor state since $H\sigma_x = \sigma_z H$. As shown in Example 5.1, these two phase errors will propagate back into the code block via the CNOT gates used for syndrome extraction. Thus one gate error during cat-state preparation has produced two errors in both the code and ancilla blocks with probability of $\mathcal{O}(p)$. This procedure for Shor state preparation is thus not fault-tolerant. To recover fault-tolerance, the preparation procedure must be modified so that the probability for this outcome becomes of $\mathcal{O}(p^2)$.

To that end, following Shor [1,6], notice that when the target for one of the CNOT gates used in cat-state preparation develops a bit-flip error, the error propagates to the target qubits of all subsequent CNOT gates in Figure 5.3. In the cat-state, the first and last ancilla qubits always have even relative parity. A single CNOT gate error that causes a bit-flip error on its target qubit causes the relative parity of the first and last ancilla qubits to become odd. Figure 5.4 gives a modified circuit for Shor state preparation that takes advantage of this observation. A new ancilla qubit is introduced in the CB-state $|0\rangle$ and it is made the target of CNOT gates from the first and fourth ancilla qubits. If these qubits have odd relative parity, the new ancilla qubit will wind up in the state $|1\rangle$. Measuring σ_z for the new qubit returns $s = 1$, telling us that an odd number of bit-flip errors occurred during cat-state preparation. In particular, to $\mathcal{O}(p)$, one bit-flip error has occurred. In this case the ancilla block is discarded, a new ancilla block is introduced, and cat-state preparation is re-tried on the new ancilla block. If measurement of the new ancilla qubit yields $s = 0$, we know its state is $|0\rangle$, and that an even number of bit-flip errors has occurred. Specifically, to $\mathcal{O}(p)$, we know that no bit-flip errors occurred. In this case the final Hadamard gates can be applied, and we are assured that to $\mathcal{O}(p)$, the output Shor state will not contain phase errors. Said another way, this modified procedure insures that the probability that two phase errors appear in the Shor state is $\mathcal{O}(p^2)$, as desired.

So far the focus has been on bit-flip errors during cat-state preparation. Phase errors can also occur when preparing this state and they will go undetected by the circuit in Figure 5.4. The final Hadamard gates in that circuit will transform these errors into bit-flip errors in the Shor state. The bit-flip

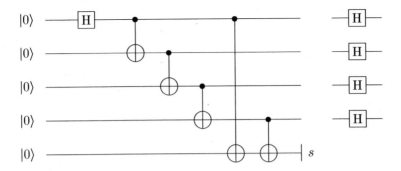

FIGURE 5.4
Modified quantum circuit for Shor state preparation for $w_i = 4$. A new ancilla qubit is introduced to record the relative parity s of the first and fourth qubits. If measurement yields $s = 1$, the ancilla block is discarded, a new ancilla block is introduced, and the circuit is applied again. The final Hadamard gates are only applied if $s = 0$. See text for full discussion. Based on Figure 11, J. Preskill, *Proc. R. Soc. Lond. A* **454**, 385 (1998); ©Royal Society; used with permission.

errors will not spread into the code block during syndrome extraction, though they can cause the measured syndrome value to be in error. As noted earlier, if blindly accepted, an erroneous syndrome value will cause the wrong recovery operator to be applied to the code block, introducing errors into that block. The following subsection will describe a fault-tolerant procedure for syndrome verification.

5.2.3 Syndrome Verification

Bit-flip errors in the Shor state can cause the measured syndrome value to be in error. This is most easily seen from eq. (5.24). If an odd number of bit-flip errors occur, $\text{wt}(\overline{A}_e)$ will be an odd rather than an even integer, and syndrome measurement will return $l_i + 1$ instead of l_i. As pointed out in Section 5.2.2, bit-flip errors in the Shor state can appear during its preparation. Other possible sources are (i) a storage error that causes a bit-flip error on an ancilla qubit, or (ii) a faulty CNOT gate during syndrome extraction that introduces errors in both the code and ancilla blocks. In the latter case, if the error in the ancilla block is a bit-flip error, then the measured syndrome value will be incorrect, and if blindly accepted, will produce a second error in the code block through application of the wrong recovery operation to that block. Thus to $\mathcal{O}(p)$, one gate error produces two errors in the code block, and the error correction procedure fails to be fault-tolerant. The modified procedure presented below insures that two errors appear in the code block only when two or more independent errors occur.

Example 5.4

Errors can occur that are not detected by the syndrome extraction circuit. Referring to Figure 5.2, suppose a storage error occurs on a qubit in the code block immediately before the final Hadamard gate is applied to it. Since this error occurs after all CNOT gates have been applied to this qubit, it will be too late for this error to affect the value of the syndrome. If this is the only error that occurs, the measured syndrome value will be $S = 0$ even though an error is present in the code block. Another example of an undetected error is a bit-flip error that occurs on a qubit in the code block immediately before the first Hadamard gate is applied to it. The Hadamard gate transforms this error into a phase error that will not be detected by the subsequent CNOT gates applied to this qubit. Supposing again that this is the only error that occurs, the measured syndrome value will be $S = 0$ even though an error is present in the code block. It is not difficult to find other examples of errors that go undetected. The reader is encouraged to look for others. The moral of this example is that obtaining $S = 0$ for the measured syndrome value does not gaurantee that the code block is error-free. The above examples show that to $\mathcal{O}(p)$, one error can be present in the code block even when $S = 0$.
□

Having seen that errors can cause the measured syndrome value to be incorrect, a procedure must be introduced to improve its reliability. To be fault-tolerant, the procedure must insure that, should a single error occur during syndrome extraction, no more than one error remains in a code block after the error correction procedure has completed. Ref. [3] introduced the following syndrome verification protocol. We state it first, then explain how it leads to fault-tolerance.

1. **Syndrome measurement returns** $S = 0$: Accept the measurement result as correct; no recovery operation applied to code block.

2. **Syndrome measurement returns** $S \neq 0$: Remeasure syndrome using new ancilla block; accept the second syndrome result as correct; apply corresponding recovery operator to code block.

We have seen in Example 5.4 that the syndrome result $S = 0$ does not gaurantee that the code block is error-free. In fact, to $\mathcal{O}(p)$, the code block can contain one error. By accepting $S = 0$ as the correct syndrome value and leaving the code block alone, the protocol does not introduce any new errors into the data. To $\mathcal{O}(p)$ then, at most an undetected error is present in the code block after error correction. The protocol is thus fault-tolerant for $S = 0$. If the syndrome measurement yields $S \neq 0$, at least one error has occurred, and to $\mathcal{O}(p)$, exactly one error has occurred. The above protocol requires that the syndrome be remeasured using a new ancilla block, and the new result be accepted as correct. This works since to $\mathcal{O}(p)$, for $S \neq 0$, one error already occurred during the first syndrome extraction, and so to first-order in p, no error can occur during the second syndrome extraction. To $\mathcal{O}(p)$, the second syndrome result will be correct. For the second syndrome result to be wrong, a new error would have to occur during the second syndrome extraction, making the overall process second-order in p. The recovery operator corresponding to the second syndrome result can be applied to the code block, and we are assured that recovery succeeds to first-order in p. The protocol is thus fault-tolerant for $S \neq 0$ as well.

Plenio et al. [3] have extended this protocol so that it only fails when $t+1$ or more errors occur during syndrome extraction. At most $t+1$ syndrome measurements are required to determine the correct syndrome to $\mathcal{O}(p^t)$.

Steane has introduced an alternative approach for fault-tolerant quantum error correction. The reader is referred to Refs. [7,8] for further discussion of this approach.

5.3 Encoded Operations in $N(\mathcal{G}_n) \cap N(\mathcal{S})$

Formally, an encoded quantum gate is a unitary operator U that maps codewords to codewords (see Section 3.1.4). As we now know, for a quantum stabilizer code $\mathcal{C}_q$ with stabilizer $\mathcal{S}$, each codeword $|c\rangle$ is fixed by every stabilizer element s: $s|c\rangle = |c\rangle$. It is easy to show that for all $|c\rangle \in \mathcal{C}_q$ and $s \in \mathcal{S}$, $U|c\rangle$ is fixed by $UsU^\dagger$:

$$\left(UsU^\dagger\right)(U|c\rangle) = Us|c\rangle = U|c\rangle \ . \tag{5.32}$$

Now if U fixes $\mathcal{S}$ under conjugation ($U\mathcal{S}U^\dagger = \mathcal{S}$), then every element of $\mathcal{S}$ can be written as $UsU^\dagger$ for some $s \in \mathcal{S}$. Then for every codeword $|c\rangle$, it follows from eq. (5.32) that $U|c\rangle$ is a codeword since it is fixed by every element of $\mathcal{S}$. Thus U maps codewords to codewords whenever $U\mathcal{S}U^\dagger = \mathcal{S}$, and so is an encoded operation. Unitary operators that fix $\mathcal{S}$ under conjugation belong to a subgroup of the unitary group $U(n)$ known as the normalizer $N_U(\mathcal{S})$. Recall that we introduced the normalizer $N_\mathcal{G}(\mathcal{S})$ in Exercise 3.2, and showed that it was equal to the centralizer $\mathcal{C}(\mathcal{S})$. In that discussion the operators that acted on $\mathcal{S}$ belonged to $\mathcal{G}_n$, while here it is operators in the unitary group $U(n)$ that act on $\mathcal{S}$. It is clear that $N_\mathcal{G}(\mathcal{S}) \subset N_U(\mathcal{S})$ since $\mathcal{G}_n \subset U(n)$. For the remainder of this chapter we will be solely interested in $N_U(\mathcal{S})$ and so will suppress the subscript U: $N_U(\mathcal{S}) \to N(\mathcal{S})$.

Having indicated why the normalizer $N(\mathcal{S})$ is central to our discussion of encoded operations, we next introduce the Clifford group $N_U(\mathcal{G}_n)$, which is the the normalizer of the Pauli group $\mathcal{G}_n$ in $U(n)$ [10–12]. We again suppress the subscript U so that $N_U(\mathcal{G}_n) \to N(\mathcal{G}_n)$. We shall see below that the groups $N(\mathcal{S})$ and $N(\mathcal{G}_n)$ intersect, though one group need not be a subset of the other. For purposes of this chapter, the Clifford group $N(\mathcal{G}_n)$ is of interest because it is generated by a simple set of quantum gates: Hadamard, phase, and CNOT gates [10, 12]. Because $N(\mathcal{G}_n)$ has such a simple generator set, we will focus on unitary operators $U \in N(\mathcal{G}_n) \cap N(\mathcal{S})$, which are thus encoded operations that can be constructed using Hadamard, phase, and CNOT gates. The examples we will consider yield encoded operations that are transversal and so are *fault-tolerant*.

Having seen that a unitary operator U that fixes $\mathcal{S}$ under conjugation is an encoded operation, we now show that such a U also maps $\mathcal{C}(\mathcal{S})/\mathcal{S} \to \mathcal{C}(\mathcal{S})/\mathcal{S}$ and so map encoded operations to encoded operations. Recall from Section 3.1.4 that the encoded Pauli operators $X_i = \xi \sigma_x^i \xi^\dagger$ and $Z_i = \xi \sigma_z^i \xi^\dagger$ label the cosets that generate $\mathcal{C}(\mathcal{S})/\mathcal{S}$, where ξ is the encoding operator, and $i = 1, \ldots, k$ labels the encoded qubits. As elements of $\mathcal{C}(\mathcal{S})$, X_i and Z_i commute with the code generators $\{g_j : j = 1, \ldots, n - k\}$. Now let $X_i' = U X_i U^\dagger$ and $Z_i' = U Z_i U^\dagger$ be the images under U of the encoded Pauli operators. It is a simple matter to show that X_i' and Z_i' also commute with the generators of $\mathcal{S}$ and so belong to $\mathcal{C}(\mathcal{S})$. We show this explicitly for X_i':

$$\begin{aligned}[][X_i', g_j] &= U X_i U^\dagger g_j - g_j U X_i U^\dagger \\ &= U X_i s_j U^\dagger - U s_j X_i U^\dagger \\ &= U [X_i, s_j] U^\dagger \\ &= 0 \ . \end{aligned} \tag{5.33}$$

In going from the first line to the second, we used that $U \mathcal{S} U^\dagger = \mathcal{S}$ so that we can write $g_j = U s_j U^\dagger$ for some $s_j \in \mathcal{S}$. Thus $X_i' \in \mathcal{C}(\mathcal{S})$, and a similar calculation shows that $Z_i' \in \mathcal{C}(\mathcal{S})$. Since U fixes $\mathcal{S}$ under conjugation, U maps the cosets $X_i \mathcal{S} \to X_i' \mathcal{S}$ and $Z_i \mathcal{S} \to Z_i' \mathcal{S}$. The following exercise shows that U maps $\mathcal{C}(\mathcal{S})/\mathcal{S} \to \mathcal{C}(\mathcal{S})/\mathcal{S}$. Since the elements of $\mathcal{C}(\mathcal{S})/\mathcal{S}$ are encoded operations (see Section 3.1.4), we see that U maps encoded operations to encoded operations.

Exercise 5.6 *Let $X(a)Z(b)\mathcal{S}$ be an arbitrary element of $\mathcal{C}(\mathcal{S})/\mathcal{S}$ with (i) $X(a) = X_1^{a_1} \cdots X_k^{a_k}$; (ii) $Z(b) = Z_1^{b_1} \cdots Z_k^{b_k}$; and (iii) $a = a_1 \cdots a_k$ and $b = b_1 \cdots b_k$ are bit strings of length k. It follows from the definition of multiplication in $\mathcal{C}(\mathcal{S})/\mathcal{S}$ (see Exercise 3.5) that*

$$X(a)Z(b)\mathcal{S} = \left(\prod_{l=1}^{k} X_l^{a_l} \mathcal{S} \right) \left(\prod_{m=1}^{k} Z_m^{b_m} \mathcal{S} \right) \ .$$

(a) *Show that U maps $X(a)Z(b)\mathcal{S} \to X'(a)Z'(b)\mathcal{S}$, where (i) $X'(a) = (X_1')^{a_1} \cdots (X_k')^{a_k}$; (ii) $Z'(b) = (Z_1')^{b_1} \cdots (Z_k')^{b_k}$; (iii) $X_l' = U X_l U^\dagger$; (iv) $Z_m' = U Z_m U^\dagger$; and*

$$X'(a)Z'(b)\mathcal{S} = \left[\prod_{l=1}^{k} (X_l')^{a_l} \mathcal{S} \right] \left[\prod_{m=1}^{k} (Z_m')^{b_m} \mathcal{S} \right] \ .$$

Fault-Tolerant Quantum Computing

(b) Argue that $X'(a)Z'(b)\mathcal{S} \in \mathcal{C}(\mathcal{S})/\mathcal{S}$ since $\mathcal{C}(\mathcal{S})/\mathcal{S}$ is a group and $X'_l\mathcal{S}$ and $Z'_m\mathcal{S}$ are elements of $\mathcal{C}(\mathcal{S})/\mathcal{S}$ for $l, m = 1, \ldots, k$. This establishes that U maps $\mathcal{C}(\mathcal{S})/\mathcal{S} \to \mathcal{C}(\mathcal{S})/\mathcal{S}$.

5.3.1 Action of $N(\mathcal{G}_n)$

An automorphism of a group G is a map $f : G \to G$ that is 1-1, onto, and preserves the multiplication table of G (viz. $f(ab) = f(a)f(b)$). An automorphism is thus an isomorphism from $G \to G$.

THEOREM 5.1
Let U be an element of $N(\mathcal{G}_n)$ which acts on $\mathcal{G}_n$ by conjugation: $f_U(g) = UgU^\dagger$ for all $g \in \mathcal{G}_n$. The map f_U is an automorphism of $\mathcal{G}_n$.

PROOF Since $U \in N(\mathcal{G}_n)$ it fixes $\mathcal{G}_n$ under conjugation and so $f_U(g) \in \mathcal{G}_n$ for all $g \in \mathcal{G}_n$. Thus f_U maps $\mathcal{G}_n \to \mathcal{G}_n$.

1-1: Let $a, b \in \mathcal{G}_n$; $a \neq b$; and assume that $f_U(a) = f_U(b)$. Then by assumption, $UaU^\dagger = UbU^\dagger$, from which it follows that $a = b$. But $a \neq b$ so we have a contradiction. Thus $f_U(a) \neq f_U(b)$ whenever $a \neq b$, and so f_U is 1-1.

Onto: Let $f_U(\mathcal{G}_n)$ denote the range of f_U. We have just seen that $f_U(\mathcal{G}_n) \subset \mathcal{G}_n$. Since f_U is 1-1, $f_U(\mathcal{G}_n)$ contains the same number of elements as $\mathcal{G}_n$ and so cannot be a proper subset of $\mathcal{G}_n$. Thus $f_U(\mathcal{G}_n) = \mathcal{G}_n$ and so f_U is onto.

Products: Let $a, b \in \mathcal{G}_n$. Then

$$\begin{aligned} f_U(ab) &= UabU^\dagger \\ &= (UaU^\dagger)(UbU^\dagger) \\ &= f_U(a) f_U(b) \ . \end{aligned} \qquad (5.34)$$

Thus f_U is an automorphism of $\mathcal{G}_n$. Note that occasionally we will say "U is an automorphism of $\mathcal{G}_n$" when we really mean f_U. ∎

Since each $U \in N(\mathcal{G}_n)$ is an automorphism of $\mathcal{G}_n$, we can determine the action of U on $\mathcal{G}_n$ once we know how the generators of $N(\mathcal{G}_n)$ act on the generators of $\mathcal{G}_n$. To that end, we now examine how the one-qubit Hadamard and phase gates act on the generators of $\mathcal{G}_1$, and the two-qubit CNOT gate acts on the generators of $\mathcal{G}_2 = \mathcal{G}_1 \otimes \mathcal{G}_1$.

- In the computational basis the Hadamard gate H has the representation

$$H = \frac{1}{\sqrt{2}} \begin{pmatrix} 1 & 1 \\ 1 & -1 \end{pmatrix} \ . \qquad (5.35)$$

Recall from Section 2.2 that the generators of $\mathcal{G}_1$ are σ_x and σ_z. Using their well-known representations in the computational basis, it is straightforward

to show that
$$\sigma_x \to H\sigma_x H^\dagger = \sigma_z$$
$$\sigma_z \to H\sigma_z H^\dagger = \sigma_x \ . \tag{5.36}$$

Since $\sigma_y = -i\sigma_z\sigma_x$, it follows that
$$\begin{aligned}\sigma_y \to H\sigma_y H^\dagger &= -i\left(H\sigma_z H^\dagger\right)\left(H\sigma_x H^\dagger\right) \\ &= -i\sigma_x\sigma_z \\ &= -\sigma_y \ . \end{aligned} \tag{5.37}$$

We see that H permutes the generators of $\mathcal{G}_1$, which insures that the map f_H is 1-1 and onto.

- The phase gate P in the computational basis is

$$P = \begin{pmatrix} 1 & 0 \\ 0 & i \end{pmatrix} \ . \tag{5.38}$$

Its action on σ_x and σ_z is
$$\sigma_x \to P\sigma_x P^\dagger = \sigma_y$$
$$\sigma_z \to P\sigma_z P^\dagger = \sigma_z \ , \tag{5.39}$$

and on σ_y is
$$\begin{aligned}\sigma_y \to P\sigma_y P^\dagger &= -i\left(\sigma_z\right)\left(\sigma_y\right) \\ &= -\sigma_x \ . \end{aligned} \tag{5.40}$$

We see that P also permutes the generators of $\mathcal{G}_1$. Notice that P is simply a $\pi/2$ rotation about the z-axis.

- As a two-qubit gate, the CNOT gate must be applied to the generators of $\mathcal{G}_2 = \mathcal{G}_1 \otimes \mathcal{G}_1$. We take the first (second) qubit as the control (target). Writing all matrices as block matrices [13], and in the computational basis, we have for the CNOT gate

$$U_{CNOT} = \begin{pmatrix} I & 0 \\ 0 & \sigma_x \end{pmatrix} \ , \tag{5.41}$$

and for the generators of $\mathcal{G}_2$

$$\sigma_x \otimes I = \begin{pmatrix} 0 & I \\ I & 0 \end{pmatrix}$$
$$I \otimes \sigma_x = \begin{pmatrix} \sigma_x & 0 \\ 0 & \sigma_x \end{pmatrix}$$
$$\sigma_z \otimes I = \begin{pmatrix} I & 0 \\ 0 & -I \end{pmatrix}$$
$$I \otimes \sigma_z = \begin{pmatrix} \sigma_z & 0 \\ 0 & \sigma_z \end{pmatrix} \ . \tag{5.42}$$

Thus

$$\sigma_x \otimes I \to U_{CNOT}[\sigma_x \otimes I]U_{CNOT}^\dagger = \begin{pmatrix} I & 0 \\ 0 & \sigma_x \end{pmatrix} \begin{pmatrix} 0 & I \\ I & 0 \end{pmatrix} \begin{pmatrix} I & 0 \\ 0 & \sigma_x \end{pmatrix}$$
$$= \begin{pmatrix} 0 & \sigma_x \\ \sigma_x & 0 \end{pmatrix}$$
$$= \sigma_x \otimes \sigma_x \ . \quad (5.43)$$

A similar calculation can be done for each of the remaining generators. Collecting together the results for all four generators gives

$$\sigma_x \otimes I \to \sigma_x \otimes \sigma_x$$
$$I \otimes \sigma_x \to I \otimes \sigma_x$$
$$\sigma_z \otimes I \to \sigma_z \otimes I$$
$$I \otimes \sigma_z \to \sigma_z \otimes \sigma_z \ . \quad (5.44)$$

As was pointed out in Example 5.1, we see that the CNOT gate propagates bit-flip errors from the control to the target, and phase errors from the target to the control.

Exercise 5.7 *Another useful two-qubit gate is the controlled-phase gate. In the computational basis it is*

$$U_{CP} = \begin{pmatrix} I & 0 \\ 0 & \sigma_z \end{pmatrix} \ . \quad (5.45)$$

(a) Show that $U_{CP} = (I \otimes H)\,U_{CNOT}\,(I \otimes H)$.

(b) Show that the controlled-phase gate has the following action on the generators of $\mathcal{G}_2$:

$$\sigma_x \otimes I \to \sigma_x \otimes \sigma_z$$
$$I \otimes \sigma_x \to \sigma_z \otimes \sigma_x$$
$$\sigma_z \otimes I \to \sigma_z \otimes I$$
$$I \otimes \sigma_z \to I \otimes \sigma_z \ . \quad (5.46)$$

We see that a controlled-phase gate does not propagate phase errors, though a bit-flip error on one qubit spreads to a phase error on the other qubit.

- We now show how to take an automorphism f_U and determine its associated unitary operator U. As an example [4], consider the automorphism of $\mathcal{G}_1$ that cyclically permutes its generators:

$$f_T : \sigma_x \to \sigma_y \to \sigma_z \to \sigma_x \ . \quad (5.47)$$

The CB state $|0\rangle$ is the eigenstate of σ_z with eigenvalue $+1$. Using eq. (5.32) we have

$$\begin{aligned} T|0\rangle &= T\sigma_z|0\rangle \\ &= \left(T\sigma_z T^\dagger\right)(T|0\rangle) \\ &= \sigma_x\left(T|0\rangle\right) \;. \end{aligned}$$

Thus $T|0\rangle$ is the eigenvector of σ_x with eigenvalue $+1$:

$$T|0\rangle = \frac{1}{\sqrt{2}}[|0\rangle + |1\rangle] \;. \tag{5.48}$$

From eq. (5.48) we read off the matrix elements $T_{00} = \langle 0|T|0\rangle = 1/\sqrt{2}$ and $T_{10} = \langle 1|T|0\rangle = 1/\sqrt{2}$. Since the CB state $|1\rangle = \sigma_x|0\rangle$, we have

$$\begin{aligned} T|1\rangle &= T\sigma_x|0\rangle \\ &= \left(T\sigma_x T^\dagger\right)(T|0\rangle) \\ &= \sigma_y\left(T|0\rangle\right) \\ &= \frac{-i}{\sqrt{2}}[|0\rangle - |1\rangle] \;, \end{aligned}$$

where we have used eq. (5.48) in going from the third line to the fourth. Thus $T_{01} = -i/\sqrt{2}$ and $T_{11} = i/\sqrt{2}$. Collecting all matrix elements gives

$$T = \frac{1}{\sqrt{2}}\begin{pmatrix} 1 & -i \\ 1 & i \end{pmatrix} \;. \tag{5.49}$$

The reader can check by direct calculation that

$$\begin{aligned} T\sigma_x T^\dagger &= \sigma_y \\ T\sigma_y T^\dagger &= \sigma_z \\ T\sigma_z T^\dagger &= \sigma_x \;. \end{aligned}$$

The following subsection examines the effect of applying the generators of $N(\mathcal{G}_n)$ transversally to qubit blocks that are encoded using CSS codes. We shall see that there are CSS codes for which this approach is sufficiently powerful to allow fault-tolerant quantum computation (when combined with the Toffoli gate construction described in Section 5.7). However, more theoretical tools are needed if fault-tolerant quantum computing is to be possible for all quantum stabilizer codes. Presentation of the new tools begins in Section 5.4.

5.3.2 CSS Codes

Recall (Section 2.3 and Problem 1.7) that a CSS code is constructed using a pair of classical binary codes C and C' with the following properties:

Fault-Tolerant Quantum Computing

1. C and C' are $[n, k, d]$ and $[n, k', d']$ codes, respectively;

2. $C' \subset C$; and

3. C and $C'_\perp$ are both t-error correcting codes.

For purposes of this subsection, two further restrictions must be made. First, we choose $C' = C_\perp$ so that condition 3 is automatically satisfied. Before stating the second restriction, recall (Problem 1.2) that for classical binary codes, either all codewords have even weight, or half have even weight and half have odd weight. Our second restriction not only assumes all codewords $c \in C$ have even weight, but that their weights are divisible by 4:

$$\text{wt}(c) \equiv 0 \pmod 4 \ . \tag{5.50}$$

Such codes are called doubly-even codes. Let $\mathbf{H}$ be the parity check matrix for C:

$$\mathbf{H} = \begin{pmatrix} \text{---} & \mathbf{H}_1^T & \text{---} \\ & \vdots & \\ \text{---} & \mathbf{H}_{n-k}^T & \text{---} \end{pmatrix} . \tag{5.51}$$

Recall (Section 1.2) that the columns $\mathbf{H}_i$ of $\mathbf{H}^T$ span $C_\perp$. Then, since $C_\perp \subset C$, it follows that for a doubly-even code C,

$$\text{wt}(\mathbf{H}_i^T) \equiv 0 \pmod 4 \tag{5.52}$$

for $i = 1, \ldots, n-k$. We will make use of this property below.

The associated CSS code is constructed by splitting the stabilizer generators into two sets $(i = 1, \ldots, n-k)$:

$$g_i = \sigma_x \left(\mathbf{H}_i^T\right)$$
$$= \left(\sigma_x^1\right)^{\mathbf{H}_{i,1}} \cdots \left(\sigma_x^n\right)^{\mathbf{H}_{i,n}} \tag{5.53}$$

and

$$g_{n-k+i} = \sigma_z \left(\mathbf{H}_i^T\right)$$
$$= \left(\sigma_z^1\right)^{\mathbf{H}_{i,1}} \cdots \left(\sigma_z^n\right)^{\mathbf{H}_{i,n}} , \tag{5.54}$$

where $\mathbf{H}_i^T$ is the ith row of $\mathbf{H}$. The encoded Pauli operators (see eq. (3.20))

$$X_j = \xi \sigma_x^j \xi^\dagger$$
$$Z_j = \xi \sigma_z^j \xi^\dagger \qquad (j = 1, \ldots, k) \tag{5.55}$$

have the form

$$X_j = \sigma_x(a_j)$$
$$= \left(\sigma_x^1\right)^{a_{j,1}} \cdots \left(\sigma_x^n\right)^{a_{j,n}} \tag{5.56}$$

and

$$Z_j = \sigma_z(a_j)$$
$$= (\sigma_z^1)^{a_{j,1}} \cdots (\sigma_z^n)^{a_{j,n}} . \tag{5.57}$$

Note that X_j and Z_j both use the same bit string a_j. It is straightforward to show that

$$X_j Z_j = (-1)^{a_j \cdot a_j} Z_j X_j$$
$$= (-1)^{\text{wt}(a_j)} Z_j X_j . \tag{5.58}$$

To insure that X_j and Z_j anticommute, we must have $\text{wt}(a_j) = 2m_j + 1$, where m_j is an integer. The encoded Pauli operator $Y_j = \xi \sigma_y^j \xi^\dagger$ is then

$$Y_j = -i Z_j X_j$$
$$= (i)^{(\text{wt}(a_j)+3)} \sigma_y(a_j) . \tag{5.59}$$

Eq. (5.59) can be inverted to give

$$\sigma_y(a_j) = (-1)^{m_j} Y_j , \tag{5.60}$$

where $m_j = [\text{wt}(a_j) - 1]/2$.

Exercise 5.8 *Derive eqs. (5.58)–(5.60).*

Suppose we apply a Hadamard gate H to each qubit in a code block. This is a transversal operation and so is fault-tolerant (Section 5.1). Since H maps $\sigma_x^j \leftrightarrow \sigma_z^j$ for $j = 1, \ldots, n$, we see from eqs. (5.53) and (5.54) that applying H bitwise maps $g_i \leftrightarrow g_{n-k+i}$ for $i = 1, \ldots, n-k$. This operation thus fixes $\mathcal{S}$ and so is an encoded operation. From eqs. (5.56) and (5.57) we see that bitwise application of H maps $X_j \leftrightarrow Z_j$ for $j = 1, \ldots, k$, and so applies an encoded Hadamard gate to all encoded qubits. We thus obtained a block encoded Hadamard gate.

Next we apply the phase gate P bitwise to a code block. This operation is transversal and so fault-tolerant. Since P maps $\sigma_z^j \to \sigma_z^j$ for $j = 1, \ldots, n$, bitwise application of P maps $g_{n-k+i} \to g_{n-k+i}$ for $i = 1, \ldots, n-k$. However P maps $\sigma_x^j \to \sigma_y^j$ and so for $i = 1, \ldots, n-k$

$$g_i \to \sigma_y\left(\mathbf{H}_i^T\right) = i^{\text{wt}(\mathbf{H}_i^T)} \sigma_z\left(\mathbf{H}_i^T\right) \sigma_x\left(\mathbf{H}_i^T\right)$$
$$= g_{n-k+i} \, g_i , \tag{5.61}$$

which is an element of $\mathcal{S}$. Note that we used eq. (5.52) so that $\text{wt}(\mathbf{H}_i^T) = 4s$ for some integer s. The images of the generators also generate $\mathcal{S}$. This can be seen by finding combinations of image products that yield the original generators. Thus bitwise application of P fixes $\mathcal{S}$ and so is an encoded operation. From eq. (5.57) we see that bitwise P maps $Z_j \to Z_j$ for $j = 1, \ldots, k$. Applied to X_j, bitwise P gives (see eq. (5.60))

$$X_j \to \sigma_y(a_j) = (-1)^{m_j} Y_j \ . \tag{5.62}$$

Thus if m_j is odd (even), bitwise application of P performs a block-encoded $P^\dagger$ (P) gate on all encoded qubits.

Finally, we apply the CNOT gate transversally from the jth qubit in one code block to the jth qubit in a second code block for $j = 1, \ldots, n$. Because of the transversality of the operation, it will be fault-tolerant. Next we examine its action on $\mathcal{S} \otimes \mathcal{S}$. From eqs. (5.44) we see that transversal CNOT maps

$$\begin{aligned} g_i \otimes I &\longrightarrow g_i \otimes g_i \\ I \otimes g_i &\longrightarrow I \otimes g_i \\ g_{n-k+i} \otimes I &\longrightarrow g_{n-k+i} \otimes I \\ I \otimes g_{n-k+i} &\longrightarrow g_{n-k+i} \otimes g_{n-k+i} \ , \end{aligned} \tag{5.63}$$

for $i = 1, \ldots, n - k$. As with the previous two cases, the generator images are also generators of $\mathcal{S}$, and so the transversal CNOT operation is an encoded operation since it fixes $\mathcal{S}$. Again, using eqs. (5.44), we see that for $j = 1, \ldots, k$

$$\begin{aligned} X_j \otimes I &\longrightarrow X_j \otimes X_j \\ I \otimes X_j &\longrightarrow I \otimes X_j \\ Z_j \otimes I &\longrightarrow Z_j \otimes I \\ I \otimes Z_j &\longrightarrow Z_j \otimes Z_j \ , \end{aligned} \tag{5.64}$$

which is the encoded version of eqs. (5.44). Thus when both code blocks are encoded using a CSS code, bitwise application of CNOT gates applies a block-encoded CNOT gate between the jth encoded qubits in the two blocks for all j. It is also possible to show that if bitwise application of CNOT gates yields an encoded operation, the code must be a CSS code (see Problem 5.3). In this case, having established that the code is a CSS code, it follows from the above analysis that the encoded operation is a block-encoded CNOT gate.

We see that for CSS codes based on doubly-even classical codes with $C_\perp \subset C$, transversal application of the generators of $N(\mathcal{G}_n)$ produces: (i) block-encoded versions of the one-qubit generators; and (ii) a block-encoded CNOT gate. Thus this procedure constructs all generators of the Clifford group for those CSS codes in this class that encode a single qubit. As noted above, combined with the Toffoli gate construction of Section 5.7, fault-tolerant quantum computing can be done using such single-qubit encoding CSS codes [1]. For CSS codes that encode multiple qubits, the procedure of this section is unable to construct an encoded Hadamard or phase gate that only acts on a subset of the encoded qubits, or an encoded CNOT gate that only acts on the encoded i^{th} and j^{th} qubits. This shortcoming will also plague the procedures developed in Sections 5.4–5.5 when applied to stabilizer codes that encode multiple qubits. Section 5.6 will show how to perform these missing encoded gates when working with such codes. Even with that blemish removed, it will still be necessary to have a quantum gate that does not belong to the Clifford group to carry out universal quantum computation. The Toffoli gate will be used for that purpose. Section 5.7 will show how to construct a fault-tolerant encoded Toffoli gate for any quantum stabilizer code. All together, the theoretical tools developed in Sections 5.3–5.7 will make encoded fault-tolerant quantum computing possible using any quantum stabilizer code.

5.4 Measurement

In this section we explain how a quantum gate can be applied to a quantum register using measurements. Transversal application of this technique, in combination with the results of Section 5.3, will eventually provide a sufficiently powerful set of tools to allow fault-tolerant quantum computation to be done using any quantum stabilizer code.

Let $|\psi_0\rangle$ be an arbitrary state in the n-qubit Hilbert space H_2^n. The set of operators in $\mathcal{G}_n$ that fix $|\psi_0\rangle$ is the stabilizer $\mathcal{S}_0$ of $|\psi_0\rangle$ in $\mathcal{G}_n$. Denoting the generators of $\mathcal{S}_0$ by $\{G_1^0, \ldots, G_m^0\}$, we have

$$G_k^0|\psi_0\rangle = |\psi_0\rangle \qquad k = 1, \ldots, m \ . \tag{5.65}$$

Note that typically $\mathcal{S}_0$ will *not* be the code stabilizer $\mathcal{S}$.

Let $\mathcal{O}$ be an Hermitian operator in $\mathcal{G}_n$ that anticommutes with one of the generators of $\mathcal{S}_0$. We choose the generators so that G_1^0 is the only anticommuting generator. This can always be done since if G_k^0 anticommutes with $\mathcal{O}$, $G_1^0 G_k^0$ will commute with it, and so we simply replace G_k^0 with $G_1^0 G_k^0$. By Theorem 3.1-(2), $\mathcal{O}^2 = I$ and so $\mathcal{O}$ has eigenvalues $\lambda_\pm = \pm 1$ and projection operators $P_\pm = (I \pm \mathcal{O})/2$ that project a state onto the $\lambda_\pm$ eigenspaces, respectively.

We are now ready to introduce the measurement protocol. *Step 1:* Measure $\mathcal{O}$. (A fault-tolerant procedure for measurement is given in Nielsen and Chuang [14].) *Step 2A:* If the measurement outcome is λ_+, take no further action. In this case the final state will be $|\psi_1\rangle = P_+|\psi_0\rangle \equiv |\psi_+\rangle$. *Step 2B:* If the outcome is λ_-, apply G_1^0 to the post-measurement state $P_-|\psi_0\rangle$. The

Fault-Tolerant Quantum Computing

final state $|\psi_1\rangle$ is then

$$\begin{aligned}|\psi_1\rangle &= G_1^0 P_- |\psi_0\rangle \\ &= G_1^0 \left[\frac{1}{2}(I - \mathcal{O})\right] |\psi_0\rangle \\ &= \frac{1}{2}(I + \mathcal{O}) G_1^0 |\psi_0\rangle \\ &= |\psi_+\rangle \;, \end{aligned} \quad (5.66)$$

where we used that G_1^0 anticommutes with $\mathcal{O}$ and fixes $|\psi_0\rangle$. Thus in *both* cases, the measurement protocol maps $|\psi_0\rangle \to |\psi_1\rangle = |\psi_+\rangle$. Following Gottesman [4] we will say "measure $\mathcal{O}$" when we actually mean "carry out the measurement protocol using $\mathcal{O}$."

Notice that the final state $|\psi_1\rangle = |\psi_+\rangle$ is fixed by the stabilizer $\mathcal{S}_1$ whose generators are $\{\mathcal{O}, G_2^1, \ldots, G_m^1\}$, where

$$G_k^1 = \begin{cases} G_k^0 & \text{if } [G_k^0, \mathcal{O}] = 0 \\ G_1^0 G_k^0 & \text{if } \{G_k^0, \mathcal{O}\} = 0 \end{cases} \quad k \neq 1 \;. \quad (5.67)$$

As with $\mathcal{S}_0$, typically $\mathcal{S}_1$ differs from the code stabilizer $\mathcal{S}$. Not surprisingly, since the measurement protocol transforms the state $|\psi_0\rangle \to |\psi_1\rangle$, it also transforms the stabilizer $\mathcal{S}_0 \to \mathcal{S}_1$.

The centralizer $\mathcal{C}(\mathcal{S}_0)$ ($\mathcal{C}(\mathcal{S}_1)$) is the set of operators in $\mathcal{G}_n$ that commute with all elements of $\mathcal{S}_0$ ($\mathcal{S}_1$). The measurement protocol induces a map from $\mathcal{C}(\mathcal{S}_0)/\mathcal{S}_0 \to \mathcal{C}(\mathcal{S}_1)/\mathcal{S}_1$. To see this, let $N \in \mathcal{C}(\mathcal{S}_0)$ and $|\psi_0\rangle$ the state fixed by $\mathcal{S}_0$. The image N' of N under measurement is determined by requiring that N' acting on the final state $P_+|\psi_0\rangle$ yield the same state as P_+ acting on the state $N|\psi_0\rangle$:

$$N'[P_+|\psi_0\rangle] = P_+[N|\psi_0\rangle] \;. \quad (5.68)$$

(1) If $[N, \mathcal{O}] = 0$, then $N' = N$ solves eq. (5.68) since

$$\begin{aligned} N' P_+ |\psi_0\rangle &= N P_+ |\psi_0\rangle \\ &= P_+ N |\psi_0\rangle \;. \end{aligned}$$

Since N commutes with $\mathcal{O}$, $N' \in \mathcal{C}(\mathcal{S}_1)$ and belongs to the coset $N\mathcal{S}_1$. In this case, measurement maps $N \to N$ and $N\mathcal{S}_0 \to N\mathcal{S}_1$. (2) On the other hand, if $\{N, \mathcal{O}\} = 0$, then $N' = G_1^0 N$ solves eq. (5.68) since $[G_1^0 N, \mathcal{O}] = 0$ and G_1^0 fixes $|\psi_0\rangle$:

$$\begin{aligned} N' P_+ |\psi_0\rangle &= G_1^0 N P_+ |\psi_0\rangle \\ &= P_+ G_1^0 N |\psi_0\rangle \\ &= P_+ N |\psi_0\rangle \;. \end{aligned}$$

Since $G_1^0 N$ commutes with $\mathcal{O}$, $N' \in \mathcal{C}(\mathcal{S}_1)$ and belongs to the coset $(G_1^0 N)\mathcal{S}_1$. In this case, measurement maps $N \to G_1^0 N$ and $N\mathcal{S}_0 \to (G_1^0 N)\mathcal{S}_1$. In the following, the analogs for $\mathcal{S}_0$ of the encoded Pauli operators X_i and Z_i will be denoted by $\mathcal{X}_i^0$ and $\mathcal{Z}_i^0$, respectively. By examining how measurement maps $\mathcal{X}_i^0$ and $\mathcal{Z}_i^0$ we will be able to identify what operation is applied to $|\psi_0\rangle$.

In summary, to formally determine the action of measuring $\mathcal{O}$ on the pre-measurement state $|\psi_0\rangle$ (with stabilizer $\mathcal{S}_0$), carry out the following procedure:

1. Identify a generator $G_1^0 \in \mathcal{S}_0$ that anticommutes with $\mathcal{O}$. Replace any generator G_k^0 ($k \neq 1$) of $\mathcal{S}_0$ that anticommutes with $\mathcal{O}$ with $G_1^0 G_k^0$.

2. Form the new stabilizer $\mathcal{S}_1$ by replacing G_1^0 in $\mathcal{S}_0$ with $\mathcal{O}$. The generators for $\mathcal{S}_1$ are then $\{\mathcal{O}, G_2^1, \ldots, G_m^1\}$, where the G_k^1 are defined in eq. (5.67).

3. Replace any $\mathcal{X}_i^0$ or $\mathcal{Z}_i^0$ that anticommute with $\mathcal{O}$ with $G_1^0 \mathcal{X}_i^0$ or $G_1^0 \mathcal{Z}_i^0$, respectively. This insures that the new operators belong to $\mathcal{C}(\mathcal{S}_1)$.

The following example due to Gottesman [4] demonstrates how to implement the generators of the Clifford group $N(\mathcal{G}_n)$ using only measurement and CNOT gates.

Example 5.5

Consider a two-qubit system in which qubit 1 is in an arbitrary state $|\psi\rangle$ and qubit 2 is in the CB-state $|0\rangle$. The stabilizer for the state $|\psi\rangle \otimes |0\rangle$ has generator $G = I^1 \otimes \sigma_z^2$. The operators $\mathcal{X} = \sigma_x^1 \otimes I^2$ and $\mathcal{Z} = \sigma_z^1 \otimes I^2$ clearly commute with G and anticommute with each other. They will serve as the analogs of σ_x and σ_z. Now apply a CNOT gate using qubit 1 as the control. From eq. (5.44) we have

$$G \to G^0 = \sigma_z^1 \otimes \sigma_z^2$$
$$\mathcal{X} \to \mathcal{X}^0 = \sigma_x^1 \otimes \sigma_x^2$$
$$\mathcal{Z} \to \mathcal{Z}^0 = \sigma_z^1 \otimes I^2 \ . \tag{5.69}$$

Finally, measure $\mathcal{O} = I^1 \otimes \sigma_y^2$. It is clear from eq. (5.69) that G^0 and $\mathcal{X}^0$ anticommute with $\mathcal{O}$. Thus measuring $\mathcal{O}$ induces the map

$$G^0 \to G^1 = I^1 \otimes \sigma_y^2 = \mathcal{O}$$
$$\mathcal{X}^0 \to \mathcal{X}^1 = -\sigma_y^1 \otimes \sigma_y^2 = G^0 \mathcal{X}^0$$
$$\mathcal{Z}^0 \to \mathcal{Z}^1 = \sigma_z^1 \otimes I^2 \ . \tag{5.70}$$

Since the measurement protocol leaves qubit 2 in the $+1$ eigenstate of σ_y^2, the parts of G^1, $\mathcal{X}^1$, and $\mathcal{Z}^1$ associated with qubit 2 fix this state and so it is conventional to discard the measured qubit. The parts of $\mathcal{X}^0$, $\mathcal{X}^1$, $\mathcal{Z}^0$, and $\mathcal{Z}^1$ associated with qubit 1 then determine what operation has been carried out. From $\mathcal{X}^0 \to \mathcal{X}^1$ we see that $\sigma_x^1 \to -\sigma_y^1$, and from $\mathcal{Z}^0 \to \mathcal{Z}^1$ that $\sigma_z^1 \to \sigma_z^1$. The above procedure has thus implemented $P^\dagger$. Note that if the measurement protocol were modified so that $|\psi_0\rangle \to P_-|\psi_0\rangle$, the above procedure would have produced P instead of $P^\dagger$.

It is also possible to follow the transformations of the state. Writing $|\psi\rangle = a|0\rangle + b|1\rangle$, we have $|\psi\rangle \otimes |0\rangle = a|00\rangle + b|10\rangle$. Applying the CNOT gate produces the state $|\psi_0\rangle = a|00\rangle + b|11\rangle$. Measuring $\mathcal{O}$ puts qubit 2 in the state $|0\rangle_y = |0\rangle + i|1\rangle$, which is the $+1$ eigenstate of σ_y^2. Since $|0\rangle = |0\rangle_y + |1\rangle_y$ and $|1\rangle = -i|0\rangle_y + i|1\rangle_y$, we see that measurement produces the state

$$|\psi_1\rangle = \{\, a|0\rangle - ib|1\rangle \,\} \otimes |0\rangle_y \ .$$

This is a non-entangled state in which qubit 1 is in the state $|\psi_f\rangle = a|0\rangle - ib|1\rangle$. This procedure has thus applied the unitary transformation $|\psi_f\rangle = U|\psi\rangle$ to qubit 1. Inserting the states $|\psi_f\rangle$ and $|\psi\rangle$ into this equation gives

$$\begin{pmatrix} a \\ -ib \end{pmatrix} = U \begin{pmatrix} a \\ b \end{pmatrix} \ ,$$

from which we find

$$U = \begin{pmatrix} 1 & 0 \\ 0 & -i \end{pmatrix} = P^\dagger \ ,$$

in agreement with the operator analysis.

It is possible to construct the Hadamard gate H using measurements and CNOT gates once we can construct the operator Q that applies a $-\pi/2$ rotation about the x-axis. Thus Q maps $\sigma_x \to \sigma_x$ and $\sigma_z \to \sigma_y$. It is simple to check using the matrix representations for Q and P that $H = PQ^\dagger P$. We now show how to produce Q from measurement and CNOT gates.

To make Q, start with qubit 1 in an arbitrary state $|\psi\rangle$ and qubit 2 in the $+1$ eigenstate of σ_x^2. Thus the two qubits begin in the state $|\psi\rangle \otimes [|0\rangle + |1\rangle]$, where we have suppressed the normalization factor. The stabilizer for this state has generator $G = I^1 \otimes \sigma_x^2$. The Pauli operator analogs are $\mathcal{X} = \sigma_x^1 \otimes I^2$ and $\mathcal{Z} = \sigma_z^1 \otimes I^2$. Now apply a CNOT gate using qubit 2 as the control. Then,

$$G \to \tilde{G} = \sigma_x^1 \otimes \sigma_x^2$$
$$\mathcal{X} \to \tilde{\mathcal{X}} = \sigma_x^1 \otimes I^2$$
$$\mathcal{Z} \to \tilde{\mathcal{Z}} = \sigma_z^1 \otimes \sigma_z^2 \ .$$

Next apply P to the second qubit (which we now know how to do using measurement and a CNOT gate). This produces the map

$$\tilde{G} \to G^0 = \sigma_x^1 \otimes \sigma_y^2$$
$$\tilde{\mathcal{X}} \to \mathcal{X}^0 = \sigma_x^1 \otimes I^2$$
$$\tilde{\mathcal{Z}} \to \mathcal{Z}^0 = \sigma_z^1 \otimes \sigma_z^2 \ . \quad (5.71)$$

Finally, measure $\mathcal{O} = I^1 \otimes \sigma_x^2$. From eq. (5.71) we see that G^0 and $\mathcal{Z}^0$ anticommute with $\mathcal{O}$. Thus measuring $\mathcal{O}$ produces the map

$$G^0 \to G^1 = I^1 \otimes \sigma_x^2 = \mathcal{O}$$
$$\mathcal{X}^0 \to \mathcal{X}^1 = \sigma_x^1 \otimes I^2$$
$$\mathcal{Z}^0 \to \mathcal{Z}^1 = \sigma_y^1 \otimes \sigma_x^2 = G^0 \mathcal{Z}^0 \ . \quad (5.72)$$

Discarding qubit 2, we see from eqs. (5.71) and (5.72) that this procedure maps $\sigma_x^1 \to \sigma_x^1$ and $\sigma_z^1 \to \sigma_y^1$, which is the action of Q. We see that the Hadamard and phase gates can be made using measurements and CNOT gates. This gives us all the generators of $N(\mathcal{G}_n)$ and so any gate in the Clifford group can be produced by combining measurements and CNOT gates (and state preparation). □

5.5 Four-Qubit Interlude

This section introduces a four-qubit unitary operation that, when applied transversally, produces an encoded version of itself for all quantum stabilizer codes. With this operation, a fault-tolerant block-encoded CNOT gate can be made, which together with measurement, allows all gates in the Clifford group to be constructed (see Example 5.5) for stabilizer codes encoding a single qubit.

We begin by specifying the action of the four-qubit operation. It produces the following transformation on the generators of $\mathcal{G}_4$:

$$
\begin{aligned}
\sigma_x^1 \otimes I^2 \otimes I^3 \otimes I^4 &\longrightarrow \sigma_x^1 \otimes \sigma_x^2 \otimes \sigma_x^3 \otimes I^4 \\
I^1 \otimes \sigma_x^2 \otimes I^3 \otimes I^4 &\longrightarrow I^1 \otimes \sigma_x^2 \otimes \sigma_x^3 \otimes \sigma_x^4 \\
I^1 \otimes I^2 \otimes \sigma_x^3 \otimes I^4 &\longrightarrow \sigma_x^1 \otimes I^2 \otimes \sigma_x^3 \otimes \sigma_x^4 \\
I^1 \otimes I^2 \otimes I^3 \otimes \sigma_x^4 &\longrightarrow \sigma_x^1 \otimes \sigma_x^2 \otimes I^3 \otimes \sigma_x^4 \\
\sigma_z^1 \otimes I^2 \otimes I^3 \otimes I^4 &\longrightarrow \sigma_z^1 \otimes \sigma_z^2 \otimes \sigma_z^3 \otimes I^4 \\
I^1 \otimes \sigma_z^2 \otimes I^3 \otimes I^4 &\longrightarrow I^1 \otimes \sigma_z^2 \otimes \sigma_z^3 \otimes \sigma_z^4 \\
I^1 \otimes I^2 \otimes \sigma_z^3 \otimes I^4 &\longrightarrow \sigma_z^1 \otimes I^2 \otimes \sigma_z^3 \otimes \sigma_z^4 \\
I^1 \otimes I^2 \otimes I^3 \otimes \sigma_z^4 &\longrightarrow \sigma_z^1 \otimes \sigma_z^2 \otimes I^3 \otimes \sigma_z^4 \quad . \quad (5.73)
\end{aligned}
$$

A quantum circuit that carries out this operation is shown in Figure 5.5 and is due to Gottesman [4].

Let $\mathcal{C}_q$ be a quantum stabilizer code that encodes k qubits in n qubits. Let

$$s = i^{\lambda_s} \sigma_x(s_x) \sigma_z(s_z)$$

be an arbitrary element of the stabilizer $\mathcal{S}$. Here λ_s is chosen to insure s is Hermitian, and s_x and s_z are bit strings of length n. Suppose we apply the four-qubit operation transversally to four code blocks, with each of the Pauli operators in s transformed according to eq. (5.73). This operation produces

Fault-Tolerant Quantum Computing

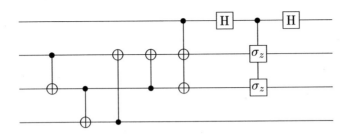

FIGURE 5.5
Quantum circuit to implement the four-qubit operation in eq. (5.73). Qubit lines are labeled from 1 to 4 in going from top to bottom. Reprinted Figure 2 with permission from D. Gottesman, *Phys. Rev. A* **57**, 127 (1998). ©American Physical Society, 1998.

the mapping:

$$\begin{aligned} s \otimes I^2 \otimes I^3 \otimes I^4 &\longrightarrow s \otimes s \otimes s \otimes I^4 \\ I^1 \otimes s \otimes I^3 \otimes I^4 &\longrightarrow I^1 \otimes s \otimes s \otimes s \\ I^1 \otimes I^2 \otimes s \otimes I^4 &\longrightarrow s \otimes I^2 \otimes s \otimes s \\ I^1 \otimes I^2 \otimes I^3 \otimes s &\longrightarrow s \otimes s \otimes I^3 \otimes s \end{aligned} \qquad (5.74)$$

The reader can show that the generator images are themselves generators of the stabilizer $S \otimes S \otimes S \otimes S$ by finding combinations of their products that yield the original set of generators. This operation thus fixes the four-block stabilizer and so is an encoded operation. Since it is transversal it is also fault-tolerant. To determine which encoded operation it is, we examine how the encoded Pauli operators $X_j = i^{\lambda_j} \sigma_x(c_j) \sigma_z(d_j)$ and $Z_j = i^{\bar{\lambda}_j} \sigma_x(e_j) \sigma_z(f_j)$ ($j = 1, \ldots, k$) are transformed. Transversal application of the four-qubit operation maps:

$$\begin{aligned} X_j \otimes I^2 \otimes I^3 \otimes I^4 &\longrightarrow X_j \otimes X_j \otimes X_j \otimes I^4 \\ I^1 \otimes X_j \otimes I^3 \otimes I^4 &\longrightarrow I^1 \otimes X_j \otimes X_j \otimes X_j \\ I^1 \otimes I^2 \otimes X_j \otimes I^4 &\longrightarrow X_j \otimes I^2 \otimes X_j \otimes X_j \\ I^1 \otimes I^2 \otimes I^3 \otimes X_j &\longrightarrow X_j \otimes X_j \otimes I^3 \otimes X_j \\ Z_j \otimes I^2 \otimes I^3 \otimes I^4 &\longrightarrow Z_j \otimes Z_j \otimes Z_j \otimes I^4 \\ I^1 \otimes Z_j \otimes I^3 \otimes I^4 &\longrightarrow I^1 \otimes Z_j \otimes Z_j \otimes Z_j \\ I^1 \otimes I^2 \otimes Z_j \otimes I^4 &\longrightarrow Z_j \otimes I^2 \otimes Z_j \otimes Z_j \\ I^1 \otimes I^2 \otimes I^3 \otimes Z_j &\longrightarrow Z_j \otimes Z_j \otimes I^3 \otimes Z_j \end{aligned} \qquad (5.75)$$

with $j = 1, \ldots, k$. Comparing eqs. (5.75) and (5.73) we see that transversal application of the four-qubit operation produces an encoded version of itself. Since the code $\mathcal{C}_q$ was chosen arbitrarily, this result is true for all quantum stabilizer codes. We see from eq. (5.75) that for multi-qubit encodings ($k > 1$), this operation simultaneously maps the encoded Pauli operators (X_j, Z_j) for all j. It cannot be used to map a subset of these operators (e. g., only X_1 and Z_1). Clearly, for single-qubit encodings ($k = 1$), this issue does not arise.

Now suppose we have two qubits in an arbitrary two-qubit state $|\psi_{12}\rangle$ and we prepare two ancilla qubits in the CB-state $|00\rangle$. The stabilizer $\mathcal{S}_0$ for the state $|\psi_{12}\rangle \otimes |00\rangle$ has generators:

$$G_1^0 = I^1 \otimes I^2 \otimes \sigma_z^3 \otimes I^4 \quad ; \quad G_2^0 = I^1 \otimes I^2 \otimes I^3 \otimes \sigma_z^4 \quad , \tag{5.76}$$

and the Pauli operator analogs are

$$\begin{aligned}
\mathcal{X}_1^0 &= \sigma_x^1 \otimes I^2 \otimes I^3 \otimes I^4 \quad ; \quad \mathcal{X}_2^0 = I^1 \otimes \sigma_x^2 \otimes I^3 \otimes I^4 \\
\mathcal{Z}_1^0 &= \sigma_z^1 \otimes I^2 \otimes I^3 \otimes I^4 \quad ; \quad \mathcal{Z}_2^0 = I^1 \otimes \sigma_z^2 \otimes I^3 \otimes I^4 \quad .
\end{aligned} \tag{5.77}$$

Applying the four-qubit operation to these qubits produces the mapping:

$$\begin{aligned}
G_1^0 \to G_1 &= \sigma_z^1 \otimes I^2 \otimes \sigma_z^3 \otimes \sigma_z^4 \quad ; \quad G_2^0 \to G_2 = \sigma_z^1 \otimes \sigma_z^2 \otimes I^3 \otimes \sigma_z^4 \\
\mathcal{X}_1^0 \to \mathcal{X}_1 &= \sigma_x^1 \otimes \sigma_x^2 \otimes \sigma_x^3 \otimes I^4 \quad ; \quad \mathcal{X}_2^0 \to \mathcal{X}_2 = I^1 \otimes \sigma_x^2 \otimes \sigma_x^3 \otimes \sigma_x^4 \\
\mathcal{Z}_1^0 \to \mathcal{Z}_1 &= \sigma_z^1 \otimes \sigma_z^2 \otimes \sigma_z^3 \otimes I^4 \quad ; \quad \mathcal{Z}_2^0 \to \mathcal{Z}_2 = I^1 \otimes \sigma_z^2 \otimes \sigma_z^3 \otimes \sigma_z^4 \quad .
\end{aligned}$$

Finally, the operators $\mathcal{O}_1 = I^1 \otimes I^2 \otimes \sigma_x^3 \otimes I^4$ and $\mathcal{O}_2 = I^1 \otimes I^2 \otimes I^3 \otimes \sigma_x^4$ are measured. They anticommute, respectively, with $\{G_1, \mathcal{Z}_1, \mathcal{Z}_2\}$ and $\{G_1, G_2, \mathcal{Z}_2\}$. Measurement thus produces the mapping:

$$\begin{aligned}
G_1, G_2 &\to G_1^1, G_2^1 \\
\mathcal{X}_1, \mathcal{X}_2 &\to \mathcal{X}_1^1, \mathcal{X}_2^1 \\
\mathcal{Z}_1, \mathcal{Z}_2 &\to \mathcal{Z}_1^1, \mathcal{Z}_2^1 \quad ,
\end{aligned}$$

where $\mathcal{Z}_1 \to \mathcal{Z}_1^1 = G_1 G_2 \mathcal{Z}_1$; $\mathcal{Z}_2 \to \mathcal{Z}_2^1 = G_1 \mathcal{Z}_2$; and

$$\begin{aligned}
G_1^1 &= \mathcal{O}_1 \quad ; \quad G_2^1 = \mathcal{O}_2 \\
\mathcal{X}_1^1 &= \sigma_x^1 \otimes \sigma_x^2 \otimes \sigma_x^3 \otimes I^4 \quad ; \quad \mathcal{X}_2^1 = I^1 \otimes \sigma_x^2 \otimes \sigma_x^3 \otimes \sigma_x^4 \\
\mathcal{Z}_1^1 &= \sigma_z^1 \otimes I^2 \otimes I^3 \otimes I^4 \quad ; \quad \mathcal{Z}_2^1 = \sigma_z^1 \otimes \sigma_z^2 \otimes I^3 \otimes I^4 \quad .
\end{aligned}$$

Discarding the measured ancilla qubits gives

$$\begin{aligned}
\mathcal{X}_1^1 &= \sigma_x^1 \otimes \sigma_x^2 \quad ; \quad \mathcal{X}_2^1 = I^1 \otimes \sigma_x^2 \\
\mathcal{Z}_1^1 &= \sigma_z^1 \otimes I^2 \quad ; \quad \mathcal{Z}_2^1 = \sigma_z^1 \otimes \sigma_z^2 \quad .
\end{aligned} \tag{5.78}$$

Discarding the ancilla qubits from eq. (5.77) we see that this procedure has mapped $\mathcal{X}_1^0, \mathcal{X}_2^0 \to \mathcal{X}_1^1, \mathcal{X}_2^1$ and $\mathcal{Z}_1^0, \mathcal{Z}_2^0 \to \mathcal{Z}_1^1, \mathcal{Z}_2^1$, which (restricted to the

Fault-Tolerant Quantum Computing

first two qubits) maps:

$$\begin{aligned}
\sigma_x^1 \otimes I^2 &\longrightarrow \sigma_x^1 \otimes \sigma_x^2 \\
I^1 \otimes \sigma_x^2 &\longrightarrow I^1 \otimes \sigma_x^2 \\
\sigma_z^1 \otimes I^2 &\longrightarrow \sigma_z^1 \otimes I^2 \\
I^1 \otimes \sigma_z^2 &\longrightarrow \sigma_z^1 \otimes \sigma_z^2 \quad,
\end{aligned} \qquad (5.79)$$

We recognize eq. (5.79) as the action of a CNOT gate. Thus the four-qubit operation can be used to construct a CNOT gate.

If we apply the procedure leading to eq. (5.79) transversally to four code blocks and discard the measured qubits in the ancilla code blocks, we produce a fault-tolerant encoded operation on the first two blocks that maps

$$\begin{aligned}
X_j \otimes I^2 &\longrightarrow X_j \otimes X_j \\
I^1 \otimes X_j &\longrightarrow I^1 \otimes X_j \\
Z_j \otimes I^2 &\longrightarrow Z_j \otimes I^2 \\
I^1 \otimes Z_j &\longrightarrow Z_j \otimes Z_j \quad,
\end{aligned} \qquad (5.80)$$

for $j = 1, \ldots, k$. We recognize this as a block-encoded CNOT gate. As with eq. (5.75), for stabilizer codes encoding more than one qubit, this procedure causes an encoded CNOT gate to be applied simultaneously to the jth encoded qubits in the two code blocks, for all j values. It cannot be used to selectively apply an encoded CNOT gate to some of the j, and not to others. This procedure is also unable to apply an encoded CNOT gate to two encoded qubits in the *same* code block. This difficulty does not occur if the stabilizer code only encodes a single qubit. For such codes, the procedure yields a fault-tolerant encoded CNOT gate, which together with measurement allows all gates in the Clifford group to be applied fault-tolerantly on encoded qubits. The following section extends this procedure so that the same can be done with stabilizer codes that encode multiple qubits.

5.6 Multi-Qubit Stabilizer Codes

Section 5.5 presented a four-qubit interaction that was used to construct a block-encoded CNOT gate. We saw that for stabilizer codes that encode multiple qubits, this block CNOT gate was unable to apply an encoded CNOT gate (i) between only the ith encoded qubits in two code blocks, or (ii) between only the ith and j^{th} encoded qubits in the same or different blocks. This section presents techniques that remove these limitations so that it will be possible to fault-tolerantly perform all gates in the Clifford group on encoded qubits for any quantum stabilizer code.

To begin, suppose we have a data qubit and an ancilla qubit. The data qubit is initially in an arbitrary state $|\psi\rangle$ and we prepare the ancilla qubit in the $+1$ eigenstate of σ_x^a. The stabilizer for this state has the generator $G^0 = I^d \otimes \sigma_x^a$, where the superscripts identify the data and ancilla qubit operators. The Pauli operator analogs are $\mathcal{X}^0 = \sigma_x^d \otimes I^a$ and $\mathcal{Z}^0 = \sigma_z^d \otimes I^a$. A CNOT gate is now applied to this state using the ancilla qubit as the control. This produces the new generator and Pauli operator analogs: $\overline{G} = \sigma_x^d \otimes \sigma_x^a$, $\overline{\mathcal{X}} = \sigma_x^d \otimes I^a$, and $\overline{\mathcal{Z}} = \sigma_z^d \otimes \sigma_z^a$. Finally we measure $\mathcal{O} = \sigma_x^d \otimes I^a$ so that the final generator and Pauli operator analogs are $G^1 = \mathcal{O}$, and $\mathcal{X}^1 = I^d \otimes \sigma_x^a$ and $\mathcal{Z}^1 = \sigma_z^d \otimes \sigma_z^a$, respectively. Discarding the measured qubit gives $\mathcal{X}^1 = \sigma_x^a$ and $\mathcal{Z}^1 = \sigma_z^a$ so that $\sigma_x^d \to \sigma_x^a$ and $\sigma_z^d \to \sigma_z^a$. Thus the data has been transferred from the data qubit to the ancilla qubit. Now suppose that we had initially prepared the ancilla qubit in the $+1$ eigenstate of σ_z^a (viz. CB-state $|0\rangle$) and then applied the CNOT gate using the ancilla as the control. Since the control qubit is in the $|0\rangle$ state, no data is transferred because the CNOT gate applies no action to the target qubit for this control state.

The two operations introduced in the preceding paragraph provide the means to move the jth encoded qubit from a data code block to an empty ancilla code block. We prepare the ancilla block so that all encoded qubits $i \neq j$ are in the $+1$ eigenstate $|\overline{0}\rangle$ of Z_i^a and the jth encoded qubit is in the $+1$ eigenstate of X_j^a. We then apply the block encoded CNOT gate from Section 5.5, using the ancilla block as the control and then measure Z_j^d. This produces the encoded version of the two operations discussed in the preceding paragraph so that the jth encoded data qubit is transferred to the jth encoded ancilla qubit, while all other encoded ancilla qubits remain in the state $|\overline{0}\rangle$. After the measurement, the jth encoded data qubit is left in the $+1$ eigenstate $|\overline{0}\rangle$ of Z_j^d, while no action is applied to the other encoded data qubits. This procedure thus provides the means to switch a specific encoded qubit from one code block to another. The procedure to transfer the jth encoded qubit back to the data block is to apply a block-encoded CNOT gate using the ancilla block as the control and then measure X_j^a. The following exercise explains how this works.

Exercise 5.9 *Suppose we have a data qubit in the $+1$ eigenstate of σ_z^d and an ancilla qubit in an arbitrary state $|\psi\rangle$. The stabilizer for this state has generator $G^0 = \sigma_z^d \otimes I^a$ and Pauli operator analogs $\mathcal{X}^0 = I^d \otimes \sigma_x^a$ and $\mathcal{Z}^0 = I^d \otimes \sigma_z^a$. Apply a CNOT gate using the ancilla qubit as the control. Finally, measure σ_x^a. Show that this procedure produces the map $\sigma_x^a \to \sigma_x^d$ and $\sigma_z^a \to \sigma_z^d$ so that the state of the ancilla qubit is transferred to the data qubit. In the encoded version of this procedure the other encoded ancilla qubits and the jth encoded data qubit must initially be in the encoded CB-state $|\overline{0}\rangle$. This will be the case in the applications we consider in this section.*

We can now overcome the first limitation mentioned above—applying an encoded CNOT gate solely to the ith encoded qubits in two data blocks. To do this we first switch the ith encoded qubits to two ancilla code blocks using

Fault-Tolerant Quantum Computing 171

the procedure presented in the previous paragraph. Upon completion of this procedure, the ith encoded qubits have been transferred to the two ancilla blocks and all other encoded ancilla qubits are in the state $|\overline{0}\rangle$. We can now apply the block-encoded CNOT gate from Section 5.5 to the ancilla blocks. This will apply an encoded CNOT gate to the ith encoded qubits while leaving unaltered the state of the other encoded ancilla qubits. At this point we have applied the desired operation to the ith encoded qubits and so can switch them back to their initial data blocks by the procedure described at the end of the previous paragraph.

The remaining limitation is how to apply an encoded CNOT gate to the ith and j^{th} encoded qubits in either the same or different data blocks. A partial solution is to transfer these encoded data qubits to two ancilla blocks in which all other encoded ancilla qubits are in the state $|\overline{0}\rangle$. If a fault-tolerant encoded SWAP gate is available to swap a given encoded qubit with the first encoded qubit in its block, we could then swap each of the two encoded data qubits with the first encoded qubit in its respective block. Applying the block encoded CNOT gate (Section 5.5) to the ancilla blocks would apply an encoded CNOT gate to the two encoded data qubits while leaving all other encoded ancilla qubits alone. Having applied the desired CNOT gate to the encoded data qubits, we could swap them back to their starting positions in the ancilla blocks, and then switch them back to their initial data blocks. Clearly, for this approach to work, a procedure must be found to apply the desired encoded SWAP gate.

If the automorphism group $Aut(\mathcal{S})$ of the stabilizer $\mathcal{S}$ is non-trivial, it may be possible to construct the desired SWAP gate from one of its elements. Let U be an automorphism of $\mathcal{S}$. Since U fixes $\mathcal{S}$, it is an encoded operation and so maps codewords to codewords. Since this map must conserve probability, U must be unitary. Unitary encoded operations were shown to map $\mathcal{C}(\mathcal{S})/\mathcal{S} \to \mathcal{C}(\mathcal{S})/\mathcal{S}$ in Section 5.3. Since $\mathcal{C}(\mathcal{S})/\mathcal{S} \cong \mathcal{G}_k$ (Section 3.1.4), $U \in N(\mathcal{G}_k)$ and its action is determined once we know how it acts on the generators of $\mathcal{G}_k$. To be of interest here, U must act on the first and j^{th} encoded qubits, though it may also act on other encoded qubits. If so, these other encoded qubits must be measured after U acts so that we wind up with an encoded operation in $N(\mathcal{G}_2)$ that only acts on the first and j^{th} encoded qubits. Gottesman [5] has shown that, to within one-qubit operations, the only two-qubit gates in $N(\mathcal{G}_2)$ are (i) the identity gate, (ii) the SWAP gate, (iii) the CNOT gate, and (iv) a CNOT gate followed by a SWAP gate. Clearly if U yields case (ii), we are done. As the following exercise shows, cases (iii) and (iv) can also produce the desired SWAP gate.

Exercise 5.10 *Parts (a) and (b) of this exercise show how to make a SWAP gate from three CNOT gates and four Hadamard gates. Part (c) works out the conditions under which a CNOT gate followed by a SWAP gate can make the desired SWAP gate.*

(a) Show that the quantum circuit in Figure 5.6 produces the following trans-

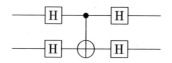

FIGURE 5.6
Quantum circuit for a CNOT gate that uses the lower qubit as control.

formation of the generators of $\mathcal{G}_2$:

$$\sigma_x^1 \otimes I^2 \longrightarrow \sigma_x^1 \otimes I^2$$
$$I^1 \otimes \sigma_x^2 \longrightarrow \sigma_x^1 \otimes \sigma_x^2$$
$$\sigma_z^1 \otimes I^2 \longrightarrow \sigma_z^1 \otimes \sigma_z^2$$
$$I^1 \otimes \sigma_z^2 \longrightarrow I^1 \otimes \sigma_z^2 \ .$$

We recognize this as the action of a CNOT gate from qubit 2 to qubit 1.

(b) Show that the quantum circuit in Figure 5.7 produces the following mapping:

$$\sigma_x^1 \otimes I^2 \longrightarrow I^1 \otimes \sigma_x^2$$
$$I^1 \otimes \sigma_x^2 \longrightarrow \sigma_x^1 \otimes I^2$$
$$\sigma_z^1 \otimes I^2 \longrightarrow I^1 \otimes \sigma_z^2$$
$$I^1 \otimes \sigma_z^2 \longrightarrow \sigma_z^1 \otimes I^2 \ .$$

We recognize this as the action of a SWAP gate.

(c) Suppose we have a CNOT gate followed by a SWAP gate that act on the first and j^{th} encoded qubits in a code block and the CNOT gate uses the first encoded qubit as the control. Explain why this combined gate reduces to a SWAP gate if qubit 1 is initially in the state $|\bar{0}\rangle$. Since the SWAP gate is to be used in a code block in which the first encoded qubit is in the state $|\bar{0}\rangle$, such a CNOT plus SWAP gate can be used to SWAP encoded qubits 1 and j.

Should the automorphism group $Aut(\mathcal{S})$ not be suitable for construction of a SWAP gate, such a gate can always be made using quantum teleportation [4, 15] (see Problem 5.5). To begin we have a codeblock in which all encoded qubits are in the state $|\bar{0}\rangle$ except for the jth encoded qubit, which is in an arbitrary state $|\psi\rangle$. We introduce a second code block that has been prepared so that all encoded qubits are in the state $|\bar{0}\rangle$ except for encoded qubits 1 and j, which are in the encoded Bell state $\left[|\bar{0}\bar{0}\rangle + |\bar{1}\bar{1}\rangle\right]/2$. Carrying out the quantum teleportation procedure between the jth encoded qubits in the two code blocks ends up transferring the state $|\psi\rangle$ from the jth encoded qubit in

Fault-Tolerant Quantum Computing

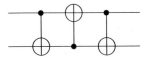

FIGURE 5.7
Quantum circuit for a SWAP gate.

the first block to the first encoded qubit in the second block. The teleportation procedure also leaves the jth encoded qubit in the second block in the state $|\overline{0}\rangle$. We then discard the first code block and keep the second. Initially, encoded qubits 1 and j in the first code block were in the states $|\overline{0}\rangle$ and $|\psi\rangle$, respectively. After teleportation, the first and j^{th} encoded qubits in the second block are in the states $|\psi\rangle$ and $|\overline{0}\rangle$, respectively. We now have a code block that has the first encoded qubit in the state $|\psi\rangle$ and the jth encoded qubit in the state $|\overline{0}\rangle$, which is what we wanted. We have in essence carried out the desired SWAP operation. Since this procedure only requires a block-encoded CNOT gate, a block Hadamard gate, and Z_j measurements, which we can do for any stabilizer code, this teleportation procedure can be used to make the desired SWAP gate for any quantum stabilizer code. Combining all the procedures presented in this section we can now fault-tolerantly apply encoded versions of all operations in the Clifford group $N(\mathcal{G}_n)$ for any quantum stabilizer code. Together with the encoded Toffoli gate introduced in the following section, we will finally be able to do fault-tolerant quantum computing on data encoded using any quantum stabilizer code.

5.7 Operations Outside $N(\mathcal{G}_n)$—Toffoli Gate

Using the techniques developed in Sections 5.3–5.6, we can now apply all gates in the Clifford group $N(\mathcal{G}_n)$ fault-tolerantly for any quantum stabilizer code. Unfortunately, this set of gates is not sufficiently powerful to perform universal quantum computation. The Gottesman-Knill theorem (see Appendix C) gives a careful statement and proof of this. Universal quantum computation becomes possible, however, if the gates in the Clifford group are supplemented with the Toffoli gate [1,16]. This section completes the theoretical framework that allows fault-tolerant quantum computing to be done using any quantum stabilizer code by showing how to make a fault-tolerant encoded Toffoli gate [1,4,5]. Note that other gates can be used instead of the Toffoli gate. A well-known alternative is the $\pi/8$ gate. We refer the reader to Refs. [17, 18] for a discussion of its fault-tolerant encoded implementation.

The Toffoli gate U_T is a controlled-controlled NOT gate that uses qubits 1 and 2 as the controls and qubit 3 as the target. Its action on the three-qubit CB-state $|i_1 \, i_2 \, i_3\rangle$ is to apply a NOT gate to qubit 3 only when $i_1 = i_2 = 1$. Otherwise, no action is applied to the qubits. Formally, the action of U_T on the three-qubit CB states is

$$U_T |i_1 \, i_2 \, i_3\rangle = |i_1 \, i_2 \, (i_3 \oplus i_1 i_2)\rangle \, , \qquad (5.81)$$

where $\oplus$ denotes binary addition and $i_1, i_2, i_3 = 0, 1$.

Let $\mathcal{O}$ be an Hermitian operator of order 2 ($\mathcal{O}^2 = I$) whose eigenvalues are thus $\lambda_\pm = \pm 1$. The operators $P_\pm(\mathcal{O})$ that project states onto the $\lambda_\pm$ eigenspaces are

$$P_\pm(\mathcal{O}) = \frac{1}{2}[I \pm \mathcal{O}] \, .$$

Using these projection operators, U_T can be written as

$$\begin{aligned} U_T &= \frac{1}{2}\left\{ P_+(\sigma_z^1) + P_+(\sigma_z^2) + P_-(\sigma_z^1 \sigma_z^2) \right\} + P_-(\sigma_z^1) P_-(\sigma_z^2) \sigma_x^3 \\ &= \frac{1}{4}\left[3I + \sigma_z^1 + \sigma_z^2 - \sigma_z^1 \sigma_z^2 + \left(I - \sigma_z^1\right)\left(I - \sigma_z^2\right) \sigma_x^3 \right] \, . \end{aligned} \qquad (5.82)$$

In an effort to make equations more readable, we will suppress the direct product symbol $\otimes$ throughout this section.

Exercise 5.11 *Show that the operator on the RHS of eq. (5.82) satisfies eq. (5.81).*

U_T acts on $\mathcal{G}_3$ by conjugating its elements: $f_T(g) = U_T \, g \, U_T^\dagger$. This action preserves the multiplication table of $\mathcal{G}_3$ since

$$\begin{aligned} f_T(gh) &= U_T \, (gh) \, U_T^\dagger \\ &= \left(U_T \, g \, U_T^\dagger\right)\left(U_T \, h \, U_T^\dagger\right) \\ &= f_T(g) \, f_T(h) \, . \end{aligned}$$

We can thus determine the action of U_T on $\mathcal{G}_3$ by examining how it maps the generators of this group. We shall see momentarily that the image of $\mathcal{G}_3$ under U_T is contained in $N(\mathcal{G}_3)$. Consequently, the Toffoli gate U_T does not fix $\mathcal{G}_3$ and so cannot belong to $N(\mathcal{G}_3)$. It cannot belong to the Clifford group $N(\mathcal{G}_n)$ either since, if it did, it would fix $\mathcal{G}_n$ and so would have to fix the subgroup $\mathcal{G}_3 \subset \mathcal{G}_n$. Since it does not, the Toffoli gate has to lie outside the Clifford group.

We begin with the mapping $\sigma_x^1 \to U_T \sigma_x^1 U_T^\dagger$:

$$\begin{aligned} U_T \sigma_x^1 U_T^\dagger &= \frac{1}{16} \left[3I + \sigma_z^1 + \sigma_z^2 - \sigma_z^1 \sigma_z^2 + \left(I - \sigma_z^1\right)\left(I - \sigma_z^2\right) \sigma_x^3 \right] \\ &\quad \times \sigma_x^1 \left[3I + \sigma_z^1 + \sigma_z^2 - \sigma_z^1 \sigma_z^2 + \left(I - \sigma_z^1\right)\left(I - \sigma_z^2\right) \sigma_x^3 \right] \\ &= \frac{1}{2} \left[\left(I + \sigma_z^2\right) + \left(I - \sigma_z^2\right) \sigma_x^3 \right] \sigma_x^1 \; . \end{aligned} \qquad (5.83)$$

The operator in square brackets on the second line of eq. (5.83) is simply a CNOT gate from qubit 2 to qubit 3, which can be verified by examining its action on the generators of $N(\mathcal{G}_2)$. Thus the image $f_T(\sigma_x^1)$ is the product of σ_x^1 with a CNOT gate, and so $f_T(\sigma_x^1) \in N(\mathcal{G}_3)$. Similar algebra determines the images of the other generators under U_T. One finds

$$\begin{aligned} \sigma_x^1 &\to \frac{1}{2} \left[\left(I + \sigma_z^2\right) + \left(I - \sigma_z^2\right) \sigma_x^3 \right] \sigma_x^1 \\ \sigma_x^2 &\to \frac{1}{2} \left[\left(I + \sigma_z^1\right) + \left(I - \sigma_z^1\right) \sigma_x^3 \right] \sigma_x^2 \\ \sigma_x^3 &\to \sigma_x^3 \\ \sigma_z^1 &\to \sigma_z^1 \\ \sigma_z^2 &\to \sigma_z^2 \\ \sigma_z^3 &\to \frac{1}{2} \left[\left(I + \sigma_z^1\right) + \left(I - \sigma_z^1\right) \sigma_z^2 \right] \sigma_z^3 \; . \end{aligned} \qquad (5.84)$$

We see that all generator images lie in $N(\mathcal{G}_3)$. Since U_T preserves the multiplication table of $\mathcal{G}_3$, all elements of $\mathcal{G}_3$ are mapped into $N(\mathcal{G}_3)$ as claimed earlier, and the Toffoli gate is seen to implement a homomorphism from $\mathcal{G}_3$ to $N(\mathcal{G}_3)$.

Suppose that we can prepare the three-qubit state

$$|A\rangle = \frac{1}{2} \left[|000\rangle + |010\rangle + |100\rangle + |111\rangle \right] \; . \qquad (5.85)$$

The stabilizer for this state has generators

$$\begin{aligned} G_1 &= \frac{1}{2} \left[\left(I + \sigma_z^2\right) + \left(I - \sigma_z^2\right) \sigma_x^3 \right] \sigma_x^1 \equiv U_{CNOT}^{23} \sigma_x^1 \\ G_2 &= \frac{1}{2} \left[\left(I + \sigma_z^1\right) + \left(I - \sigma_z^1\right) \sigma_x^3 \right] \sigma_x^2 \equiv U_{CNOT}^{13} \sigma_x^2 \\ G_3 &= \frac{1}{2} \left[\left(I + \sigma_z^1\right) + \left(I - \sigma_z^1\right) \sigma_z^2 \right] \sigma_z^3 \equiv U_{phase}^{12} \sigma_z^3 \; , \end{aligned} \qquad (5.86)$$

where U_{CNOT}^{ij} is a CNOT gate from qubit i to j, and U_{phase}^{12} is a controlled-phase gate acting on qubits 1 and 2 (see Exercise 5.7).

Now introduce another three-qubit block (with qubits labeled $4, 5, 6$) prepared in an arbitrary three-qubit state $|\psi\rangle$. The stabilizer for the composite state $|A\rangle|\psi\rangle$ has generators

$$G_1^0 = G_1 I^4 I^5 I^6$$
$$G_2^0 = G_2 I^4 I^5 I^6$$
$$G_3^0 = G_3 I^4 I^5 I^6 \tag{5.87}$$

and Pauli operator analogs

$$\begin{aligned}\mathcal{X}_1^0 &= \sigma_x^4 &;& \quad \mathcal{Z}_1^0 = \sigma_z^4 \\ \mathcal{X}_2^0 &= \sigma_x^5 &;& \quad \mathcal{Z}_2^0 = \sigma_z^5 \\ \mathcal{X}_3^0 &= \sigma_x^6 &;& \quad \mathcal{Z}_3^0 = \sigma_z^6 \end{aligned} \tag{5.88}$$

Next apply CNOT gates from qubits $1 \to 4$, $2 \to 5$, and $6 \to 3$. This produces the mappings

$$\begin{aligned} G_1^0 &\longrightarrow \overline{G}_1 = C_{NOT}^{23} \sigma_x^1 \sigma_x^4 \\ G_2^0 &\longrightarrow \overline{G}_2 = C_{NOT}^{13} \sigma_x^2 \sigma_x^5 \\ G_3^0 &\longrightarrow \overline{G}_3 = C_{phase}^{12} \sigma_z^3 \sigma_z^6 \end{aligned} \tag{5.89}$$

and

$$\begin{aligned} \mathcal{X}_1^0 &\longrightarrow \overline{\mathcal{X}}_1 = \sigma_x^4 &;& \quad \mathcal{Z}_1^0 \to \overline{\mathcal{Z}}_1 = \sigma_z^1 \sigma_z^4 \\ \mathcal{X}_2^0 &\longrightarrow \overline{\mathcal{X}}_2 = \sigma_x^5 &;& \quad \mathcal{Z}_2^0 \to \overline{\mathcal{Z}}_2 = \sigma_z^2 \sigma_z^5 \\ \mathcal{X}_3^0 &\longrightarrow \overline{\mathcal{X}}_3 = \sigma_x^3 \sigma_x^6 &;& \quad \mathcal{Z}_3^0 \to \overline{\mathcal{Z}}_3 = \sigma_z^6 \end{aligned} \tag{5.90}$$

Finally, measure $\mathcal{O}_1 = \sigma_z^4$, $\mathcal{O}_2 = \sigma_z^5$, and $\mathcal{O}_3 = \sigma_x^6$. Here $\mathcal{O}_1$ anticommutes with $\{\overline{G}_1, \overline{\mathcal{X}}_1\}$; $\mathcal{O}_2$ with $\{\overline{G}_2, \overline{\mathcal{X}}_2\}$; and $\mathcal{O}_3$ with $\{\overline{G}_3, \overline{\mathcal{Z}}_3\}$. This set of measurements produces the mappings

$$\begin{aligned} \overline{G}_1 &\longrightarrow G_1^1 = \mathcal{O}_1 \\ \overline{G}_2 &\longrightarrow G_2^1 = \mathcal{O}_2 \\ \overline{G}_3 &\longrightarrow G_3^1 = \mathcal{O}_3 \end{aligned} \tag{5.91}$$

and

$$\begin{aligned} \overline{\mathcal{X}}_1 \to \mathcal{X}_1^1 &= \overline{G}_1 \overline{\mathcal{X}}_1 & & \overline{\mathcal{Z}}_1 \to \mathcal{Z}_1^1 = \overline{\mathcal{Z}}_1 \\ &= G_1 &;& \qquad\qquad = \sigma_z^1 \sigma_z^4 \\ \overline{\mathcal{X}}_2 \to \mathcal{X}_2^1 &= \overline{G}_2 \overline{\mathcal{X}}_2 & & \overline{\mathcal{Z}}_2 \to \mathcal{Z}_2^1 = \overline{\mathcal{Z}}_2 \\ &= G_2 &;& \qquad\qquad = \sigma_z^2 \sigma_z^5 \\ \overline{\mathcal{X}}_3 \to \mathcal{X}_3^1 &= \overline{\mathcal{X}}_3 & & \overline{\mathcal{Z}}_3 \to \mathcal{Z}_3^1 = \overline{G}_3 \overline{\mathcal{Z}}_3 \\ &= \sigma_x^3 \sigma_x^6 &;& \qquad\qquad = G_3 \end{aligned} \tag{5.92}$$

Fault-Tolerant Quantum Computing

This procedure has mapped eqs. (5.88) onto (5.92) which, discarding the measured qubits in eqs. (5.92), becomes

$$\sigma_x^4 \to G_1 = \frac{1}{2}\left[(I+\sigma_z^2) + (I-\sigma_z^2)\sigma_x^3\right]\sigma_x^1$$
$$\sigma_x^5 \to G_2 = \frac{1}{2}\left[(I+\sigma_z^1) + (I-\sigma_z^1)\sigma_x^3\right]\sigma_x^2$$
$$\sigma_x^6 \to \sigma_x^3$$
$$\sigma_z^4 \to \sigma_z^1$$
$$\sigma_z^5 \to \sigma_z^2$$
$$\sigma_z^6 \to G_3 = \frac{1}{2}\left[(I+\sigma_z^1) + (I-\sigma_z^1)\sigma_z^2\right]\sigma_z^3 \ . \quad (5.93)$$

We recognize eq. (5.93) as the homomorphism produced by the Toffoli gate (see eq. (5.84)) in combination with the transfer of all data from qubits 4–6 to qubits 1–3. Thus if we can produce an encoded $|A\rangle$ state, we can apply the above procedure using encoded CNOT gates and measurements (which we know how to do) to implement an encoded Toffoli gate. To simplify the discussion of making an encoded $|A\rangle$ state, we focus on a quantum stabilizer code that encodes a single qubit. To treat a multi-qubit code we would start with three ancilla blocks with all encoded qubits in the state $|\bar{0}\rangle$ and use the X_1 and Z_1 operators for each block to implement the procedure described below to make the encoded $|A\rangle$ state. At the end of this procedure, the first encoded qubits in these three blocks would be in the encoded $|A\rangle$ state and all other encoded qubits would remain in the state $|\bar{0}\rangle$. The three encoded data qubits that are to be acted on by the Toffoli gate are switched to three ancilla blocks in which all encoded qubits are initially in the state $|\bar{0}\rangle$. The three encoded data qubits are then swapped with the first encoded qubits in their respective blocks. The just-described Toffoli gate procedure is then applied to the six ancilla blocks using block-encoded CNOT gates and measurements. Afterwards the encoded data qubits are transferred back to their original code blocks. There is thus no loss of generality in focusing on single-qubit encodings.

Let $|\overline{A}\rangle$ represent the encoded version of eq. (5.85) and introduce $|\overline{B}\rangle = X_3|\overline{A}\rangle$, where X_3 is the encoded-σ_x^3 operator. Let the stabilizer code $\mathcal{C}_q$ that encodes each of the three code blocks be an $[n,1,d]$ code. We will at times below refer to the collection of three code blocks as simply the code block to unclutter the language. It should be clear from the context when this is being done. Suppressing normalization factors in the remainder of this section, the states $|\overline{A}\rangle$ and $|\overline{B}\rangle$ are

$$|\overline{A}\rangle = |\overline{000}\rangle + |\overline{010}\rangle + |\overline{100}\rangle + |\overline{111}\rangle$$
$$|\overline{B}\rangle = |\overline{001}\rangle + |\overline{011}\rangle + |\overline{101}\rangle + |\overline{110}\rangle \ . \quad (5.94)$$

Notice that

$$|\overline{A}\rangle + |\overline{B}\rangle = \sum_{i_1,i_2,i_3=0}^{1} |\overline{i_1 i_2 i_3}\rangle$$
$$= [\,|\overline{0}\rangle + |\overline{1}\rangle\,][\,|\overline{0}\rangle + |\overline{1}\rangle\,][\,|\overline{0}\rangle + |\overline{1}\rangle\,] \quad . \tag{5.95}$$

The state $|\overline{A}\rangle + |\overline{B}\rangle$ can thus be prepared by measuring X_i on each code block $i = 1, 2, 3$. Now introduce an ancilla block containing n qubits and prepare it in the cat-state $|\psi_{cat}\rangle = |0\cdots 0\rangle + |1\cdots 1\rangle$. Next suppose the following operation can be applied to the code and ancilla blocks:

$$\begin{aligned}
|0\cdots 0\rangle|\overline{A}\rangle &\longrightarrow |0\cdots 0\rangle|\overline{A}\rangle \\
|1\cdots 1\rangle|\overline{A}\rangle &\longrightarrow |1\cdots 1\rangle|\overline{A}\rangle \\
|0\cdots 0\rangle|\overline{B}\rangle &\longrightarrow |0\cdots 0\rangle|\overline{B}\rangle \\
|1\cdots 1\rangle|\overline{B}\rangle &\longrightarrow -|1\cdots 1\rangle|\overline{B}\rangle \quad .
\end{aligned} \tag{5.96}$$

With this operation the state $|\overline{A}\rangle$ can be constructed as follows. Let the ancilla block be prepared in the state $|\psi_{cat}\rangle$ and the code block in the state $|\overline{A}\rangle + |\overline{B}\rangle$. Apply the operation in eq. (5.96) to this composite state. The result is

$$[\,|0\cdots 0\rangle + |1\cdots 1\rangle\,][\,|\overline{A}\rangle + |\overline{B}\rangle\,] \longrightarrow$$
$$[\,|0\cdots 0\rangle + |1\cdots 1\rangle\,]|\overline{A}\rangle + [\,|0\cdots 0\rangle - |1\cdots 1\rangle\,]|\overline{B}\rangle \quad . \tag{5.97}$$

Finally, measure $\mathcal{O} = \sigma_x^1 \cdots \sigma_x^n$ on the ancilla block. Since

$$\mathcal{O}[\,|0\cdots 0\rangle \pm |1\cdots 1\rangle\,] = \pm[\,|0\cdots 0\rangle \pm |1\cdots 1\rangle\,] \quad ,$$

if the measurement outcome is $+1$, we know from eq. (5.97) that the code block will be in the state $|\overline{A}\rangle$. If the outcome is -1, then the code block will be in the state $|\overline{B}\rangle$. In this case, applying the operator X_3 to the code block puts it in the state $X_3|\overline{B}\rangle = |\overline{A}\rangle$. In either case, at the end of all operations, the code block will be in the state $|\overline{A}\rangle$ as desired.

As we know from Section 5.2, cat-state preparation is not fault-tolerant and so a candidate cat-state must be carefully verified before it is used in the production of the state $|\overline{A}\rangle$. The quantum circuit in Figure 5.4 (minus the final four Hadamard gates) produces the cat-state if no errors occur. It also verifies to $\mathcal{O}(p)$ that no bit-flip errors are present in the state at the end of the preparation protocol. As noted there, this circuit does not detect phase errors in the output state. A phase error in the cat-state will cause an error in the construction of $|\overline{A}\rangle$. To see this, note that

$$\sigma_z^i[\,|0\cdots 0\rangle + |1\cdots 1\rangle\,][\,|\overline{A}\rangle + |\overline{B}\rangle\,] = [\,|0\cdots 0\rangle - |1\cdots 1\rangle\,][\,|\overline{A}\rangle + |\overline{B}\rangle\,]. \tag{5.98}$$

Applying the operation in eq. (5.96) to the state in eq. (5.98) yields

$$\sigma_z^i[\,|0\cdots 0\rangle + |1\cdots 1\rangle\,][\,|\overline{A}\rangle + |\overline{B}\rangle\,] \longrightarrow$$
$$[\,|0\cdots 0\rangle - |1\cdots 1\rangle\,]|\overline{A}\rangle + [\,|0\cdots 0\rangle + |1\cdots 1\rangle\,]|\overline{B}\rangle \quad . \tag{5.99}$$

Measuring $\mathcal{O} = \sigma_x^1 \cdots \sigma_x^n$ on the ancilla block yields $|\overline{B}\rangle$ ($|\overline{A}\rangle$) when the measurement result is $+1$ (-1) instead of $|\overline{A}\rangle$ ($|\overline{B}\rangle$). Thus a phase error during cat-state preparation causes the above protocol to misidentify the code block state. A procedure that to $\mathcal{O}(p)$ verifies that the code block is in the state $|\overline{A}\rangle$ is to repeat the cat-state preparation plus $\mathcal{O}$-measurement sequence until the same measurement outcome occurs twice in a row. Each run must use a fresh ancilla block so that the runs will be statistically independent. To see why this procedure works, notice that if no phase error occurs during cat-state preparation on the first two runs, then both runs will yield the same measurement result, and this result will properly identify the code block state. If a phase error occurs (during cat-state preparation) on the first run, then to $\mathcal{O}(p)$, no phase error will occur during the second and third runs. Hence these runs will yield the same measurement outcome, and again the code block state will be properly identified. Finally if a phase error occurs during the second run, then to $\mathcal{O}(p)$, no phase error occurs on the third and fourth runs, their measurement outcomes agree, and the code block state is correctly identified. To $\mathcal{O}(p)$, these are the only possibilities since if no phase error occurs on the first two runs, there will be no need to do a third run. Thus, to $\mathcal{O}(p)$, we are certain that the code block is in the state $|\overline{A}\rangle$ upon completion of the procedure. We leave it to the interested reader to work out extensions of this procedure to higher order in p.

One loose end remains—how can the operation in eq. (5.96) be applied? This operation applies the map $|\overline{A}\rangle \to |\overline{A}\rangle$ and $|\overline{B}\rangle \to -|\overline{B}\rangle$ when the ancilla block is in the state $|1 \cdots 1\rangle$, and does nothing to the three code blocks when the ancilla are in the state $|0 \cdots 0\rangle$. Let $\overline{G}_3$ be the encoded version of G_3 (see eq. (5.86)). Notice that

$$\overline{G}_3 |\overline{A}\rangle = |\overline{A}\rangle$$
$$\overline{G}_3 |\overline{B}\rangle = -|\overline{B}\rangle \ .$$

Thus the desired operation is to apply $\overline{G}_3$ only when the ancilla are in the state $|1 \cdots 1\rangle$. As pointed out earlier, $\overline{G}_3 \in N(\mathcal{G}_3)$ and so can be constructed using encoded Hadamard, phase, and CNOT gates. We know how to apply these encoded one- and two-qubit gates by combining transversal application of the four-qubit operation introduced in Section 5.5 with measurement. Thus $\overline{G}_3$ can be applied using a sequence of transversal operations and measurements. Each transversal operation or measurement T only involves corresponding qubits in the three code blocks. If we can use the ith ancilla qubit to control whether T is applied to the i-th qubits in the code blocks, we will be able to control the application of all the transversal operations and measurements used to implement $\overline{G}_3$. Then when the ancilla block is in the state $|1 \cdots 1\rangle$ ($|0 \cdots 0\rangle$), all T operations and measurements will (not) be applied transversally, and $\overline{G}_3$ will (not) be fault-tolerantly applied to the code blocks. This is the desired operation. It is assumed that a universal set of quantum gates is available that can be applied to *unencoded* qubits. Then the controlled-T

operations can be implemented using gates from this set. With this assumption, the above remarks describe how to implement the operation in eq. (5.96). With it we can construct $|\overline{A}\rangle$, which can then be used to perform an encoded Toffoli gate for any quantum stabilizer code. This, together with the results of Sections 5.3–5.6, allows fault-tolerant quantum computing to be done on encoded data using any quantum stabilizer code.

5.8 Example: [5,1,3] Code

This section constructs a universal set of encoded fault-tolerant quantum gates for the [5,1,3] code [4]. The generators and encoded Pauli operators for this code were given in Section 3.2.1:

$$g_1 = \sigma_x \otimes \sigma_z \otimes \sigma_z \otimes \sigma_x \otimes I$$
$$g_2 = I \otimes \sigma_x \otimes \sigma_z \otimes \sigma_z \otimes \sigma_x$$
$$g_3 = \sigma_x \otimes I \otimes \sigma_x \otimes \sigma_z \otimes \sigma_z$$
$$g_4 = \sigma_z \otimes \sigma_x \otimes I \otimes \sigma_x \otimes \sigma_z \qquad (5.100)$$

and

$$X = \sigma_x \otimes \sigma_x \otimes \sigma_x \otimes \sigma_x \otimes \sigma_x$$
$$Z = \sigma_z \otimes \sigma_z \otimes \sigma_z \otimes \sigma_z \otimes \sigma_z \quad . \qquad (5.101)$$

To unclutter the notation we will usually suppress the superscripts labeling the qubits throughout this section. Qubits are implicitly labeled from 1 to 5 in going from left to right.

The operator T that cyclically permutes the Pauli matrices (see Section 5.3.1),

$$T : \sigma_x \to \sigma_y \to \sigma_z \to \sigma_x \quad , \qquad (5.102)$$

applied transversally to the code block leaves the stabilizer S invariant. To see this notice that

$$g_1 \longrightarrow \sigma_y \otimes \sigma_x \otimes \sigma_x \otimes \sigma_y \otimes I = g_3\, g_4$$
$$g_2 \longrightarrow I \otimes \sigma_y \otimes \sigma_x \otimes \sigma_x \otimes \sigma_y = g_1\, g_2\, g_3$$
$$g_3 \longrightarrow \sigma_y \otimes I \otimes \sigma_y \otimes \sigma_x \otimes \sigma_x = g_2\, g_3\, g_4$$
$$g_4 \longrightarrow \sigma_x \otimes \sigma_y \otimes I \otimes \sigma_y \otimes \sigma_x = g_1\, g_2. \qquad (5.103)$$

It is a straightforward exercise to show that the generator images are also a set of generators (e. g. find combinations of products that yield $g_1, \ldots, g_4$). Thus T applied transversally fixes the stabilizer S and so is an encoded fault-tolerant operation. To see which operation it is, we examine its action on X

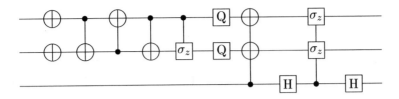

FIGURE 5.8
Quantum circuit to apply the three-qubit operator T_3. Qubit lines are labeled from 1 to 3 in going from top to bottom. Reprinted Figure 1 with permission from D. Gottesman, *Phys. Rev. A* **57**, 127 (1998). ©American Physical Society, 1998.

and Z. From eq. (5.101) we see that

$$X \longrightarrow \sigma_y \otimes \sigma_y \otimes \sigma_y \otimes \sigma_y \otimes \sigma_y = (-i)^5 \, ZX = Y$$
$$Z \longrightarrow \sigma_x \otimes \sigma_x \otimes \sigma_x \otimes \sigma_x \otimes \sigma_x = X$$
$$Y \longrightarrow -iXY = Z \ . \tag{5.104}$$

Thus T applied transversally to the code block performs an encoded T operation.

Although the four-qubit operation introduced in Section 5.5 is available, there exists a three-qubit operation T_3 that serves equally as well for the [5,1,3] code. Its action on unencoded qubits is

$$\sigma_x \otimes I \otimes I \longrightarrow \sigma_x \otimes \sigma_y \otimes \sigma_z$$
$$\sigma_z \otimes I \otimes I \longrightarrow \sigma_z \otimes \sigma_x \otimes \sigma_y$$
$$I \otimes \sigma_x \otimes I \longrightarrow \sigma_y \otimes \sigma_x \otimes \sigma_z$$
$$I \otimes \sigma_z \otimes I \longrightarrow \sigma_x \otimes \sigma_z \otimes \sigma_y$$
$$I \otimes I \otimes \sigma_x \longrightarrow \sigma_x \otimes \sigma_x \otimes \sigma_x$$
$$I \otimes I \otimes \sigma_z \longrightarrow \sigma_z \otimes \sigma_z \otimes \sigma_z \ . \tag{5.105}$$

Gottesman [4] gives an 8×8 matrix representation for T_3 and a quantum circuit that implements it (see Figure 5.8).

We now show that applying T_3 transversally fixes $\mathcal{S} \otimes \mathcal{S} \otimes \mathcal{S}$. Transversal application of T_3 produces the following map on the generators:

$$g_1 \otimes I \otimes I \longrightarrow g_1 \otimes g_3 \, g_4 \otimes g_1 \, g_3 \, g_4$$
$$g_2 \otimes I \otimes I \longrightarrow g_2 \otimes g_1 \, g_2 \, g_3 \otimes g_1 \, g_3$$
$$g_3 \otimes I \otimes I \longrightarrow g_3 \otimes g_2 \, g_3 \, g_4 \otimes g_2 \, g_4$$
$$g_4 \otimes I \otimes I \longrightarrow g_4 \otimes g_1 \, g_2 \otimes g_1 \, g_2 \, g_4 \ , \tag{5.106}$$

$$I \otimes g_1 \otimes I \longrightarrow g_3 g_4 \otimes g_1 \otimes g_1 g_3 g_4$$
$$I \otimes g_2 \otimes I \longrightarrow g_1 g_2 g_3 \otimes g_2 \otimes g_1 g_3$$
$$I \otimes g_3 \otimes I \longrightarrow g_2 g_3 g_4 \otimes g_3 \otimes g_2 g_4$$
$$I \otimes g_4 \otimes I \longrightarrow g_1 g_2 \otimes g_4 \otimes g_1 g_2 g_4 \quad , \tag{5.107}$$

and

$$I \otimes I \otimes g_i \to g_i \otimes g_i \otimes g_i \qquad (i = 1, \ldots, 4) \quad . \tag{5.108}$$

For the two cases described by eqs. (5.106) and (5.107), we see that T_3 applies T and T^2 to the other two code blocks. Since T and T^2 fix $\mathcal{S}$, it follows that T_3 fixes $\mathcal{S} \otimes \mathcal{S} \otimes \mathcal{S}$ for these two cases. Inspection of eq. (5.108) indicates that T_3 fixes $\mathcal{S} \otimes \mathcal{S} \otimes \mathcal{S}$ in this case as well. Thus T_3 applied transversally is an encoded fault-tolerant operation for the [5,1,3] code. We leave it as an exercise for the reader to show that T_3 applied transversally produces the following map on the encoded Pauli operators:

$$X \otimes I \otimes I \to X \otimes Y \otimes Z$$
$$Z \otimes I \otimes I \to Z \otimes X \otimes Y$$
$$I \otimes X \otimes I \to Y \otimes X \otimes Z$$
$$I \otimes Z \otimes I \to X \otimes Z \otimes Y$$
$$I \otimes I \otimes X \to X \otimes X \otimes X$$
$$I \otimes I \otimes Z \to Z \otimes Z \otimes Z \quad . \tag{5.109}$$

Thus for the [5,1,3] code, T_3 applied transversally performs an encoded version of itself.

By combining T, T_3, and measurements, it is possible to fault-tolerantly perform all operations in the Clifford group $N(\mathcal{G}_n)$ for the [5, 1, 3] code. It is enough to describe the construction procedures for unencoded qubits since transversal application of these procedures will promote the unencoded T and T_3 operators to encoded versions of themselves. The procedures also use measurement of one-qubit Hermitian operators of order 2. A quantum circuit to measure such an operator $\mathcal{O}$ is given in Figure 5.9 (see Problem 5.6). Since the qubits in Figure 5.9 are unencoded qubits, the circuit constructions in Ref. [19] can be used to apply the controlled-$\mathcal{O}$ operation using only single-qubit gates and CNOT gates. In the procedures described below, $\mathcal{O}$ will be one of the Pauli operators. For such $\mathcal{O}$, we see from eqs. (5.101) that transversal application of the circuit in Figure 5.9 causes the encoded operator $\overline{\mathcal{O}}$ to be measured. For example, if $\mathcal{O} = \sigma_x$, transversal application of Figure 5.9 causes $\sigma_x \otimes \sigma_x \otimes \sigma_x \otimes \sigma_x \otimes \sigma_x$ to be measured. From eq. (5.101), this is X, which is precisely the encoded version of $\mathcal{O} = \sigma_x$.

We begin with the one-qubit gates in $N(\mathcal{G}_n)$. To perform the phase gate P, take two ancilla qubits prepared in the state $|00\rangle$ and a third data qubit in an arbitrary state $|\psi\rangle$. The stabilizer for this three-qubit state has generators $G_1^0 = \sigma_z \otimes I \otimes I$ and $G_2^0 = I \otimes \sigma_z \otimes I$, and the Pauli operator analogs are

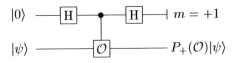

FIGURE 5.9
Quantum circuit to measure a one-qubit Hermitian operator $\mathcal{O}$ of order 2. The control qubit is measured in the computational basis $\{|0\rangle, |1\rangle\}$. When the measurement yields $m = +1$, the control qubit is in the final state $|0\rangle$ and the target qubit is in the final state $P_+(\mathcal{O})|\psi\rangle$. The projection operator $P_+(\mathcal{O}) = (I+\mathcal{O})/2$ insures that the target qubit's final state is the eigenstate of $\mathcal{O}$ with eigenvalue $\mathcal{O}_+ = +1$. For the applications considered in this section, $|\psi\rangle$ is never an eigenstate of $\mathcal{O}$. Based on Figure 10.26, Ref. [14]; ©Cambridge University Press (2000), used with permission.

$\mathcal{X}^0 = I \otimes I \otimes \sigma_x$ and $\mathcal{Z}^0 = I \otimes I \otimes \sigma_z$. Now apply T_3 so that

$$\begin{aligned} G_1^0 \to G_1 &= \sigma_z \otimes \sigma_x \otimes \sigma_y \\ G_2^0 \to G_2 &= \sigma_x \otimes \sigma_z \otimes \sigma_y \\ \mathcal{X}^0 \to \mathcal{X} &= \sigma_x \otimes \sigma_x \otimes \sigma_x \\ \mathcal{Z}^0 \to \mathcal{Z} &= \sigma_z \otimes \sigma_z \otimes \sigma_z \end{aligned} \quad (5.110)$$

Next measure $\mathcal{O}_1 = I \otimes \sigma_z \otimes I$ and $\mathcal{O}_2 = I \otimes I \otimes \sigma_z$. $\mathcal{O}_1$ anticommutes with $\{G_1, \mathcal{X}\}$ and $\mathcal{O}_2$ anticommutes with $\{G_1, G_2, \mathcal{X}\}$. The measurement then maps

$$\begin{aligned} G_1 \to G_1^1 &= \mathcal{O}_1 \\ G_2 \to G_2^1 &= \mathcal{O}_2 \\ \mathcal{X} \to \mathcal{X}^1 &= G_1 \mathcal{X} = \sigma_y \otimes I \otimes \sigma_z \\ \mathcal{Z} \to \mathcal{Z}^1 &= \sigma_z \otimes \sigma_z \otimes \sigma_z \end{aligned} \quad (5.111)$$

Discarding the measured qubits gives $\mathcal{X}^1 = \sigma_y^1$ and $\mathcal{Z}^1 = \sigma_z^1$. This procedure has thus mapped $\sigma_x^3 \to \sigma_y^1$ and $\sigma_z^3 \to \sigma_z^1$, which is P together with the transfer of the data from qubit 3 to qubit 1. The operator Q is now available since $Q = T^\dagger P$, and from it, the Hadamard gate becomes available since $H = PQ^\dagger P$. Thus we can make all one-qubit gates in the Clifford group using T, T_3, and measurements.

T_3 can be used to produce a two-qubit gate U_2 that, combined with one-qubit gates in $N(\mathcal{G}_n)$, allows a CNOT gate to be constructed. To produce U_2, begin with two qubits in an arbitrary state $|\psi_{12}\rangle$ and a third qubit in the state $|0\rangle$. The initial generator is $G_1^0 = I \otimes I \otimes \sigma_z$, and the Pauli operator analogs are $\mathcal{X}_1^0 = \sigma_x \otimes I \otimes I$; $\mathcal{X}_2^0 = I \otimes \sigma_x \otimes I$; $\mathcal{Z}_1^0 = \sigma_z \otimes I \otimes I$; and $\mathcal{Z}_2^0 = I \otimes \sigma_z \otimes I$.

Now apply T_3 so that

$$G_1^0 \to G_1 = \sigma_z \otimes \sigma_z \otimes \sigma_z$$
$$\mathcal{X}_1^0 \to \mathcal{X}_1 = \sigma_x \otimes \sigma_y \otimes \sigma_z$$
$$\mathcal{X}_2^0 \to \mathcal{X}_2 = \sigma_y \otimes \sigma_x \otimes \sigma_z$$
$$\mathcal{Z}_1^0 \to \mathcal{Z}_1 = \sigma_z \otimes \sigma_x \otimes \sigma_y$$
$$\mathcal{Z}_2^0 \to \mathcal{Z}_2 = \sigma_x \otimes \sigma_z \otimes \sigma_y \quad . \tag{5.112}$$

Next measure $\mathcal{O} = I \otimes \sigma_x \otimes I$, which anticommutes with G_1, $\mathcal{X}_1$, and $\mathcal{Z}_2$. The measurement maps

$$G_1 \to G_1^1 = \mathcal{O}$$
$$\mathcal{X}_1 \to \mathcal{X}_1^1 = \sigma_y \otimes \sigma_x \otimes I$$
$$\mathcal{X}_2 \to \mathcal{X}_2^1 = \sigma_y \otimes \sigma_x \otimes \sigma_z$$
$$\mathcal{Z}_1 \to \mathcal{Z}_1^1 = \sigma_z \otimes \sigma_x \otimes \sigma_y$$
$$\mathcal{Z}_2 \to \mathcal{Z}_2^1 = \sigma_y \otimes I \otimes \sigma_x \quad . \tag{5.113}$$

Discarding the measured qubit gives $\mathcal{X}_1^1 = \sigma_y^1 \otimes I^3$; $\mathcal{X}_2^1 = \sigma_y^1 \otimes \sigma_z^3$; $\mathcal{Z}_1^1 = \sigma_z^1 \otimes \sigma_y^3$; and $\mathcal{Z}_2^1 = \sigma_y^1 \otimes \sigma_x^3$. This procedure has thus mapped

$$\sigma_x^1 \otimes I^2 \to \sigma_y^1 \otimes I^3$$
$$I^1 \otimes \sigma_x^2 \to \sigma_y^1 \otimes \sigma_z^3$$
$$\sigma_z^1 \otimes I^2 \to \sigma_z^1 \otimes \sigma_y^3$$
$$I^1 \otimes \sigma_z^2 \to \sigma_y^1 \otimes \sigma_x^3 \quad . \tag{5.114}$$

Changing qubit label $3 \to 2$ on the RHS, Gottesman [4] has pointed out that U_2 and U_{CNOT}^{21} are related:

$$U_2 = \left(I \otimes T^2\right)(P \otimes I) U_{CNOT}^{21} (I \otimes Q) \quad . \tag{5.115}$$

Solving for U_{CNOT}^{21} gives

$$U_{CNOT}^{21} = \left(P^\dagger \otimes I\right)(I \otimes T) U_2 \left(I \otimes Q^\dagger\right) \quad , \tag{5.116}$$

where we used that T is an operator of order 3. Exercise 5.10 showed how to obtain U_{CNOT}^{12} from U_{CNOT}^{21}. Thus we can make all generators of the Clifford group for the [5,1,3] code. Our remaining task is to show how to construct the Toffoli gate for this code.

As we saw in Section 5.7, to make an encoded Toffoli gate we must be able to apply a controlled-$\overline{G}_3$ operation using an ancilla block as the control. From the encoded version of eq. (5.86), $\overline{G}_3$ is the product of an encoded controlled-phase gate on encoded qubits 1 and 2 and the encoded Pauli operator Z_3. From Exercise 5.7,

$$\overline{U}_{phase}^{12} = \left(I \otimes \overline{H}\right) \overline{U}_{CNOT}^{12} \left(I \otimes \overline{H}\right) \quad . \tag{5.117}$$

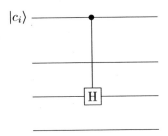

FIGURE 5.10
Quantum circuit that applies a Hadamard gate to qubit 2 conditioned on the ancilla qubit being in the state $|c_i\rangle = |1\rangle$. The lower three lines, going from top to bottom, correspond to qubits 1–3.

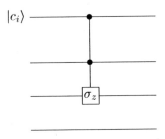

FIGURE 5.11
Quantum circuit that applies a controlled-phase gate to qubits 1 and 2 conditioned on the ancilla qubit being in the state $|c_i\rangle = |1\rangle$. The lower three lines, going from top to bottom, correspond to qubits 1–3.

Now that we know how to apply $\overline{H}$ and $\overline{U}_{CNOT}^{12}$ using transversal operations, eq. (5.117) shows how to construct $\overline{U}_{phase}^{12}$ transversally. Finally, since $Z_3 = i^{\lambda_3}\sigma_x(a_3)\sigma_z(b_3)$ is a product of one-qubit operators, its application is automatically transversal. Putting all these remarks together, we see that $\overline{G}_3$ can be applied through a sequence of transversal operations and measurements. The situation for the [5,1,3] code thus corresponds to that described in Section 5.7: the ith ancilla qubit can be used to control each operation T applied to the ith qubits in the three code blocks. As examples, Figures 5.10 and 5.11 show how to apply a controlled-T operation when T is a Hadamard gate on qubit 2 and a controlled-phase gate on qubits 1 and 2, respectively. The state of the ancilla qubit controls whether T is applied. The circuit constructions in Ref. [19] allow the controlled operations to be implemented using only single qubit gates and CNOT gates. As described in Section 5.7, by applying operations such as these transversally, we are able to apply $\overline{G}_3$ conditioned on an ancilla block being in the state $|1\cdots 1\rangle$. This allows us to construct the state $|\overline{A}\rangle$, which in turn allows us to apply an encoded Toffoli gate. Since all operations are transversal, the implementation is fault-tolerant. We thus have a universal set of fault-tolerant encoded quantum gates so that fault-tolerant quantum computing can be done using the [5,1,3] code.

5.9 Example: [4,2,2] Code

Let C_q be a [4,2,2] quantum stabilizer code with generators

$$g_1 = \sigma_x^1 \sigma_x^2 \sigma_x^3 \sigma_x^4$$
$$g_2 = \sigma_z^1 \sigma_z^2 \sigma_z^3 \sigma_z^4 \ . \tag{5.118}$$

These generators can be transformed to the generators of the [4,2,2] code introduced in Section 3.2.2 by applying Hadamard gates to qubits 2 and 3 and Q gates to qubits 1 and 4. The encoded Pauli operators X_i and Z_i are

$$X_1 = \sigma_x^1 \sigma_x^2 \quad ; \quad Z_1 = \sigma_z^2 \sigma_z^4$$
$$X_2 = \sigma_x^1 \sigma_x^3 \quad ; \quad Z_2 = \sigma_z^3 \sigma_z^4 \ . \tag{5.119}$$

It is a simple matter to check that (i) $[g_1, g_2] = 0$; (ii) $\{X_i, Z_i\} = 0$; and (iii) $[X_i, Z_j] = 0$ for $i \neq j$.

5.9.1 Fault-Tolerant Quantum Computing

Problem 5.7 establishes that C_q is a CSS code and that transversal application of CNOT gates applies a block-encoded CNOT gate. The operation is fault-tolerant since it is applied transversally. As described in Section 5.6, we can use the block-encoded CNOT gate and measurement to fault-tolerantly switch an encoded qubit from a code block to an ancilla block. After an encoded qubit has been switched, the procedure presented in Section 5.4 to make all one-qubit gates in the Clifford group using a CNOT gate and measurement can be applied transversally to implement fault-tolerant encoded versions of these one-qubit gates. To complete the generators of the Clifford group, we need to make a fault-tolerant encoded CNOT gate. As explained in Section 5.6, switching allows a fault-tolerant encoded CNOT gate to be applied to the ith encoded qubits in two code blocks, however, to apply this operation to i^{th} and j^{th} encoded qubits when $i \neq j$ requires the ability to swap encoded qubits within in a block. We shall see below that C_q allows such an encoded SWAP gate to be applied fault-tolerantly. Thus for C_q, it will be possible to apply all gates in the Clifford group fault-tolerantly on encoded qubits.

Fault-Tolerant Quantum Computing

Section 5.7 outlined how the Toffoli gate construction must be modified to handle a stabilizer code that encodes multiple qubits. The modification requires being able to switch encoded qubits to ancilla blocks and swap encoded qubits within a block. Thus the one remaining task is to explain how to make an intra-block SWAP gate. Once this is done, all the ingredients needed to make a universal set of fault-tolerant encoded quantum gate will be in place, and we will have established that fault-tolerant quantum computing is possible using a [4,2,2] quantum stabilizer code.

5.9.2 Fault-Tolerant Encoded Operations

To complete the discussion we show how to perform a fault-tolerant SWAP gate on encoded qubits belonging to the same block. We also show how to produce a number of other fault-tolerant encoded gates.

A number of useful encoded operations can be produced by swapping a pair of *unencoded* qubits. If we use the fault-tolerant SWAP procedure described in Example 5.2, the resulting encoded operations will also be fault-tolerant. As our first example, we swap unencoded qubits 2 and 3. This produces the map

$$\begin{aligned} X_1 &\longrightarrow \sigma_x^1 \sigma_x^3 = X_2 \\ X_2 &\longrightarrow \sigma_x^1 \sigma_x^2 = X_1 \\ Z_1 &\longrightarrow \sigma_z^3 \sigma_z^4 = Z_2 \\ Z_2 &\longrightarrow \sigma_z^2 \sigma_z^4 = Z_1 \ . \end{aligned} \qquad (5.120)$$

This simple operation has applied a fault-tolerant SWAP gate to the two encoded qubits. As explained above, with this gate, fault-tolerant quantum computing can be done using a [4,2,2] code. Another useful gate comes from swapping unencoded qubits 1 and 2. This produces the transformation

$$\begin{aligned} X_1 &\longrightarrow \sigma_x^2 \sigma_x^1 = X_1 \\ X_2 &\longrightarrow \sigma_x^2 \sigma_x^3 = X_1 X_2 \\ Z_1 &\longrightarrow \sigma_z^1 \sigma_z^4 = g_2 Z_1 Z_2 \sim Z_1 Z_2 \\ Z_2 &\longrightarrow \sigma_z^3 \sigma_z^4 = Z_2 \ . \end{aligned} \qquad (5.121)$$

In the third line we used that $g_2 Z_1 Z_2$ and $Z_1 Z_2$ produce the same encoded operation since they belong to the same coset in $\mathcal{C}(\mathcal{S})/\mathcal{S}$. We recognize eq. (5.121) as a CNOT gate from encoded qubit 2 to encoded qubit 1. Since the swapping of unencoded qubits was done fault-tolerantly, the encoded CNOT gate is also fault-tolerant. In a similar manner, the reader can show that swapping unencoded qubits 1 and 3 produces a fault-tolerant CNOT gate from encoded qubit 1 to encoded qubit 2.

Transversal operations are also possible. We already mentioned applying CNOT gates transversally between two code blocks to obtain a fault-tolerant block encoded CNOT gate. We consider two more possibilities. Applying Hadamard gates transversally to a code block maps

$$\begin{aligned} X_1 &\longrightarrow \sigma_z^1 \sigma_z^2 = g_2 Z_2 \sim Z_2 \\ X_2 &\longrightarrow \sigma_z^1 \sigma_z^3 = g_2 Z_1 \sim Z_1 \\ Z_1 &\longrightarrow \sigma_x^2 \sigma_x^4 = g_1 X_2 \sim X_2 \\ Z_2 &\longrightarrow \sigma_x^3 \sigma_x^4 = g_1 X_1 \sim X_1 \end{aligned} \quad (5.122)$$

Thus transversal application of Hadamard gates performs Hadamard gates on both encoded qubits followed by an encoded SWAP gate. Since the operation was done transversally, it is also fault-tolerant. Finally, applying a phase gate transversally produces the map

$$\begin{aligned} X_1 &\longrightarrow \sigma_y^1 \sigma_y^2 = -X_1 (g_2 Z_2) \sim -X_1 Z_2 \\ X_2 &\longrightarrow \sigma_y^1 \sigma_y^3 = -X_2 (g_2 Z_1) \sim -Z_1 X_2 \\ Z_1 &\longrightarrow \sigma_z^2 \sigma_z^4 = Z_1 \\ Z_2 &\longrightarrow \sigma_z^3 \sigma_z^4 = Z_2 \end{aligned} \quad (5.123)$$

We recognize this as a fault-tolerant encoded controlled-phase gate (see Exercise 5.7) followed by $Z_1 Z_2$.

Problems

5.1 We consider modifications to the syndrome extraction protocol that arise when the code generators contain factors of σ_y. Let $g_i = \sigma_x(a_i)\sigma_z(b_i)$. As noted in Section 5.2.1, if g_i contains σ_y^k, then the Hadamard gate H_k used in constructing $\bar{g}_i$ must be replaced by $\tilde{H}_k$.

(a) Let $I_i = a_i * b_i$ and $d_i = a_i * \not{b}_i$. With the substitution $H_k \to \tilde{H}_k$ made for each index k associated with a factor of σ_y^k in g_i, show that the operator $\mathbf{H}_i \to \tilde{\mathbf{H}}_i$, where:

$$\tilde{\mathbf{H}}_i = \prod_{j=1}^n \left(\tilde{H}_j\right)^{I_{i,j}} (H_j)^{d_{i,j}}$$

and that

$$\bar{g}_i = \tilde{\mathbf{H}}_i g_i \tilde{\mathbf{H}}_i = \sigma_z(a_i + b_i + a_i * b_i) \ .$$

(b) Let $E = \sigma_x(e_x)\sigma_z(e_z)$. Show that

$$\begin{aligned} \bar{E} &= \tilde{\mathbf{H}}_i E \tilde{\mathbf{H}}_i \\ &= (-1)^\phi \sigma_z(e_x * d_i + e_z * \not{d}_i)\sigma_x(e_x * \not{d}_i + e_z * a_i) \ , \end{aligned}$$

where $\phi = (e_x * \not{d}_i) \cdot (e_z * \not{d}_i)$.

Fault-Tolerant Quantum Computing

(c) As in Section 5.2.1, write $|\bar{c}\rangle = \sum_k \bar{c}(k)|k\rangle$. Show that

$$\overline{E}|\bar{c}\rangle = (-1)^\phi \sigma_z(e_x * d_i + e_z * \not{d}_i) \sum_k \bar{c}(k)|k + e_x * \not{d}_i + e_z * a_i\rangle .$$

(d) Show that $\bar{g}_i|\bar{c}\rangle = |\bar{c}\rangle$ requires

$$k \cdot (a_i + b_i + a_i * b_i) \equiv 0 \pmod{2} .$$

(e) Defining $U_{CNOT}(a_i + b_i + a_i * b_i)$ by letting $a_i + b_i \to a_i + b_i + a_i * b_i$ in eq. (5.19), and suppressing the normalization factor on the Shor state, show that

$$U_{CNOT}(a_i + b_i + a_i * b_i)\overline{E}|\bar{c}\rangle \otimes |A\rangle =$$
$$(-1)^\phi \sigma_z(e_x * d_i + e_z * \not{d}_i) \sum_k \bar{c}(k)|k + e_x * \not{d}_i + e_z * a_i\rangle$$
$$\otimes \sum_{A_e} |A_e + (a_i + b_i + a_i * b_i) * (k + e_x * \not{d}_i + e_z * a_i)\rangle .$$

(f) Show that

$$(a_i + b_i + a_i * b_i) * (e_x * \not{d}_i + e_z * a_i) = e_z * a_i + e_x * b_i .$$

Combine this with parts (d) and (e) to show that

$$U_{CNOT}(a_i + b_i + a_i * b_i)\overline{E}|\bar{c}\rangle \otimes |A\rangle =$$
$$(-1)^\phi \sigma_z(e_x * d_i + e_z * \not{d}_i) \sigma_x(e_x * \not{d}_i + e_z * a_i) |\bar{c}\rangle$$
$$\otimes \sum_{A_e} |A_e + e_z * a_i + e_x * b_i\rangle$$
$$= \overline{E}|\bar{c}\rangle \otimes \sum_{A_e} |A_e + e_z * a_i + e_x * b_i\rangle .$$

At this point the discussion following eq. (5.22) can be taken up.

5.2 The generators for the [4,2,2] quantum stabilizer code are given in eq. (3.37).

(a) Writing the generators as $g_i = \sigma_x(a_i)\sigma_z(b_i)$ for $i = 1, 2$, show that

$$a_1 = (1001) \quad ; \quad b_1 = (0110)$$
$$a_2 = (1111) \quad ; \quad b_2 = (1001) .$$

(b) Using the results of Problem 5.1 to handle g_2, show that

$$\mathbf{H}_1 = H_1 H_4 \quad ; \quad \bar{g}_1 = \sigma_z(1111)$$
$$\tilde{\mathbf{H}}_2 = \tilde{H}_1 \tilde{H}_4 H_2 H_3 \quad ; \quad \bar{g}_2 = \sigma_z(1111) .$$

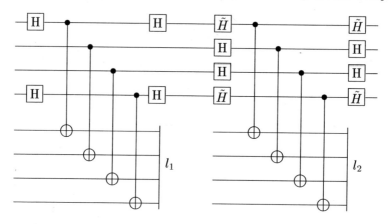

FIGURE 5.12
Quantum circuit for syndrome extraction for the [4,2,2] quantum stabilizer code.

(c) Show that the quantum circuit for syndrome extraction for the [4,2,2] code is that given in Figure 5.12.

5.3 Let C_q be a quantum error correcting code. Consider two blocks of qubits, each encoded using C_q. Show that if transversal application of CNOT gates to this pair of code blocks yields an encoded operation, then C_q must be a CSS code. (Since the CNOT gates are applied transversally, the encoded operation will be fault-tolerant.)

5.4 The one-qubit operator T that cyclically permutes the Pauli operators was introduced in Section 5.3. This problem shows how T can be implemented using measurements.

(a) Consider a two-qubit system with qubit 1 in an arbitrary state $|\psi\rangle$ and qubit 2 in the $+1$ eigenstate of σ_y^2. Apply a CNOT gate using qubit 2 as the control; then measure the operator $\mathcal{O} = \sigma_y^1 \otimes I^2$. Show that this procedure maps

$$\sigma_x^1 \to \sigma_y^2$$
$$\sigma_y^1 \to \sigma_z^2$$
$$\sigma_z^1 \to \sigma_x^2 \ .$$

This is T, except that the data has been transferred from qubit 1 to qubit 2.

(b) Follow the transformation of the intial state

$$|\psi\rangle \otimes |0\rangle_y = [a|0\rangle + b|1\rangle] \otimes [|0\rangle + i|1\rangle]/\sqrt{2}$$

Fault-Tolerant Quantum Computing

under the procedure described in part (a). Show that the state of qubit 2 after the measurement is

$$|\psi_f\rangle = a|0\rangle_x - ib|1\rangle_x = \begin{pmatrix} a - ib \\ a + ib \end{pmatrix},$$

where $|k\rangle_x = [|0\rangle + (-1)^k|1\rangle]/\sqrt{2}$ and $k = 0, 1$. The procedure in part (a) has applied a unitary transformation U to the data: $|\psi_f\rangle = U|\psi\rangle$. Inserting the states $|\psi_f\rangle$ and $|\psi\rangle$ into this equation gives

$$\frac{1}{\sqrt{2}} \begin{pmatrix} a - ib \\ a + ib \end{pmatrix} = U \begin{pmatrix} a \\ b \end{pmatrix}.$$

Show that

$$U = \frac{1}{\sqrt{2}} \begin{pmatrix} 1 & -i \\ 1 & i \end{pmatrix},$$

which is recognized as T (see eq. (5.49)).

5.5 In quantum teleportation [15], the entanglement in an Einstein-Podolsky-Rosen (EPR) pair is used to transport an unknown quantum state $|\psi\rangle$ between two distant observers A and B. Qubit 1 is initially in an arbitrary state $|\psi\rangle$ and qubits 2 and 3 are in the Bell state $[|00\rangle + |11\rangle]/2$. Observer A has possession of qubits 1 and 2 and observer B has qubit 3. The two observers also share a classical communication channel.

(a) The initial state is $|\psi\rangle \otimes [|00\rangle + |11\rangle]/2$, which has a stabilizer with generators $G_1^0 = I^1 \otimes \sigma_x^2 \otimes \sigma_x^3$ and $G_2^0 = I^1 \otimes \sigma_z^2 \otimes \sigma_z^3$. The Pauli operator analogs are $\mathcal{X}^0 = \sigma_x^1 \otimes I^2 \otimes I^3$ and $\mathcal{Z}^0 = \sigma_z^1 \otimes I^2 \otimes I^3$. Observer A applies a CNOT gate from qubit 1 to qubit 2 and then applies a Hadamard gate to qubit 1. Show that the stabilizer generators G_1^0 and G_2^0 and the Pauli operator analogs $\mathcal{X}^0$ and $\mathcal{Z}^0$ map to

$$G_1^* = I^1 \otimes \sigma_x^2 \otimes \sigma_x^3 \quad ; \quad G_2^* = \sigma_x^1 \otimes \sigma_z^2 \otimes \sigma_z^3$$
$$\mathcal{X}^* = \sigma_z^1 \otimes \sigma_x^2 \otimes I^3 \quad ; \quad \mathcal{Z}^* = \sigma_x^1 \otimes I^2 \otimes I^3.$$

(b) Observer A now measures the operators $\mathcal{O}_1 = \sigma_z^1 \otimes I^2 \otimes I^3$ and $\mathcal{O}_2 = I^1 \otimes \sigma_z^2 \otimes I^3$. Show that the generators and Pauli operator analogs are transformed to

$$G_1^1 = \mathcal{O}_2 \quad ; \quad G_2^1 = \mathcal{O}_1$$
$$\mathcal{X}^1 = G_1^* \mathcal{X}^* \quad ; \quad \mathcal{Z}^1 = G_2^* \mathcal{Z}^*$$
$$= \sigma_z^1 \otimes I^2 \otimes \sigma_x^3 \quad ; \quad = I^1 \otimes \sigma_z^2 \otimes \sigma_z^3.$$

Discarding the measured qubits gives $\mathcal{X}^1 = \sigma_x^3$ and $\mathcal{Z}^1 = \sigma_z^3$ so that this procedure has mapped $\sigma_x^1 \to \sigma_x^3$ and $\sigma_z^1 \to \sigma_z^3$. The state $|\psi\rangle$ has thus been transported from qubit 1 to qubit 3.

5.6 Let $\mathcal{O}$ be an Hermitian operator of order 2. Show that the quantum circuit in Figure 5.9 will leave the target qubit in the state $P_+(\mathcal{O})|\psi\rangle$ whenever the measurement of the control qubit in the computational basis yields $m = +1$ and leaves it in the state $|0\rangle$.

5.7 Consider two four-qubit blocks, each encoded using the [4,2,2] stabilizer code presented in Section 5.9.

(a) Show that transversal application of CNOT gates using the qubits in block 1 as controls produces the following map on the code generators:

$$g_1 \otimes I \longrightarrow g_1 \otimes g_1$$
$$I \otimes g_1 \longrightarrow I \otimes g_1$$
$$g_2 \otimes I \longrightarrow g_2 \otimes I$$
$$I \otimes g_2 \longrightarrow g_2 \otimes g_2 \ .$$

The images of the map form a generator set for $\mathcal{S} \otimes \mathcal{S}$, which can be shown by finding combinations of products that give the original generator set. This establishes that transversal application of CNOT gates is an encoded operation for this code. By Problem 5.3, this code must be a CSS code; and so as shown in Section 5.3.2, the encoded operation must be a block-encoded CNOT gate. This is confirmed in part (b) of this problem.

(b) Show that transversal application of CNOT gates from block 1 to block 2 produces the following map:

$$X_i \otimes I \longrightarrow X_i \otimes X_i$$
$$I \otimes X_i \longrightarrow I \otimes X_i$$
$$Z_i \otimes I \longrightarrow Z_i \otimes I$$
$$I \otimes Z_i \longrightarrow Z_i \otimes Z_i \ ,$$

for $i = 1, 2$. As anticipated, this is the action of a block-encoded CNOT gate.

References

[1] Shor, P. W., Fault-tolerant quantum computation, in *Proceedings, 37th Annual Symposium on Fundamentals of Computer Science*, IEEE Press, Los Alamitos, CA, 1996, pp. 56–65.

[2] DiVincenzo, D. P. and Shor, P. W., Fault-tolerant error correction with efficient quantum codes, *Phys. Rev. Lett.* **77**, 3260, 1996.

[3] Plenio, M. B., Vedral, V., and Knight, P. L., Conditional generation of error syndromes in fault-tolerant error correction, *Phys. Rev. A* **55**, 4593, 1997.

[4] Gottesman, D., Theory of fault-tolerant quantum computation, *Phys. Rev. A* **57**, 127, 1998.

[5] Gottesman, D., Stabilizer codes and quantum error correction, Ph. D. thesis, California Institute of Technology, Pasadena, CA, 1997.

[6] Preskill, J., Fault-tolerant quantum computation, in *Introduction to Quantum Computation and Information*, H.-K. Lo, S. Popescu, and T. Spiller, Eds., World Scientific, Singapore, 1998.

[7] Steane, A. M., Active stabilization, quantum computation, and quantum state synthesis, *Phys. Rev. Lett.* **78**, 2252, 1997.

[8] Steane, A. M., Efficient fault-tolerant quantum computing, *Nature* **399**, 124, 1999.

[9] Schrodinger, E., The present situation in quantum mechanics, in *Quantum Theory and Measurement*, J. A. Wheeler and W. H. Zurek, Eds., Princeton University Press, Princeton, NJ, 1983.

[10] Calderbank, A. R. et al., Quantum error correction and orthogonal geometry, *Phys. Rev. Lett.* **78**, 405, 1997.

[11] Calderbank, A. R. et al., Quantum error correction via codes over GF(4), *IEEE Trans. Inf. Theor.* **44**, 1369, 1998.

[12] Bennett, C. H. et al., Mixed-state entanglement and quantum error correction, *Phys. Rev. A* **54**, 3824, 1996.

[13] Shilov, G. E., *Linear Algebra*, Dover Publications, New York, 1977, 89.

[14] Nielsen, M. A. and Chuang, I. L., *Quantum Computation and Quantum Information*, Cambridge University Press, London, 2000, Section 10.6.3.

[15] Bennett, C. H. et al., Teleporting an unknown quantum state via dual classical and EPR channels, *Phys. Rev. Lett.* **70**, 1895, 1993.

[16] Kitaev, A. Y., Quantum computations: algorithms and error correction, *Russ. Math. Surv.* **52**, 1191, 1997.

[17] Gottesman, D. and Chuang, I. L., Quantum teleportation is a universal computational primitive, *Nature* **402**, 390, 1999.

[18] Zhou, X., Leung, D. W., and Chuang, I. L., Methodology for quantum logic gate construction, *Phys. Rev. A* **62**, 052316, 2000.

[19] Barenco, A. et al., Elementary gates for quantum computation, *Phys. Rev. A* **52**, 3457, 1995.

6

Accuracy Threshold Theorem

In Chapters 2–5 we gave a detailed presentation of the theory of quantum error correction and fault-tolerant quantum computing. The stage is finally set for a proof of the accuracy threshold theorem [2–9]. As pointed out in Chapter 1, this theorem spells out the conditions under which quantum computing can be done reliably using imperfect quantum gates and in the presence of noise. The proof presented in this chapter follows Ref. [6]. Section 6.1 addresses a number of preliminary topics. It explains why a concatenated quantum error correcting code (QECC) should be used to protect the computational data when fault-tolerance is an issue; it states the principal assumptions underlying the threshold calculations presented in Section 6.2; and finally, it closes with a statement of the accuracy threshold theorem. The proof of this theorem is taken up in Section 6.2. For the error model introduced in Section 6.1, the accuracy threshold is calculated for gates in the Clifford group $N(\mathcal{G}_n)$, storage registers, and the Toffoli gate. The calculations make explicit use of the recursive structure of a concatenated QECC and the fault-tolerant procedures introduced in Chapter 5. The theorem is proved by showing that all three cases yield non-zero threshold values.

6.1 Preliminaries

We have learned how to protect data using a QECC and how to control the proliferation of errors through the use of fault-tolerant procedures for encoded quantum gates, error correction, and measurement. The question that faces us now is whether this substantial theoretical investment provides us with the ability to do a quantum computation of arbitrary duration with arbitrarily small error probability. Ultimately we will find the answer is yes, though only under appropriate conditions. First, it is clear that active harm will be done to the data if an error correction step introduces more errors than it removes. Thus one anticipates that if error correction is to be useful, the error probabilities for quantum gates and storage registers used for quantum computation and error correction cannot be too large. Second, and perhaps a little less obvious, is that not all QECCs are suitable for fault-tolerant quantum computing. The introduction of fault-tolerant procedures increases both

196   *Quantum Error Correction and Fault-Tolerant Quantum Computing*

the circuit complexity and time needed to do syndrome extraction. Ironically, the desire to control error proliferation through fault-tolerant procedures provides more opportunities for errors to be introduced into the data. Again, if error probabilities are too large, more errors will appear during error correction than are removed. Concatenated QECCs provide a way around this second difficulty. Their recursive structure allows them to correct a large number of errors, while at the same time making it possible to do syndrome extraction in a sufficiently efficient manner.

6.1.1 Concatenated QECCs

The reader may wish to review the discussion of concatenated QECCs given in Section 3.4 before reading this subsection. There, a detailed discussion is given of codes that contain two layers of concatenation. In the present chapter our interest is in codes that contain l layers of concatenation. Fortunately, code construction will not require as detailed a description as in Section 3.4.

In situations where fault-tolerance is an issue, it is highly convenient when constructing a concatenated QECC to base the construction on a CSS code that uses a doubly even classical code with $C_\perp \subset C$ (Section 5.3.2). The reason such a code is so convenient is that encoded gates in the Clifford group can be applied fault-tolerantly by applying the unencoded gates transversally to code blocks. Consequently, for the remainder of this chapter, we restrict ourselves to a concatenated QECC that is based on the [7,1,3] CSS code (Section 2.3).

Concatenated QECCs can be given two complementary descriptions: (i) top-down; and (ii) bottom-up.

(i) In the top-down description we begin with an unencoded qubit and encode it into a block of seven qubits using the [7,1,3] CSS code. Next each qubit in this block is itself encoded into a block of seven qubits. Repeating this process l times produces our concatenated QECC in which one data qubit is recursively encoded into the state of 7^l physical qubits. By using induction on Theorem 3.4 (see Problem 6.1), it can be shown that our concatenated code is a $[7^l,1,d]$ code with $d \geq 3^l$. Recall that a distance d QECC can correct t errors, where $t = [(d-1)/2]$ and $[x]$ is the integer part of x. Since $d \geq 3^l$, we see that the number of errors that a concatenated QECC can correct grows *exponentially* with the number of layers of concatenation l.

(ii) In the remainder of this chapter we will use the bottom-up description of a concatenated QECC. In the bottom-up description we begin with 7^l physical qubits. This is the zeroth-level of the (concatenated) code. The first-level of the code is obtained by collecting the physical qubits into non-overlapping blocks of seven qubits. Each block encodes a single level-1 qubit using the [7,1,3] CSS code. There are thus a total of 7^{l-1} such level-1 qubits. The second-level of the code is obtained by collecting the level-1 qubits into (non-overlapping) blocks of seven qubits so that there are 7^{l-2} such level-2 qubits. This blocking/coarse-graining of level-$(j-1)$ qubits to form level-j qubits is repeated l times so that at level-l there is only one level-l qubit, which is the unencoded data qubit we are trying to protect. From the construction procedure it follows that a level-j qubit is made up of 7^j physical qubits. To measure the error syndrome for this code we begin at level-1, measuring the syndrome for each level-1 code block. To simplify the analysis of Section 6.2 it is assumed that the syndrome for all these code blocks can be measured in parallel. Then we do not need to worry about storage errors accumulating on code blocks while they wait around for other code blocks to have their syndrome measured. As long as l is not too large, this assumption is reasonable, and so the time to measure the level-1 error syndrome is simply the time it takes to measure and verify the syndrome of a level-1 (code) block. Once the syndrome for all level-1 blocks has been determined, the appropriate recovery operation is applied in parallel to each block. Next, the error syndrome is measured in parallel for all level-2 blocks. As with the level-1 blocks, the time needed to do this is the time it takes to measure and verify the error syndrome for a level-2 block. Once the syndrome has been determined for all level-2 blocks, error recovery is done in parallel on each of these blocks. Repeating this process for all l code-levels implements one error correction cycle on our concatenated QECC. The time required to carry out one error correction cycle is then the sum of the times needed to do error correction on a code block at each code-level. We can now state why concatenated QECCs are so essential to the proof of the accuracy threshold theorem: (i) the number of errors these codes can correct grows exponentially with the number of layers of concatenation, and (ii) the time needed to do error correction is linear in the time it takes to error correct a code block in each layer.

6.1.2 Principal Assumptions

As noted in Chapter 1, the specific value obtained for the accuracy threshold depends on the detailed assumptions made concerning the error model and the types of operations that can be applied to blocks of data and ancilla qubits. This subsection will state the principal assumptions on which the threshold calculations of Section 6.2 will be based. Further secondary assumptions will be made there when appropriate/convenient to simplify analysis of the calculational results.

Two assumptions were already stated in Section 6.1.1. We restate them here so that all principal assumptions can be found in this subsection.

Assumption 6.1 *For a concatenated QECC, it is possible to measure the error syndrome for all code blocks in a given layer of concatenation simultaneously (viz. in parallel).*

Assumption 6.2 *Each layer of a concatenated QECC will be encoded using the same CSS code. In this Chapter the [7,1,3] CSS code will be used for this purpose.*

The following three assumptions describe the nature of the errors that are assumed to occur.

Assumption 6.3 *The physical process responsible for errors is such that independent errors form exclusive alternatives [10] and so the probabilities for independent errors add.*

Errors can occur on qubits as they wait around in storage registers, are operated on with quantum gates, and during measurements and state preparation.

Assumption 6.4 *Storage errors occur independently on different qubits with an error probability* **rate** p_{stor}.

Thus if a qubit is in storage for a time Δt, the probability that a storage error occurs is $p_{stor} \Delta t$. The error probability for quantum gates in the Clifford group $N(\mathcal{G}_n)$ is p_g, and for the Toffoli gate is p_{Tof}. Note that the gate error probability is defined to *include* the storage error probability that accrues on qubits while they are being acted on by the gate.

Assumption 6.5 *Quantum gates can only produce errors on the qubits upon which they act. Qubits that are not the object of a gate operation do however accumulate storage errors while they wait for a gate to act on other qubits.*

The error probability for state preparation and measurement are denoted p_{prep} and p_{meas}, respectively. The next assumption is a procedural one.

Assumption 6.6 *After any operation at code level-j, error correction is applied to all level-j code blocks. We denote the error probabilities for level-j Clifford group gates, Toffoli gates, and storage registers by* $p_g^{(j)}$, $p_{Tof}^{(j)}$, *and* $p_{stor}^{(j)}$, *respectively.*

The time required to implement an operation will depend on the code-level at which it is being applied.

Assumption 6.7 *The unit of time is chosen so that the time needed to apply an operation on a physical qubit at level-0 is 1.*

Clearly not all operations on level-0 qubits will take exactly one time-unit. It is being assumed that these operations will have application times that are of the same order of magnitude. This representative time then defines the unit of time, and we ignore the small differences in application time for the different gates. We denote the time it takes to apply a Toffoli gate at level-j by $t_{Tof}^{(j)}$, and the time to prepare an encoded state at level-j by $t_{prep}^{(j)}$. Our final assumption involves the times to do state preparation and measurement on physical qubits at level-0.

Assumption 6.8 *The time required to prepare the state of a level-0 qubit* $t_{prep}^{(0)}$ *is assumed to be negligible compared to the time to prepare a level-j* encoded state for $j \geq 1$ and so it is set to zero: $t_{prep}^{(0)} = 0$. The time to measure the state of a level-0 qubit $t_{meas}^{(0)}$ is assumed to be of the same order of magnitude as the time to apply an operation to it and so $t_{meas}^{(0)} = 1$.

As noted earlier, since we have restricted ourselves to a concatenated QECC built up from the [7,1,3] CSS code, all gates in the Clifford group can be applied at level-j by applying it transversally to qubits in the level-j code blocks. Because of the recursive structure of a concatenated QECC, this ultimately means that we apply the gate transversally to level-0 qubits. The time thus needed to implement a gate in the Clifford group at any code-level is simply the time it takes to apply the gate to a level-0 qubit and so $t_g^{(j)} = 1$.

Having listed our principal assumptions, we close this section with a statement of the Accuracy Threshold Theorem.

THEOREM 6.1 Accuracy Threshold Theorem
A quantum computation of arbitrary duration can be done with arbitrarily small error probability in the presence of noise and using imperfect quantum gates if the following conditions hold:

1. *the computational data is protected using a concatenated QECC;*

2. *fault-tolerant procedures are used for encoded quantum gates, error correction, and measurement; and*

3. *storage registers and a universal set of quantum gates are available that have error probabilities P_e that are less than a threshold value P_a known as the accuracy threshold: $P_e < P_a$.*

The proof of this theorem forms the content of Section 6.2. The threshold calculations will use a concatenated QECC to encode the data, and all procedures will be fault-tolerant. Thus, proof of the theorem will reduce to showing that the accuracy threshold takes on non-zero values for (i) a storage register, (ii) all gates in the Clifford group, and (iii) the Toffoli gate. The final value for the accuracy threshold for our set of assumptions will then be the smallest of these three values.

6.2 Threshold Analysis

It proves convenient to think of a storage register as applying a quantum operation to the qubits that are stored in it. Ideally the operation is the identity. However, because of unwanted interactions with the environment, the operation actually applied will differ from the identity with a probability per unit

200   *Quantum Error Correction and Fault-Tolerant Quantum Computing*

time p_{stor}. In this section, when speaking of a generic quantum operation, we will have in mind either a gate in the Clifford group, a storage register "gate", or a Toffoli gate. Given an l-layered concatenated QECC, we refer to the error probability for level-l operations as effective error probabilities since they give the probability that a real error appears in the data qubit we are trying to protect. We refer to the error probability for level-0 operations as primitive error probabilities since they will be found to recursively determine the error probability for all higher-level operations. To establish this connection, we derive a recursion-relation that connects the error probability $p_{op}^{(j)}$ for a level-j operation to the error probability of the level-$(j-1)$ operations that are used to apply it. This is done in Section 6.2.1. The recursion-relation is then used in Sections 6.2.2 and 6.2.3 to determine the accuracy threshold for, respectively, gates in the Clifford group and storage registers, and Toffoli gates.

6.2.1   Recursion-Relation for $p_{op}^{(j)}$

Recall that error correction is applied to every level-j code block after the application of a level-j quantum operation (Assumption 6.6). Let $p_{EC}^{(j)}$ denote the probability that error correction applied to a level-j code block produces an error on one of the level-$(j-1)$ qubits that belong to the block. By Assumption 6.2, we restrict ourselves to a concatenated QECC based on the [7,1,3] CSS code. Thus for a quantum operation to produce an error at level-j, at least two errors must occur at level-$(j-1)$ during the two-step procedure of quantum operation plus follow-up error correction. In the remainder of Section 6.2 we focus on the most probable case where only two errors occur, and to simplify the analysis, assume that all code blocks acted on by a quantum operation begin the operation error-free.

Consider a level-j code block that has developed an error due to a level-j quantum operation. Two of the level-$(j-1)$ qubits belonging to the block must contain errors. There are $\binom{7}{2} = 21$ different ways in which two qubits can be chosen from a block of seven qubits to act as error sites; and for a given choice of qubits, the two errors can appear in one of three ways: (i) both during the quantum operation; (ii) one during the quantum operation and the second during error correction; or (iii) both during error correction. For case (i), the quantum operation produces a level-j error if two of the level-$(j-1)$ operations used to apply it produce errors. This occurs with probability $(p_{op}^{(j-1)})^{2*}$. For case (ii), the syndrome verification protocol of Section 5.2.3 requires that

---
*This is true for gates in the Clifford group and storage registers. For a Toffoli gate, $p_{Tof}^{(j-1)}$ is *not* equal to the probability E that a level-$(j-1)$ qubit develops an error during a level-j Toffoli gate. As will be shown in Section 6.2.3, E can be upper bounded by the largest of the E_i appearing in eq. (6.30), and is seen to be a linear inhomogeneous function of $p_{Tof}^{(j-1)}$.

at most two syndrome extractions be carried out. Since an error has occurred during the quantum operation, the second error must occur during one of these two syndrome extractions. Let qubits A and B denote the two qubits on which the errors occur. The first error can occur on qubit A or B, and the second error can occur during the first or the second syndrome extraction. Thus the total probability for a case (ii) error is $4p_{op}^{(j-1)}p_{EC}^{(j)}$. For case (iii), both errors appear in the code block during error correction. Since the syndrome verification procedure requires at most two syndrome extraction operations, this can happen in two ways: either both errors occur during a single syndrome extraction; or when two syndrome extraction operations are applied, each produces an error. For the first scenario, the probability is $2\left(p_{EC}^{(j)}\right)^2$ since both errors occur during either the first or second syndrome extraction. For the second scenario, there are only two ways in which this scenario can occur. Either the first error appears on qubit A and the second on qubit B, or vice versa. The probability for the second scenario is thus $2\left(p_{EC}^{(j)}\right)^2$, and so the total probability for a case (iii) error is $4\left(p_{EC}^{(j)}\right)^2$. [Note that Ref. [6] used a different syndrome verification procedure and found $8\left(p_{EC}^{(j)}\right)^2$ for the case (iii) error probability.] Putting together all these results we obtain the probability $p_{op}^{(j)}$ that a level-j quantum operation followed by error correction will produce a level-j error:

$$p_{op}^{(j)} = 21\left[\left(p_{op}^{(j-1)}\right)^2 + 4p_{op}^{(j-1)}p_{EC}^{(j)} + 4\left(p_{EC}^{(j)}\right)^2\right]. \qquad (6.1)$$

Eq. (6.1) allows us to understand why concatenated QECCs are so important to establishing an accuracy threshold. As we have seen, since our concatenated QECC is based on the [7,1,3] CSS code, an error at level-j requires that at least two errors occur at level-$(j-1)$, and so $p_{op}^{(j)}$ will be $\mathcal{O}((p_{op}^{(j-1)})^2)$. This is (loosely speaking) what eq. (6.1) is telling us. Thus for some value C,

$$p_{op}^{(l)} \leq C\left(p_{op}^{(l-1)}\right)^2$$
$$\leq C^{2^l-1}\left(p_{op}^{(0)}\right)^{2^l}. \qquad (6.2)$$

If we define C_{th} to be the smallest C for which eq. (6.2) is true, then C_{th} yields the tightest upper bound for $p_{op}^{(l)}$:

$$p_{op}^{(l)} \leq C_{th}^{2^l-1}\left(p_{op}^{(0)}\right)^{2^l}$$
$$\leq p_{op}^{(0)}\left(\frac{p_{op}^{(0)}}{p_{th}}\right)^{2^l-1}, \qquad (6.3)$$

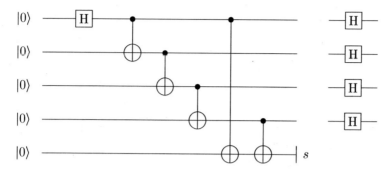

FIGURE 6.1
Quantum circuit for Shor state preparation for $w_i = 4$. A fifth ancilla qubit is introduced to record the relative parity s of the first and fourth qubits. If measurement yields $s = 1$, the ancilla block is discarded, a new ancilla block is introduced, and the circuit is applied again. The final Hadamard gates are only applied if $s = 0$. See Section 5.2.2 for a full discussion.

where $p_{th} = 1/C_{th}$. Thus if $p_{op}^{(0)}/p_{th} < 1$, the effective error probability $p_{op}^{(l)}$ decreases exponentially with the number of concatenation layers l. We see that p_{th} is the accuracy threshold for the quantum operation since, whenever the primitive error probability $p_{op}^{(0)}$ is less than p_{th}, the effective error probability can be made arbitrarily small by using a suitable number of concatenation layers. In fact, suppose we want $p_{op}^{(l)} = \epsilon$. Solving eq. (6.3) for l gives

$$l \geq \log_2 \left[1 + \frac{\log_2(\epsilon/p_{op}^{(0)})}{\log_2(p_{op}^{(0)}/p_{th})} \right] . \qquad (6.4)$$

If we want $\epsilon/p_{op}^{(0)} = 0.2$, and assume $p_{op}^{(0)}/p_{th} = 0.5$ so that $p_{op}^{(l)}$ will be an order of magnitude less than p_{th}, eq. (6.4) gives $l = 1.73$. Thus two layers of concatenation should deliver a quantum operation with the desired effective error probability.

To obtain our final expression for $p_{op}^{(j)}$ we must evaluate $p_{EC}^{(j)}$. We assume that error correction is being carried out using the fault-tolerant procedure presented in Section 5.2. The [7,1,3] CSS code being used to construct our concatenated QECC has six generators, each of weight $w_i = 4$ (see eq. (2.50)). Syndrome extraction will thus require six ancilla blocks containing four qubits, all prepared in the Shor state $|A\rangle$. The quantum circuit used to prepare the Shor state appears in Figure 6.1. The sequence of quantum gates is the same as in Figure 5.4, although the input states are the $|0\rangle$ computational basis states for level-$(j-1)$ of our concatenated QECC. Recall that the portion of the circuit preceding the final Hadamard gates: (i) places the four top-most ancilla qubits in the cat-state in the absence of errors, and (ii) to $\mathcal{O}(p)$ insures that no bit-flip errors are present in the cat-state. If the measurement of the fifth ancilla qubit returns $s = 0$, the final Hadamard gates are applied. This transforms the cat-state into the Shor state, and to $\mathcal{O}(p)$, we are sure that the Shor state is free of phase errors.

As discussed in Section 5.2.2, a phase error produced during Shor state preparation will spread to the data block during syndrome extraction. The probability that this occurs is thus the probability that a phase error is present in the ancilla block after Shor state preparation. We need not worry about the presence of a bit-flip error in the Shor state since such an error cannot directly spread into the data block, and its effects can be managed using the syndrome verification protocol presented in Section 5.2.3. Shor state preparation can leave a phase error in the ancilla block if a bit-flip error occurs during cat-state preparation, or a phase error occurs during one of the final Hadamard gates used to transform the cat-state into the Shor state. Recall (Section 5.2.2) that the measurement of the fifth ancilla qubit detects any bit-flip error that occurs prior to the fourth CNOT gate in Figure 6.1. Thus a bit-flip error must occur during the fourth or fifth CNOT gate if it is to produce a phase error in the Shor state. For qubit 1 (4) to wind up with a phase error at the end of Shor state preparation, it must pick up either (i) a bit-flip error during the fourth (fifth) CNOT gate in Figure 6.1, (ii) a bit-flip error while it waits for the fifth (fourth) CNOT gate to be applied to qubit 4 (1), or (iii) a phase error when one of the final Hadamard gates is applied to it. An upper bound for the probability that one of these errors occurs is then $2p_g^{(j-1)} + p_{stor}^{(j-1)}$. This is an upper bound since $p_g^{(j-1)}$ and $p_{stor}^{(j-1)}$ are the probabilities that *any* one-qubit error occurs during a Clifford group gate or during one time-unit of storage, respectively. Thus $p_g^{(j-1)}$ overestimates the probability that *only* a bit-flip error occurs during (i) or (ii), and $p_{stor}^{(j-1)}$ overestimates the probability that only a phase error occurs during (iii). We conservatively identify this upper bound with the error probability that either qubit 1 or 4 winds up with a phase error at the end of Shor state preparation. For qubit 2 (3) to have a phase error after Shor state preparation, it must pick up either (i) a bit-flip error while it waits for the fourth and fifth CNOT gates in Figure 6.1 to be applied, or (ii) a phase error when the final Hadamard gate is applied to it. Thus $p_g^{(j-1)} + 2p_{stor}^{(j-1)}$ upper bounds the probability that either of these errors occurs, and we again identify this probability with the upper bound. Summing over the four ancilla qubits gives $6p_g^{(j-1)} + 6p_{stor}^{(j-1)}$ for the probability that one of the ancilla qubits will have a phase error after Shor state preparation, and this in turn gives the probability that a phase error spreads to the data block during syndrome extraction. We will use this result below.

Next we determine the probability that an error appears on a data qubit during syndrome extraction and measurement. Problem 6.2 puts the generators of the [7,1,3] CSS code into standard form (Section 4.1), and uses the approach of Plenio et al. [11] to construct a quantum circuit for error correction. The resulting circuit is shown in Figure 6.2. Examination of this circuit indicates that each data qubit is involved in at most four syndrome bit

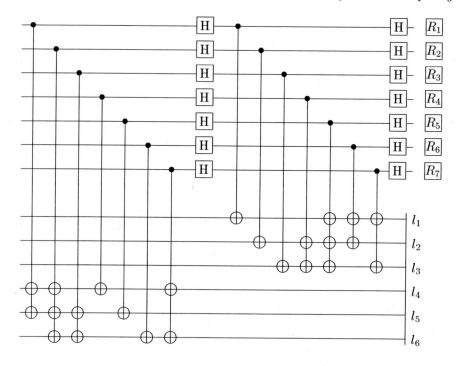

FIGURE 6.2
Error correction circuit for the [7,1,3] CSS code in standard form. The seven top-most lines correspond to the data block and each of the six lower lines corresponds to a block of four ancilla qubits. The four qubit lines in each ancilla block have been compressed into a single line to make the circuit easier to examine, and the four CNOT gates applied to each ancilla block are applied transversally. The measured syndrome value $S(E) = l_1 \cdots l_6$ determines the recovery operation $\mathbf{R} = R_1 \cdots R_7$ that is applied to the data block. If no errors occur during ancilla block preparation, each ancilla block enters the circuit in the Shor state $|A\rangle$.

extractions, each time acting as the control for a CNOT gate. To simplify the analysis, we suppose that all data qubits are involved in exactly four such extractions, and thus take part in four CNOT gates. Each data qubit must then wait around for completion of the two syndrome extractions in which it does not participate. Since each CNOT gate takes only one time-step to execute (see remarks following Assumption 6.8), each data qubit waits a total of two time-steps while the two syndrome extractions in which it does not take part finish. The error correction circuit also applies two Hadamard gates to each data qubit. It follows from these remarks that each data qubit is acted on by four CNOT gates, two Hadamard gates, and waits in storage for two time-steps. The probability that an error occurs on a data qubit during syndrome extraction is thus $6p_g^{(j-1)} + 2p_{stor}^{(j-1)}$. Storage errors can also appear on a data qubit while it waits for (i) the cat-states to be prepared, (ii) the cat-states to be converted to Shor states, and (iii) the extracted syndrome bit values to be measured. We assume the six cat-states can be prepared in parallel so that a data block only has to wait for the time $t_{prep}^{(j-1)}$ it takes to prepare a single cat-state. The data block must also wait while the Hadamard gates that convert a cat-state to a Shor state are applied. This takes one time-step per cat-state, and we assume that this operation is applied immediately before an ancilla block is used to extract a syndrome bit. Thus by assumption, cat-state conversion is done serially, and a data block will thus wait a total of six time-steps for all six conversions to be done. Finally, we assume that measurement of the syndrome bit that was transferred to ancilla block i can be measured at the same time that the next syndrome bit is being extracted to ancilla block $i + 1$. Thus a data block only has to wait for the time $t_{meas}^{(j-1)}$ it takes to measure the last syndrome bit l_6. The probability that a storage error occurs on a data qubit during these three stages of error correction is thus $(t_{prep}^{(j-1)} + 6 + t_{meas}^{(j-1)})p_{stor}^{(j-1)}$.

We can now determine $p_{EC}^{(j)}$. It is the sum of the probabilities that (i) a phase error in the Shor state spreads to the data block during syndrome extraction, (ii) an error occurs during syndrome extraction, and (iii) a storage error appears during cat-state preparation and conversion, and during measurement of the error sydrome $S(E) = l_1 \cdots l_6$:

$$p_{EC}^{(j)} = \left(6p_g^{(j-1)} + 6p_{stor}^{(j-1)}\right)$$
$$+ \left(6p_g^{(j-1)} + 2p_{stor}^{(j-1)}\right) + \left(t_{prep}^{(j-1)} + 6 + t_{meas}^{(j-1)}\right) p_{stor}^{(j-1)}$$
$$= 12p_g^{(j-1)} + \left(14 + t_{prep}^{(j-1)} + t_{meas}^{(j-1)}\right) p_{stor}^{(j-1)} \quad . \quad (6.5)$$

To complete our calculation of $p_{EC}^{(j)}$, we must determine $t_{prep}^{(j-1)}$ and $t_{meas}^{(j-1)}$. We begin with $t_{meas}^{(j-1)}$. Recall that at the zeroth-level of our concatenated code $(j = 0)$, $t_{meas}^{(0)} = 1$ by Assumption 6.8.

(i) At the first level of concatenation $(j = 1)$, a code block contains seven level-0 qubits and each ancilla block used to extract a syndrome bit-value contains four level-0 qubits. Recall that $t_{meas}^{(1)}$ is the time it takes to measure the last syndrome bit l_6. Thus it is the time needed to measure the four level-0 qubits in the sixth ancilla block in the level-0 computational basis (CB). It is assumed that these four qubits can be measured in parallel so that $t_{meas}^{(1)} = t_{meas}^{(0)} = 1$.

(ii) A new complication enters at the second level of concatenation $(j = 2)$. A level-2 code block contains seven level-1 qubits, and an ancilla block contains four level-1 qubits. Measuring a syndrome bit-value at level-2 thus requires measurement of four level-1 ancilla qubits in the level-1 CB. Since each level-1 ancilla qubit is itself a block of seven level-0 qubits, measuring a level-1 ancilla qubit requires measuring seven level-0 qubits (in the level-0 CB). In the absence of errors, the measurement result $c = c_1 \cdots c_7$ is a codeword in the [7,4,3] classical Hamming code (Section 2.3). If the parity of c is even (odd), then the level-1 qubit is in the level-1 CB state $|\bar{0}\rangle$ ($|\bar{1}\rangle$) (see eq. (2.49)). To $\mathcal{O}(p^{(0)})$, a single bit-flip error e can be present in the bit-string c. Calculation of the classical syndrome $S(e)$ determines the most probable error e_p, and decoding returns $c + e_p$ as the most probable codeword (Section 1.2.3). The decoding procedure fails with probability $\mathcal{O}((p^{(0)})^2)$. Assuming all $4 \cdot 7 = 28$ level-0 qubits can be measured in parallel, the time needed to measure all these qubits is $t_{meas}^{(0)} = 1$. Let $T^{(2)}$ be the total time needed to (i) calculate $S(e)$ and correct c for each level-1 ancilla qubit; and (ii) reconstruct the level-2 syndrome bit-value from the four level-1 ancilla qubit measurements. It is assumed that these classical computations can be done sufficiently quickly that $T^{(2)}$ is negligible compared to $t_{meas}^{(0)} = 1$. It then follows that

$$t_{meas}^{(2)} = t_{meas}^{(0)} + T^{(2)}$$
$$= 1 + T^{(2)}$$
$$= 1 \; .$$

(iii) To measure a level-$(j-1)$ syndrome bit-value we must measure four level-$(j-2)$ ancilla qubits. Measurement of each level-$(j-2)$ ancilla qubit in turn requires measurement of seven level-$(j-3)$ qubits, and so on until we reach level-0. It is assumed (as above) that all $4 \cdot 7^{(j-2)}$ level-0 qubits can be measured in parallel so that the time it takes to do these measurements is $t_{meas}^{(0)} = 1$. It is also assumed that the time $T^{(j-1)}$ needed to reconstruct the level-$(j-1)$ syndrome bit-value from all the level-0 measurement results is negligible compared to $t_{meas}^{(0)} = 1$. This last assumption will eventually fail for $j-1$ sufficiently large. However, since the number of concatenation layers l has a log-log dependence on the effective error probability $p_{op}^{(l)} = \epsilon$ (eq. (6.4)), we assume in the remainder of this chapter that l is sufficiently small that $T^{(j-1)} \ll t_{meas}^{(0)}$, and so for $0 \leq j - 1 \leq l$,

$$t_{meas}^{(j-1)} = t_{meas}^{(0)} + T^{(j-1)}$$
$$= 1 + T^{(j-1)}$$
$$= 1 \; . \tag{6.6}$$

Finally we determine $t_{prep}^{(j-1)}$. The quantum circuit that prepares the Shor state (Figure 6.1) requires each of the four level-$(j-1)$ ancilla qubits to enter the circuit in the level-$(j-1)$ CB-state $|0\rangle$. Applying the following two-step procedure to an ancilla qubit will leave it in this state. First apply an error correction cycle to the qubit. This projects its state onto the level-$(j-1)$ code

Accuracy Threshold Theorem

space. Then measure the level-$(j-1)$ encoded Pauli operator $Z^{(j-1)}$. This leaves the ancilla qubit in the desired $|0\rangle$ CB-state (recall the measurement discussion in Section 5.4). Note that measuring $Z^{(j-1)}$ corresponds to doing a measurement in the computational basis. We have just seen that the time needed to do this is $t^{(j-1)}_{meas} = 1$. The effective time $t^{(j-1)}_{EC}$ to do an error correction cycle on a level-$(j-1)$ code block can be read off from eq. (6.5):

$$t^{(j-1)}_{EC} = 14 + t^{(j-2)}_{prep} + t^{(j-2)}_{meas}$$
$$= 15 + t^{(j-2)}_{prep} , \qquad (6.7)$$

where we have used $t^{(j-2)}_{meas} = 1$, and let $j \to j-1$ in eq. (6.5). Thus, the time needed to carry out this two-step procedure is $t^{(j-1)}_{EC} + 1 = 16 + t^{(j-2)}_{prep}$. Since our syndrome verification protocol (Section 5.2.3) requires at most two syndrome extractions, the time $t^{(j-1)}_0$ needed to prepare the state $|0\rangle$ at level-$(j-1)$ satisfies:

$$t^{(j-1)}_0 \leq 2\left(t^{(j-1)}_{EC} + 1\right)$$
$$\leq 32 + 2\, t^{(j-2)}_{prep} . \qquad (6.8)$$

Following Ref. [6], we assume that both Shor states used for syndrome verification can be prepared in advance. Then the second syndrome extraction will not have to wait for the second Shor state to be prepared. The effective time needed to do the second syndrome extraction will then be $t^{(j-1)}_{EC} - (6 + t^{(j-2)}_{prep}) = 9$. If we conservatively assume that both syndrome measurements are always needed, then

$$t^{(j-1)}_0 = \left(16 + t^{(j-2)}_{prep}\right) + 9$$
$$= 25 + t^{(j-2)}_{prep} . \qquad (6.9)$$

Having prepared the $|0\rangle$ states, the circuit in Figure 6.1 then applies a Hadamard gate, five CNOT gates, and measures the fifth ancilla qubit to put the top four ancilla qubits in the cat-state. Since each of these operation takes one time-unit to complete, the time $t^{(j-1)}_{prep}$ to prepare a cat-state is

$$t^{(j-1)}_{prep} = \left(25 + t^{(j-2)}_{prep}\right) + 7$$
$$= 32 + t^{(j-1)}_{prep} . \qquad (6.10)$$

Since $t^{(0)}_{prep} = 0$ (Assumption 6.8), eq. (6.10) gives

$$t^{(j-1)}_{prep} = 32\,(j-1) . \qquad (6.11)$$

[Note that Ref. [6] used a different syndrome verification protocol and found $t^{(j-1)}_{prep} = 43(j-1)$.] Using eqs. (6.6) and (6.11) in eq. (6.5) gives

$$p^{(j)}_{EC} = 12 p^{(j-1)}_g + [15 + 32\,(j-1)]\, p^{(j-1)}_{stor} . \qquad (6.12)$$

In the following subsections we will use eqs. (6.12) and (6.1) to determine the accuracy threshold for (i) gates in the Clifford group and storage registers (Section 6.2.2), and (ii) Toffoli gates (Section 6.2.3).

6.2.2 Clifford Group Gates and Storage Registers

Substituting eq. (6.12) into eq. (6.1) allows us to determine the recursion relation for the level-j error probabilities $p_g^{(j)}$ and $p_{stor}^{(j)}$ for gates in the Clifford group and for storage registers, respectively. One finds

$$p_g^{(j)} = 13125 \left(p_g^{(j-1)}\right)^2 + p_g^{(j-1)} p_{stor}^{(j-1)} [31500 + 67200(j-1)]$$
$$+ \left(p_{stor}^{(j-1)}\right)^2 [18900 + 80640(j-1) + 86016(j-1)^2], \quad (6.13)$$

and

$$p_{stor}^{(j)} = 12096 \left(p_g^{(j-1)}\right)^2 + p_g^{(j-1)} p_{stor}^{(j-1)} [31248 + 64512(j-1)]$$
$$+ \left(p_{stor}^{(j-1)}\right)^2 [20181 + 83328(j-1) + 86016(j-1)^2]. \quad (6.14)$$

Exercise 6.1 *Verify eqs. (6.13) and (6.14).*

We examine three cases that illustrate how to obtain the accuracy threshold from eqs. (6.13) and (6.14).

(i) Consider the case where level-0 storage errors occur with negligible probability: $p_{stor}^{(0)} = 0$. There is then no need to implement error correction on qubits that are not acted on by a quantum gate. For such a qubit, $p_{EC}^{(j)} = 0$ in eq. (6.1), and so for storage registers: $p_{stor}^{(j)} = 0$ for $0 \le j \le l$. Thus, for reliable storage registers, we only need to worry about the accuracy threshold for gates in the Clifford group $N(\mathcal{G}_n)$, and for the Toffoli gate. We use eq. (6.13) to find the accuracy threshold for gates in $N(\mathcal{G}_n)$, and will examine the Toffoli gate in Section 6.2.3. Using $p_{stor}^{(j-1)} = 0$ in eq. (6.13) gives

$$p_g^{(j)} = 13125 \left(p_g^{(j-1)}\right)^2$$
$$= p_g^{(0)} \left(\frac{p_g^{(0)}}{p_{th}}\right)^{2^j - 1}, \quad (6.15)$$

where $p_{th} = 1/13125 = 7.62 \times 10^{-5}$ is the accuracy threshold for gates in the Clifford group when reliable storage registers are available.

(ii) We next consider the case where storage registers and gates in the Clifford group have equal level-0 error probabilities: $p_{stor}^{(0)} = p_g^{(0)}$. Since the

Accuracy Threshold Theorem

numerical coefficients in eqs. (6.13) and (6.14) are nearly equal, we make the approximation that $p_{stor}^{(j)} = p_g^{(j)}$ for $0 \leq j \leq l$. It follows that the accuracy threshold for storage registers and for gates in $N(\mathcal{G}_n)$ will be the same, and so it is sufficient to calculate the threshold for (say) a storage register. Setting $p_g^{(j-1)} = p_{stor}^{(j-1)}$ in eq. (6.14) gives

$$p_{stor}^{(j)} = \left(p_{stor}^{(j-1)}\right)^2 \left[63525 + 147840(j-1) + 86016(j-1)^2\right]. \tag{6.16}$$

Thus

$$p_{stor}^{(1)} = 63525 \left(p_{stor}^{(0)}\right)^2 \tag{6.17}$$

$$p_{stor}^{(2)} = 297381 \left(p_{stor}^{(1)}\right)^2 \tag{6.18}$$

$$p_{stor}^{(3)} = 703269 \left(p_{stor}^{(2)}\right)^2, \tag{6.19}$$

and for $j \geq 4$ we make the approximation

$$p_{stor}^{(j)} = 86016(j-1)^2 \left(p_{stor}^{(j-1)}\right)^2. \tag{6.20}$$

Concatenation will succeed in driving down the effective error probability if, for example, $p_{stor}^{(j)}/p_{stor}^{(j-1)} < (j-1)^2/j^2$ for $j \geq 4$. Then, from eq. (6.20),

$$86016(j-1)^2 \left(p_{stor}^{(j-1)}\right) < \frac{(j-1)^2}{j^2},$$

or

$$p_{stor}^{(j-1)} < \left(86016 j^2\right)^{-1}. \tag{6.21}$$

For $j = 4$ this gives

$$p_{stor}^{(3)} < 7.27 \times 10^{-7},$$

and eqs. (6.19)–(6.17) give, respectively,

$$p_{stor}^{(2)} < 1.02 \times 10^{-6}$$
$$p_{stor}^{(1)} < 5.85 \times 10^{-6}$$
$$p_{stor}^{(0)} < 9.59 \times 10^{-6} \sim 1 \times 10^{-5}. \tag{6.22}$$

Thus $p_{th} \sim 10^{-5}$ for the case where storage registers and Clifford group gates have error probabilities that are approximately equal.

(iii) Finally, we consider the case where ancilla blocks in the Shor state are always available when needed so that storage errors no longer accumulate over the time $t_{prep}^{(j-1)} + 6$ needed to prepare a level-$(j-1)$ Shor state. Then eq. (6.5) becomes

$$p_{EC}^{(j)} = 12 p_g^{(j-1)} + 9 p_{stor}^{(j-1)}.$$

Using this in eq. (6.1) gives

$$p_g^{(j)} = 13125 \left(p_g^{(j-1)}\right)^2 + 18900 p_g^{(j-1)} p_{stor}^{(j-1)} + 6804 \left(p_{stor}^{(j-1)}\right)^2$$

$$p_{stor}^{(j)} = 12096 \left(p_g^{(j-1)}\right)^2 + 19152 p_g^{(j-1)} p_{stor}^{(j-1)} + 7581 \left(p_{stor}^{(j-1)}\right)^2 .$$

As in case (ii) above, since the numerical coefficients in these two equations are nearly equal, we approximate $p_g^{(j)} = p_{stor}^{(j)}$ for $j \geq 1$. Then

$$p_g^{(j)} = 38829 \left(p_g^{(j-1)}\right)^2$$

$$= p_g^{(0)} \left(\frac{p_g^{(0)}}{p_{th}}\right)^{2^j - 1} ,$$

where $p_{th} = 1/38829 = 2.58 \times 10^{-5}$.

6.2.3 Toffoli Gate

One task remains—to estimate the Toffoli gate accuracy threshold. Recall how this gate is implemented (pp. 175–177):

1. We begin with six level-j qubits. Qubits 1–3 are prepared in the state

$$|A\rangle = (1/2)\left[|000\rangle + |010\rangle + |100\rangle + |111\rangle\right],$$

   and qubits 4–6 in an arbitrary three-qubit state $|\psi\rangle$.

2. Level-j CNOT gates are applied from qubits $1 \to 4$; $2 \to 5$; and $6 \to 3$.

3. The encoded operators $Z_4^{(j)}$, $Z_5^{(j)}$, and $X_6^{(j)} = H_6^{(j)} Z_6^{(j)}$ are measured yielding outcomes m_4, m_5, and m_6, respectively, and $H_6^{(j)}$ is the level-j Hadamard gate applied to qubit 6.

4. The operator $G_1^{(j)} = X_1^{(j)} U_{CNOT}^{23}$ is applied to qubits 1–3 when $m_4 = -1$. Similarly, $G_2^{(j)} = U_{CNOT}^{13} X_2^{(j)}$ ($G_3^{(j)} = U_{phase}^{12} Z_3^{(j)}$) is applied when m_5 (m_6) is -1. No operation is applied when a measurement outcome is $+1$.

In an effort to keep the notation from getting too cluttered, the level-j superscript has been suppressed on $U_{CNOT}^{\mu\nu}$ and $U_{phase}^{\mu\nu}$. Notice that in Figure 6.3 the final three CNOT gates, three measurements, and measurement-conditioned operations on qubits 1–3 implement steps 2–4 in the above Toffoli gate procedure. We now show that the remainder of the circuit in Figure 6.3 places qubits 1–3 in the state $|A\rangle$ (to $\mathcal{O}(p)$) so that the complete circuit implements a level-j Toffoli gate.

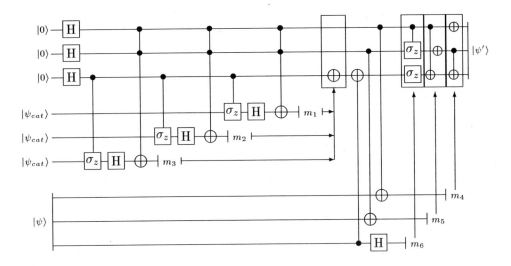

FIGURE 6.3
Quantum circuit to implement a Toffoli gate U_T on level-j qubits: $|\psi'\rangle = U_T|\psi\rangle$. The top (bottom) three lines correspond to qubits 1–3 (4–6). Three ancilla blocks (denoted A_1–A_3 from top to bottom) are prepared in cat-states $|\psi_{cat}\rangle$, which are used to insure that (to $\mathcal{O}(p)$) qubits 1–3 are in the state $|A\rangle$ after the boxed NOT gate. Majority voting on the measurement outcomes $\{m_1, m_2, m_3\}$ determines whether the boxed NOT gate is applied. See text for further discussion. Based on Figure 12, J. Preskill, *Proc. R. Soc. Lond. A* **454**, 385 (1998). ©Royal Society, used with permission.

To prepare the state $|A\rangle$ (pp. 177–178), we start with three level-j qubits labeled 1–3, and an ancilla block of seven level-$(j-1)$ qubits that have been prepared in the cat-state $|\psi_{cat}\rangle$. Qubits 1–3 each begin in the $|0\rangle$ CB-state. Hadamard gates are then applied to qubits 1–3. Next the operator $G_3 = U_{phase}^{12} Z_3^{(j)}$ is applied transversally to qubits 1–3 conditioned on the state of the qubits in the ancilla block. Next the operator

$$\mathcal{O}^{(j-1)} = \prod_{i=1}^{7} X_i^{(j-1)}$$
$$= \prod_{i=1}^{7} H_i^{(j-1)} Z_i^{(j-1)}$$

is measured on the ancilla block. In the absence of errors, the measurement outcome $m = +1$ indicates that qubits 1–3 are in the state $|A\rangle$, as desired. On the other hand, the outcome $m = -1$ tells us that the three-qubit state is $|B\rangle = X_3^{(j)}|A\rangle$. In this case, applying a NOT gate to qubit 3 transforms $|B\rangle \to$

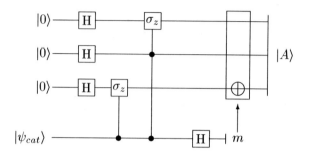

FIGURE 6.4
Quantum circuit to place qubits 1–3 (top three lines) in the state $|A\rangle$. The ancilla block contains seven level-$(j-1)$ qubits prepared in the cat-state $|\psi_{cat}\rangle$. The seven ancilla qubit lines have been collapsed into a single line to make the circuit easier to examine. The operator $\mathcal{O}^{(j-1)} = X_1^{(j-1)} \cdots X_7^{(j-1)}$ is measured on the ancilla block. The measurement outcome m determines whether a NOT gate is applied to qubit 3. See text for a discussion of how this circuit must be modified when phase errors are possible during cat-state preparation.

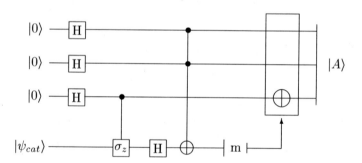

FIGURE 6.5
Alternative quantum circuit for preparation of $|A\rangle$.

$|A\rangle$. Figure 6.4 shows the quantum circuit that implements this construction procedure for $|A\rangle$.

The circuit appearing in Figure 6.4 is shown to be equivalent to the one that appears in Problem 6.3(c) if all operations in the latter circuit are applied transversally. For easy reference, the latter circuit has been redrawn in Figure 6.5. The circuit in Figure 6.5 proves more convenient for our calculation of the accuracy threshold and so we bid farewell to the circuit in Figure 6.4. Continuing with our discussion of $|A\rangle$-state preparation, the circuit in Figure 6.5 must be modified to protect against phase errors that might occur during cat-state preparation. Recall that the quantum circuit used for cat-state preparation tests for bit-flip errors, but does not test for phase errors. We saw (pp. 178–179) that a phase error in the cat-state will cause the measurement outcome m to misidentify the state $|A\rangle$. To insure that m is correct to $\mathcal{O}(p)$ it is sufficient to expand the circuit in Figure 6.5 to (i) include three ancilla blocks, each prepared in the cat-state; (ii) interact each ancilla block with qubits 1–3 as in Figure 6.5; and then (iii) measure $\mathcal{O}^{(j-1)}$ for each ancilla block. A majority vote on the measurement outcomes m_1, m_2, and m_3 then correctly determines (to $\mathcal{O}(p)$) whether the NOT gate should be applied to qubit 3. The modified circuit for $|A\rangle$-state preparation appears in Figure 6.3 where it precedes the three CNOT gates that begin the Toffoli gate implementation.

To determine the Toffoli gate accuracy threshold we use eq. (6.1) to obtain a recursion relation for $p_{op}^{(j)} = p_{Tof}^{(j)}$. As input for that relation, we must find the probability E that a single error appears on a level-$(j-1)$ qubit during the level-j Toffoli gate. This probability is then identified with $p_{op}^{(j-1)}$ in eq. (6.1). To that end, since qubits 1–3 in Figure 6.3 carry the data at the end of the circuit, we determine the probability E_i that an error appears during the circuit on one of the level-$(j-1)$ qubits that constitute the ith qubit ($i = 1, 2, 3$). We then use the largest of the E_i to upper bound E. To simplify the analysis, we assume the initial state $|0\rangle$ for each of these qubits is error-free.

To begin, let t_{cat} be the time needed to prepare a cat-state, and p_{cat} the probability that the cat-state has a bit-flip error at the end of its preparation. The probability that qubit i ($i = 1, 2, 3$) develops a storage error during cat-state preparation is then $t_{cat} p_{stor}^{(j-1)}$. Once the cat-states are ready, Hadamard gates are applied to qubits 1–3. The probability that an error occurs during one of these gates is $p_g^{(j-1)}$. Next, qubit 3 is acted on by a controlled-phase gate. The probability that this gate fails and produces an error on qubit 3 is $p_g^{(j-1)}$. Recall (Exercise 5.7) that a controlled-phase gate does not spread a phase error from one qubit to the other, though a bit-flip error on one qubit does produce a phase error on the other. Since p_{cat} is the probability that ancilla block A_3 in Figure 6.3 contains a bit-flip error prior to applying the controlled-phase gate, it is also the probability that qubit 3 picks up a phase error (through this gate) due to a bit-flip error on A_3. Following the controlled-phase gate, qubit 3 must wait for a Hadamard gate to be applied to A_3, and a Toffoli gate to be applied to A_3 and qubits 1 and 2. This takes a time $1 + t_{Tof}^{(j-1)}$ and so the probability that qubit 3 develops a storage error during this time is $(1 + t_{Tof}^{(j-1)}) p_{stor}^{(j-1)}$. We see that by the time A_3 is measured, qubit 3 has accumulated an error probability of

$$E_3 = \left(t_{cat} + t_{Tof}^{(j-1)} + 1\right) p_{stor}^{(j-1)} + 2 p_g^{(j-1)} + p_{cat} \ . \tag{6.23}$$

For qubits 1 and 2, after the initial Hadamard gates are applied, each must wait for two time-steps while the controlled-phase gate is applied to qubit 3 and A_3, and a Hadamard gate is applied to A_3. The probability a storage error occurs during this time is $2p_{stor}^{(j-1)}$. Next, a Toffoli gate is applied to these qubits and A_3. The probability that this gate fails and produces a level-$(j-1)$ error on either qubit 1 or 2 is $p_{Tof}^{(j-1)}$. It is also possible for an error in the ancilla A_3 to profilerate/spread onto qubits 1 and 2 through the Toffoli gate. The probability that A_3 has such an error can be upper bounded by the probability that A_3 has an error prior to applying the Toffoli gate. From Figure 6.3 we see that this probability is $p_{cat} + 2p_g^{(j-1)}$. Adding up all these error probabilities gives E_1 and E_2 up to the time A_3 is measured:

$$E_1 = E_2 = (t_{cat} + 2) p_{stor}^{(j-1)} + p_{cat} + 3p_g^{(j-1)} + p_{Tof}^{(j-1)} . \qquad (6.24)$$

As noted earlier, to insure $|A\rangle$-state preparation is correct to $\mathcal{O}(p)$, qubits 1–3 must repeat the above sequence of operations with two more ancilla blocks (A_2, A_1). It is assumed the three cat-states in Figure 6.3 can be prepared in parallel so that the final two cat-states are ready when needed. It is also assumed that measurement of ancilla A_{i+1} can be done while ancilla A_i is interacting with qubits 1–3. Then the qubit trio only waits for $t_{meas}^{(j-1)}$ while the final ancilla A_1 is measured. Finally, when estimating error probabilities, it will be assumed that the (boxed) NOT gate is always applied to qubit 3. Since the operations involving qubits 1–3 and the ancilla A_2 and A_1 parallel those with A_3, we will be brief when describing the sources of error for these qubits. The error probability for each error source will appear in square brackets following the description of the source. For qubit 3, errors can arise:

1. during application of the two controlled-phase gates and the (boxed) NOT gate $[3p_g^{(j-1)}]$;

2. by spreading from the ancilla during the two controlled-phase gates $[2p_{cat}]$;

3. while waiting for the two Hadamard gates to be applied to the ancilla, the two Toffoli gates, and the measurement of A_1 $[(t_{meas}^{(j-1)} + 2t_{Tof}^{(j-1)} + 2) p_{stor}^{(j-1)}]$.

Combining these error probabilities with eq. (6.23) gives the probability qubit 3 has an error at the end of $|A\rangle$-state preparation:

$$E_3 = \left(t_{cat} + t_{meas}^{(j-1)} + 3t_{Tof}^{(j-1)} + 3\right) p_{stor}^{(j-1)} + 3p_{cat} + 5p_g^{(j-1)} . \qquad (6.25)$$

For qubits 1 and 2, the error sources are:

1. the two Toffoli gates $[2p_{Tof}^{(j-1)}]$;

Accuracy Threshold Theorem

2. the spreading of errors from the ancilla during the two Toffoli gates $[2(p_{cat} + 2p_g^{(j-1)})]$;

3. the time spent waiting for the two Hadamard gates on the ancilla, the two controlled-phase gates, the measurement of A_1, and the application of the (boxed) NOT gate $[(5 + t_{meas}^{(j-1)})p_{stor}^{(j-1)}]$.

Combining these error probabilities with eq. (6.24) gives

$$E_1 = E_2 = \left(t_{cat} + t_{meas}^{(j-1)} + 7\right) p_{stor}^{(j-1)} + 3p_{cat} + 7p_g^{(j-1)} + 3p_{Tof}^{(j-1)} . \quad (6.26)$$

At this point, qubits 1–3 have been placed in the state $|A\rangle$ and they are ready to be interacted with qubits 4–6 via the three CNOT gates. The latter three qubits have been waiting for a time $(t_{cat} + t_{meas}^{(j-1)} + 3t_{Tof}^{(j-1)} + 7)$ while the state $|A\rangle$ was being prepared†. The probability that a storage error appears on any one of these qubits is then $(t_{cat} + t_{meas}^{(j-1)} + 3t_{Tof}^{(j-1)} + 7) p_{stor}^{(j-1)}$. This probability then gives an upper bound for the probability that an error spreads from qubit $(3+i)$ to qubit i via the ith CNOT gate ($i = 1, 2, 3$). We identify this upper bound with the probability that such an error occurs. The ith CNOT gate can itself introduce an error on the ith qubit. This occurs with probability $p_g^{(j-1)}$. Finally, qubits 1–3 must wait for a time $1 + t_{meas}^{(j-1)}$ while a Hadamard gate is applied to qubit 6 and qubits 4–6 are measured (in parallel). Combining these error probabilities with eqs. (6.25) and (6.26) gives:

$$E_1 = \left(2t_{cat} + 3t_{meas}^{(j-1)} + 3t_{Tof}^{(j-1)} + 15\right) p_{stor}^{(j-1)} + 3p_{cat} + 8p_g^{(j-1)} + 3p_{Tof}^{(j-1)}$$
$$E_2 = E_1$$
$$E_3 = \left(2t_{cat} + 3t_{meas}^{(j-1)} + 6t_{Tof}^{(j-1)} + 11\right) p_{stor}^{(j-1)} + 3p_{cat} + 6p_g^{(j-1)} . \quad (6.27)$$

The measurement outcomes m_4, m_5, and m_6 in Figure 6.3 determine whether the final boxed operations are applied to qubits 1–3. For purposes of estimating the final error probabilities E_i ($i = 1, 2, 3$), we make the conservative assumption that these boxed operations are always applied. The sources of error during these operations are (i) failure of a Clifford group gate, and (ii) the spreading of errors from one qubit to another via a 2-qubit gate. For the boxed operation associated with m_6, the probability an error spreads from qubit 1 (2) to qubit 2 (1) is given by the probability E_1 (E_2) in eq. (6.27). The probability that either qubit 1 or 2 has an error after the first boxed operation is thus the sum of these expressions for E_1 and E_2, plus the probability

---

†It is assumed that qubits 1–3 and the ancilla A_1–A_3 are prepared in their initial states in parallel, and that $t_{cat} \geq t_{prep}^{(j)} + 1$. Then qubits 4–6 do not have to wait while qubits 1–3 are prepared in their initial states and have the initial Hadamard gates applied to them. This assumption is not essential, nor is it entirely consistent with what was assumed in deriving eqs. (6.23) and (6.24). Its sole purpose is to slightly simplify the following formulas.

$p_g^{(j-1)}$ that the controlled-phase gate fails. For qubit 3 we only need to add the probability $p_g^{(j-1)}$ that the σ_z gate fails to E_3 in eq. (6.27). Thus,

$$E_1 = \left(4t_{cat} + 6t_{meas}^{(j-1)} + 6t_{Tof}^{(j-1)} + 30\right) p_{stor}^{(j-1)} + 6p_{cat} + 17p_g^{(j-1)} + 6p_{Tof}^{(j-1)}$$
$$E_2 = E_1$$
$$E_3 = \left(2t_{cat} + 3t_{meas}^{(j-1)} + 6t_{Tof}^{(j-1)} + 11\right) p_{stor}^{(j-1)} + 3p_{cat} + 7p_g^{(j-1)} \ . \qquad (6.28)$$

For the boxed operation associated with m_5, the error sources are (i) failure of a Clifford group gate, and (ii) the spreading of errors through the CNOT gate. The probability that an error spreads from qubit 1 (3) to qubit 3 (1) is given by E_1 (E_3) in eq. (6.28). Thus the probability that qubit 1 or 3 has an error after the second boxed operation is the sum of E_1 and E_3 in eq. (6.28), plus the probability $p_g^{(j-1)}$ that the CNOT gate fails. For qubit 2 we simply add $p_g^{(j-1)}$ to E_2 in eq. (6.28). Thus we find:

$$E_1 = \left(6t_{cat} + 9t_{meas}^{(j-1)} + 12t_{Tof}^{(j-1)} + 41\right) p_{stor}^{(j-1)} + 9p_{cat} + 25p_g^{(j-1)} + 6p_{Tof}^{(j-1)}$$
$$E_2 = \left(4t_{cat} + 6t_{meas}^{(j-1)} + 6t_{Tof}^{(j-1)} + 30\right) p_{stor}^{(j-1)} + 6p_{cat} + 18p_g^{(j-1)} + 6p_{Tof}^{(j-1)}$$
$$E_3 = E_1 \ . \qquad (6.29)$$

Analysis of the final boxed operation requires some care to avoid double-counting the probability that an error spreads to qubit 2 or 3 from qubit 1. Qubit 2 picks up this probability during the controlled-phase gate in the first of these three boxed operations, while qubit 3 picks it up during the CNOT gate in the second boxed operation. Thus E_2 and E_3 in eq. (6.29) both contain the qubit 1 contribution. The probability that qubit 2 contains an error after the last boxed operation is the probability that it had one before this operation (E_2 in eq. (6.29)) plus the probability the CNOT gate applied to it fails [$p_g^{(j-1)}$], and that an error spreads to it from qubit 3 during the CNOT gate. We should not use E_3 in eq. (6.29) to determine the probability that an error spreads from qubit 3 to 2 as this would double-count the qubit 1 contribution. Instead we must use E_3 in eq. (6.27) and add to it $2p_g^{(j-1)}$ to account for possible failure of the σ_z and CNOT gates that are applied to qubit 3 during the first two boxed operations. Similar remarks apply to qubit 3. The probability that it has an error after the final boxed operation is the probability that it had one before the operation (E_3 in eq. (6.29)) plus the probability the CNOT gate applied to it fails [$p_g^{(j-1)}$], and that an error spreads to it from qubit 2 during the CNOT gate. E_2 in eq. (6.29) should not be used to determine the probability that an error spreads from qubit 2 to 3 as this would double-count the qubit 1 contribution. One should use instead the probability $p_g^{(j-1)}$ that the NOT gate applied to qubit 2 during the second box operation fails. Finally, for qubit 1, all that is needed is to add $p_g^{(j-1)}$ to

E_1 in eq. (6.29) to account for possible failure of the NOT gate applied to it in the final boxed operation. Putting together all these remarks gives

$$E_1 = \left(6t_{cat} + 9t_{meas}^{(j-1)} + 12t_{Tof}^{(j-1)} + 41\right) p_{stor}^{(j-1)} + 9p_{cat} + 26p_g^{(j-1)} + 6p_{Tof}^{(j-1)}$$
$$E_2 = \left(6t_{cat} + 9t_{meas}^{(j-1)} + 12t_{Tof}^{(j-1)} + 41\right) p_{stor}^{(j-1)} + 9p_{cat} + 27p_g^{(j-1)} + 6p_{Tof}^{(j-1)}$$
$$E_3 = E_2 . \qquad (6.30)$$

We upper bound the probability E that a level-$(j-1)$ qubit (belonging to one of the data qubits) develops an error during a level-j Toffoli gate with the largest of (E_1, E_2, E_3). We see that E is not equal to $p_{Tof}^{(j-1)}$, but is instead a linear inhomogeneous function of this probability. It is E that must be plugged in for $p_{op}^{(j-1)}$ in eq. (6.1) (see footnote on page 200). Doing so gives

$$p_{Tof}^{(j)} =$$
$$21 \left\{ \left(\left[6t_{cat} + 9t_{meas}^{(j-1)} + 12t_{Tof}^{(j-1)} + 41\right] p_{stor}^{(j-1)} + 9p_{cat} + 27p_g^{(j-1)} + 6p_{Tof}^{(j-1)} \right)^2 \right.$$
$$+ 4 \left(\left[6t_{cat} + 9t_{meas}^{(j-1)} + 12t_{Tof}^{(j-1)} + 41\right] p_{stor}^{(j-1)} + 9p_{cat} + 27p_g^{(j-1)} + 6p_{Tof}^{(j-1)} \right) p_{EC}^{(j)}$$
$$\left. + 4 \left(p_{EC}^{(j)}\right)^2 \right\}. \qquad (6.31)$$

We now work out the Toffoli gate accuracy threshold for the case where storage errors are negligible: $p_{stor}^{(0)} = 0$. As we saw in Section 6.2.1 and 6.2.2, this implies

$$p_{stor}^{(j)} = 0; \qquad (6.32)$$
$$p_{EC}^{(j)} = 12p_g^{(j-1)}; \qquad (6.33)$$
$$p_g^{(j)} = 13125 \left(p_g^{(j-1)}\right)^2. \qquad (6.34)$$

Eq. (6.34) reduces to $(j \to j-1)$:

$$p_g^{(j-1)} = p_{th} \left(\frac{p_g^{(0)}}{p_{th}}\right)^{2^{(j-1)}}, \qquad (6.35)$$

where $p_{th} = 1/13125 = 7.62 \times 10^{-5}$ is the accuracy threshold for gates in the Clifford group. Since storage errors are assumed to be negligible, it makes sense to verify the cat-state until we are sure its error probability is arbitrarily small (viz. $p_{cat} = 0$). Then eq. (6.31) becomes

$$p_{Tof}^{(j)} = 21 \left[\begin{array}{c} \left(27p_g^{(j-1)} + 6p_{Tof}^{(j-1)}\right)^2 \\ + 4\left(27p_g^{(j-1)} + 6p_{Tof}^{(j-1)}\right) \cdot \left(12p_g^{(j-1)}\right) \\ + 4\left(12p_g^{(j-1)}\right)^2 \end{array} \right]$$
$$= 54621 \left[p_g^{(j-1)}\right]^2 + 12852 \left[p_g^{(j-1)} p_{Tof}^{(j-1)}\right] + 756 \left[p_{Tof}^{(j-1)}\right]^2. \quad (6.36)$$

Now define $\epsilon = p_g^{(0)}/p_{th}$. Combining this with eqs. (6.35) and (6.36) gives

$$p_{Tof}^{(j)} = (3.17 \times 10^{-4})\,\epsilon^{2^j} + \left[(0.98)\epsilon^{2^{(j-1)}} + 756 p_{Tof}^{(j-1)}\right] p_{Tof}^{(j-1)} \,. \qquad (6.37)$$

(i) Consider the limit where $\epsilon \ll 1$. In this case, eq. (6.37) becomes $p_{Tof}^{(j)} = 756 \left(p_{Tof}^{(j-1)}\right)^2$. It follows that (in this limit) the Toffoli gate accuracy threshold is $p_a^{Tof} = 1/756 = 1.3 \times 10^{-3}$.

(ii) Suppose the level-0 Toffoli gate and Clifford group gates have comparable accuracy $p_{Tof}^{(0)} = p_g^{(0)}$, and that $\epsilon = p_g^{(0)}/p_{th} \lesssim 1$. Then $p_{Tof}^{(0)} = \epsilon p_{th}$, and from eq. (6.37) we have

$$p_{Tof}^{(1)} = (3.17 \times 10^{-4})\epsilon^2 + \left[0.98\epsilon + 756(7.62 \times 10^{-5})\epsilon\right](7.62 \times 10^{-5})\epsilon$$
$$= (3.96 \times 10^{-4})\epsilon^2 \,; \qquad (6.38)$$

$$p_{Tof}^{(2)} = (3.17 \times 10^{-4})\epsilon^4 + \left[0.98\epsilon^2 + 756(3.96 \times 10^{-4})\epsilon^2\right](3.96 \times 10^{-4})\epsilon^2$$
$$= (8.24 \times 10^{-4})\epsilon^4 \,; \qquad (6.39)$$

$$p_{Tof}^{(3)} = (3.17 \times 10^{-4})\epsilon^8 + \left[0.98\epsilon^4 + 756(8.24 \times 10^{-4})\epsilon^4\right](8.24 \times 10^{-4})\epsilon^4$$
$$= (1.64 \times 10^{-3})\epsilon^8 \,. \qquad (6.40)$$

Suppose that $p_{Tof}^{(3)} = p_{Tof}^{(2)}$. Then further concatenation will begin to reduce the effective error probability for the Toffoli gate. Equating eqs. (6.39) and (6.40) gives $\epsilon^4 = 8.24/16.4 = 0.502$ and so $\epsilon = 0.84$. In this case, the Toffoli gate accuracy threshold is:

$$P_a^{Tof} = \epsilon p_{th} = 6.4 \times 10^{-5} \,. \qquad (6.41)$$

Since the Toffoli gate threshold is smaller than the threshold p_{th} for gates in the Clifford group, it determines the accuracy threshold for reliable quantum computation for our set of assumptions. This section has thus shown that if (i) computational data is protected using a sufficiently layered concatenated QECC; (ii) fault-tolerant procedures are used for quantum computation, error correction, and measurement; and (iii) storage registers and a universal set of quantum gates are available whose primitive error probabilities are less than $P_a = 6.4 \times 10^{-5}$, then the effective error probabilities for these registers and quantum gates can be made arbitrarily small. This in turn allows an arbitrarily long quantum computation to be done with arbitrarily small error probability, and so establishes the accuracy threshold theorem for the set of assumptions made in this chapter. In principle, adaptation of the above arguments will allow the accuracy threshold theorem to be established for other sets of assumptions.

Problems

6.1 Use Theorem 3.4 and mathematical induction to show that a concatenated QECC based on the [7,1,3] CSS code and containing l layers of concatenation is a $[7^l, 1, d]$ QECC with $d \geq 3^l$.

6.2 Given the generators for the [7,1,3] CSS code (eq. (2.50)), construct the associated matrix $\mathcal{H}$ (eq. (4.4)).

(a) Apply the following operations to $\mathcal{H}$: (i) interchange qubits 1 and 4; (ii) add row 3 to row 2; (iii) interchange row 4→row 5→row 6→row 4; (iv) add row 5 to row 4; and (v) finally, add row 6 to row 5. Show that these operations give

$$\mathcal{H} = \begin{pmatrix} 1 & 0 & 0 & 0 & 1 & 1 & 1 & 0 & 0 & 0 & 0 & 0 & 0 & 0 \\ 0 & 1 & 0 & 1 & 1 & 1 & 0 & 0 & 0 & 0 & 0 & 0 & 0 & 0 \\ 0 & 0 & 1 & 1 & 1 & 0 & 1 & 0 & 0 & 0 & 0 & 0 & 0 & 0 \\ 0 & 0 & 0 & 0 & 0 & 0 & 0 & 1 & 1 & 0 & 1 & 0 & 0 & 1 \\ 0 & 0 & 0 & 0 & 0 & 0 & 0 & 1 & 1 & 1 & 0 & 1 & 0 & 0 \\ 0 & 0 & 0 & 0 & 0 & 0 & 0 & 0 & 1 & 1 & 0 & 0 & 1 & 1 \end{pmatrix}.$$

We see that $\mathcal{H}$ is now in standard form.

(b) Using the approach of Ref. [11], show that the quantum circuit in Figure 6.2 is the error correction circuit for the [7,1,3] CSS code (in standard form).

6.3 The reader may wish to review Section 5.7 before trying this problem.

(a) Show that the action of the level-0 two-qubit controlled-phase gate (Example 5.7) is independent of which of the two qubits acts as control (viz. prove the following gate identity).

Since transversal application of this gate gives the encoded operation, the independence of gate action on the choice of control qubit is also true of the encoded controlled-phase gate.

(b) Prove the following level-0 gate identity:

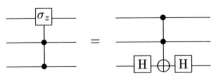

Note that this gate identity will be used in Section 6.2.3 for three level-j code blocks. It is understood that all operations are applied transversally, first from level-0 to level-1, then from level-1 to level-2, and so on up to level-j.

(c) Show that in the absence of phase errors in the cat-state $|\psi_{cat}\rangle$, the following quantum circuit produces the state $|A\rangle = (1/2)[|000\rangle + |010\rangle + |100\rangle + |111\rangle]$. Note that the qubits are level-0 qubits, measurement is in the computational basis, and the final NOT gate is only applied when the measurement outcome is $m = -1$.

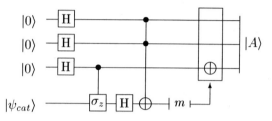

As in part (b), this circuit will be used in Section 6.2.3 for level-j code blocks. The operations are understood to be applied transversally, starting at level-0, and then (as in described in part (b)) going recursively up to level-j.

(d) When phase errors can occur in the cat-state, the measurement result m in part (c) must be verified. If the cat-state appearing in part (c) is replaced with three cat-states, then majority voting (Section 6.2.3) can be used to determine the most probable measurement result. Each cat-state interacts with the three qubits as in part (c) and is then measured. The final NOT gate is only applied to the third qubit when the majority of the measurement outcomes is -1. This insures the measurement result is correct to $\mathcal{O}(p)$. As in part (c), transversal application of all operations allows this procedure to be extended to level-j code blocks. With this modification to the preparation of $|A\rangle$, show that the quantum circuit in Figure 6.3 applies the Toffoli gate to the state $|\psi\rangle$.

References

[1] Aharonov, D. and Ben-Or, M., Fault-tolerant quantum computation with constant error, in *Proceedings of the Twenty-Ninth ACM Symposium on the Theory of Computing*, 1997, pp. 176–188.

[2] Knill, E. and Laflamme, R., Concatenated quantum codes, download at http://arXiv.org/abs/quant-ph/9608012, 1996.

[3] Knill, E., Laflamme, R., and Zurek, W. H., Resilient quantum computation, *Science* **279**, 342, 1998.

[4] Knill, E., Laflamme, R., and Zurek, W. H., Resilient quantum computation: error models and thresholds, *Proc. R. Soc. Lond. A* **454**, 365, 1998.

[5] Zalka, C., Threshold estimate for fault-tolerant quantum computation, download at http://arXiv.org/abs/quant-ph/9612028, 1996.

[6] Gottesman, D., Stabilizer codes and quantum error correction, Ph. D. thesis, California Institute of Technology, Pasadena, CA, 1997.

[7] Preskill, J., Reliable quantum computers, *Proc. R. Soc. Lond. A* **454**, 385, 1998.

[8] Kitaev, A. Y., Quantum computation: algorithms and error correction, *Russ. Math. Surv.* **52**, 1191, 1997.

[9] Kitaev, A. Y., Quantum error correction with imperfect gates, in *Quantum Communication, Computing, and Measurement*, Plenum Press, New York, 1997, pp. 181–188.

[10] Feynman, R. P. and Hibbs, A. R., *Quantum Mechanics and Path Integrals*, McGraw-Hill, New York, 1965.

[11] Plenio, M. B., Vedral, V., and Knight, P. L., Conditional generation of error syndromes in fault-tolerant error correction, *Phys. Rev. A* **55**, 4593, 1997.

7

Bounds on Quantum Error Correcting Codes

In this final chapter we present a number of bounds on the minimum distance of a quantum error correcting code (QECC). Sections 7.1–7.3 derive the quantum Hamming, Gilbert-Varshamov, and Singleton bounds, respectively. The corresponding classical bounds were derived in Section 1.2.5. Section 7.4 introduces quantum generalizations of the classical MacWilliams identities, and of weight and shadow enumerators [1, 2]. With these tools in hand, the problem of finding an optimal upper bound on the code distance is reduced to solving a problem in linear programming [3–5]. Finally, we close in Section 7.5 by showing how QECCs can be connected to the process of entanglement purification [6, 7]. This final section allows our study of QECCs to segue into the developing field of quantum information theory [8, 9], and provides a good place to bring the discussion in this book to a close.

7.1 Quantum Hamming Bound

Let C_q be an $[n, k, d]$ QECC that corrects the set of errors $\{E_a\}$ and has basis codewords $|\bar{i}\rangle$. It is assumed that C_q corrects all errors E_a in the Pauli group $\mathcal{G}_n$ with weight less than or equal to t: $\text{wt}(E_a) \leq t$. Thus the code distance $d = 2t + 1$. One preliminary result is needed before deriving the quantum Hamming bound. Recall (see p. 74) that for a degenerate QECC, the Hermitian matrix C_{ab} is singular and has at least one vanishing eigenvalue c_l. The operator

$$F_l = \sum_a U_{la} E_a \qquad (7.1)$$

associated with c_l was shown to annihilate all basis codewords $|\bar{i}\rangle$:

$$F_l |\bar{i}\rangle = 0 \ . \qquad (7.2)$$

Combining eqs. (7.1) and (7.2) gives

$$\sum_a U_{la} \left[E_a |\bar{i}\rangle \right] = 0 \qquad (7.3)$$

for all basis codewords $|\bar{i}\rangle$. Eq. (7.3) indicates that, for a degenerate QECC, the set of states $\{E_a |\bar{i}\rangle\}$ form a linearly dependent set. On the other hand, for a non-degenerate QECC, C_{ab} is non-singular and so all eigenvalues are non-vanishing: $c_l \neq 0$. It follows from eq. (2.45) that for non-degenerate codes, the set of states $\{F_l|\bar{i}\rangle\}$ are orthogonal and F_l does not annihilate basis codewords. In this case, the set of states $\mathcal{E} = \{E_a|i\rangle\}$ form a linearly independent set. The following derivation of the quantum Hamming bound requires that $\mathcal{E}$ form a linearly independent set of states and so we restrict ourselves to non-degenerate QECCs in the remainder of this section.

Let $E_a \in \mathcal{G}_n$ be an error of weight j: $\text{wt}(E_a) = j$. There are a total of

$$3^j \binom{n}{j}$$

such errors. The binomial coefficient counts the different ways that j errors can be assigned to n qubits, and each error can be either σ_x, σ_y, or σ_z. Since there are 2^k basis codewords $|\bar{i}\rangle$, there are

$$3^j \binom{n}{j} 2^k$$

possible states $E_a|\bar{i}\rangle$. Since we are restricting ourselves to non-degenerate QECCs, this set of states is linearly independent. Since the code is assumed to correct all errors in $\mathcal{G}_n$ of weight less than or equal to t, it follows that

$$\sum_{j=0}^{t} 3^j \binom{n}{j} 2^k \leq 2^n \tag{7.4}$$

since the total number of linearly independent states $\{E_a|\bar{i}\rangle\}$ cannot exceed the dimension 2^n of the n-qubit Hilbert space H_2^n. Eq. (7.4) is the quantum Hamming bound [10]. It places an upper bound on the number of errors t that a non-degenerate QECC can correct for given n and k. Since the code distance $d = 2t + 1$, it also places an upper bound on the code distance d. As noted above, the derivation just presented only applies to non-degenerate QECCs. At present it is unknown whether a degenerate QECC might allow a violation of the quantum Hamming bound. The linear programming bounds introduced in Section 7.4 have been used to show that no degenerate QECC with $n \leq 30$ violates this bound [11].

It proves useful to obtain the large n limit of the quantum Hamming bound. In this limit we must have t/n finite and non-zero. We thus write $t = \mu n$, where $\mu < 1$ and $n \gg 1$. We also introduce $P = 1/4$ and $Q = 3/4$. It is then straightforward to show that

$$\sum_{j=0}^{t} 3^j \binom{n}{j} = 4^n \sum_{j=0}^{\mu n} \binom{n}{j} P^{n-j} Q^j$$

$$= 4^n \sum_{j=\lambda n}^{n} \binom{n}{j} P^j Q^{n-j}, \tag{7.5}$$

where $\lambda = 1 - \mu$. The following two inequalities are, respectively, Eqs. (A3) and (A5) from Appendix A of Ref. [12]:

$$\frac{1}{2\sqrt{2n\lambda\mu}} \lambda^{-\lambda n} \mu^{-\mu n} \leq \binom{n}{\lambda n} \tag{7.6}$$

$$\binom{n}{\lambda n} P^{\lambda n} Q^{\mu n} < \sum_{j=\lambda n}^{n} \binom{n}{j} P^j Q^{n-j} . \tag{7.7}$$

Combining eqs. (7.5) and (7.7) gives

$$\binom{n}{\lambda n} 3^{\mu n} < \sum_{j=0}^{t} 3^j \binom{n}{j} . \tag{7.8}$$

Eqs. (7.4), (7.6), and (7.8) then give ($n \gg 1$)

$$2^{n-k} \geq \frac{\lambda^{-\lambda n} \mu^{-\mu n} 3^{\mu n}}{2\sqrt{2n\lambda\mu}} . \tag{7.9}$$

Taking the base-2 logarithm of both sides of eq. (7.9) gives

$$n - k \geq -n\left(\lambda \log \lambda + \mu \log \mu\right) + \mu n \log 3 - \log\left(2\sqrt{2n\lambda\mu}\right). \tag{7.10}$$

Introducing the binary entropy function $H(\mu) = -\mu \log \mu - (1-\mu) \log(1-\mu)$ and letting $n \to \infty$ allows eq. (7.10) to be reduced to

$$\frac{k}{n} \leq 1 - \frac{t}{n} \log 3 - H(t/n) , \tag{7.11}$$

where we used that $\mu = t/n$. This is the asymptotic form of the quantum Hamming bound [10].

7.2 Quantum Gilbert-Varshamov Bound

Consider an [n,k,d] QECC $\mathcal{C}_q$ that corrects t errors and encodes the maximum number of qubits k for given n and $d = 2t + 1$. Since $\mathcal{C}_q$ has distance d, eq. (2.46) must be satisfied by all basis codewords $|\bar{i}\rangle$ and all errors E in the Pauli group $\mathcal{G}_n$ with weight $\text{wt}(E) < d$. There are a total of

$$N = \sum_{j=0}^{d-1} 3^j \binom{n}{j} \tag{7.12}$$

such errors. Now consider a state $|\psi_1\rangle$ that satisfies all N instances of eq. (2.46) and so can be used as a basis codeword for C_q. Since C_q has 2^k basis codewords, such a state exists if $2^k \geq 1$. Let S_1 be the set of states in the n-qubit Hilbert space H_2^n,

$$S_1 = \{E|\psi_1\rangle : E \in \mathcal{G}_n \text{ and wt}(E) < d\} \;,$$

and introduce the dual subspace $S_1^\perp \subset H_2^n$ that is orthogonal to S_1. The dimension of $S_1^\perp$ is $2^n - N$. Now choose an element $|\psi_2\rangle \in S_1^\perp$ that satisfies eq. (2.46). Such a state will exist if $2^k \geq 2$. Let

$$S_2 = \{E|\psi\rangle : (|\psi\rangle = |\psi_1\rangle \text{ or } |\psi_2\rangle) \text{ and } (E \in \mathcal{G}_n \text{ and wt}(E) < d)\} \quad (7.13)$$

and define the dual subspace $S_2^\perp \subset H_2^n$ orthogonal to S_2. The dimension of $S_2^\perp$ is $2^n - 2N$. Iterating this procedure i times, where $i < 2^k$, leads to the dual space $S_i^\perp \subset H_2^n$ orthogonal to the set of states

$$S_i = \{E|\psi\rangle : (|\psi\rangle = |\psi_1\rangle, \ldots, |\psi_i\rangle) \text{ and } (E \in \mathcal{G}_n \text{ and wt}(E) < d)\}. \quad (7.14)$$

The subspace $S_i^\perp$ has dimension $2^n - iN$, which is positive since $i < 2^k$. Thus $2^n - iN > 0$, which together with eq. (7.12), gives

$$\sum_{j=0}^{d-1} 3^j \binom{n}{j} i < 2^n \;. \quad (7.15)$$

Since C_q encodes the maximum number of qubits k, $i = 2^k$ will be the first value of i at which eq. (7.15) is violated. Otherwise we could iterate the procedure another time to get $2^k + 1$ basis codewords, which contradicts the assumption that k is maximal. Thus for $i = 2^k$ we must have

$$\sum_{j=0}^{d-1} 3^j \binom{n}{j} 2^k \geq 2^n \;. \quad (7.16)$$

Eq. (7.16) is the quantum Gilbert-Varshamov bound [10]. It provides a lower bound on the code distance d. Since our starting point eq. (2.46) is simply a rewriting of eq. (2.44), which is valid for all QECCs, eq. (7.16) applies to both degenerate and non-degenerate codes.

The asymptotic limit of the quantum Gilbert-Varshamov bound can be found by following the procedure used for the quantum Hamming bound. We will sketch out the derivation and leave the details to Problem 7.2. For $n \gg 1$, we must have $2t = d - 1 = \mu n$, where $\mu < 1$. It is possible to show that

$$\sum_{j=0}^{2t} 3^j \binom{n}{j} = 4^n \sum_{j=\lambda n}^{n} \binom{n}{j} P^j Q^{n-j} \;, \quad (7.17)$$

where $\lambda = 1 - \mu$; $P = 1/4$; and $Q = 3/4$. Eqs. (7.17), (7.6), and (7.7) allow eq. (7.16) to be rewritten as

$$2^{n-k} \le \frac{\lambda^{-\lambda n}\mu^{-\mu n}3^{\mu n}}{2\sqrt{2n\lambda\mu}} . \tag{7.18}$$

Taking the base-2 logarithm of eq. (7.18) followed by simple algebra gives

$$1 - \frac{k}{n} \le \mu \log 3 - H(\mu) ,$$

or

$$\frac{k}{n} \ge 1 - \frac{2t}{n} \log 3 - H(2t/n) , \tag{7.19}$$

where $H(\mu)$ is the binary entropy and we recall that $\mu = 2t/n$. Eq. (7.19) is the desired asymptotic form of the quantum Gilbert-Varshamov bound [10]. Combining it with eq. (7.11) gives

$$1 - \frac{2t}{n} \log 3 - H(2t/n) \le \frac{k}{n} \le 1 - \frac{t}{n} \log 3 - H(t/n) . \tag{7.20}$$

Eq. (7.20) provides asymptotic upper and lower bounds on the code rate k/n of an [n,k,d] QECC that corrects t errors. As noted previously, the upper bound only applies for non-degenerate QECCs.

7.3 Quantum Singleton Bound

The quantum analog of the Singleton bound is due to Knill and Laflamme [13]. The following derivation is due to Preskill [14].

Let Q be an n-qubit register whose computational data is encoded using an [n,k,d] QECC $\mathcal{C}_q$ that corrects t errors ($d = 2t + 1$). Let $|\bar{i}\rangle$ denote the basis codewords for $\mathcal{C}_q$. Now introduce an auxiliary quantum system A whose Hilbert space $\mathcal{H}_A$ is also 2^k-dimensional. In a slight abuse of notation, let $|\bar{i}\rangle$ also denote a basis set for $\mathcal{H}_A$. Suppose the composite system AQ is placed in the pure state

$$|AQ\rangle = \frac{1}{\sqrt{2^k}} \sum_{\bar{i}} |\bar{i}\rangle|\bar{i}\rangle . \tag{7.21}$$

It proves useful to partition Q into three subsystems (Q_1, Q_2, Q_3), such that Q_1 and Q_2 each contain $d-1$ qubits, and Q_3 contains the remaining $n-2(d-1)$ qubits. Since $\mathcal{C}_q$ has distance d, any error $E \in \mathcal{G}_n$ with weight $\text{wt}(E) < d$ is a correctable error (see page 74). In particular, if E only affects the qubits in Q_1, it must have $\text{wt}(E) \le d - 1$ and so is a correctable error. The same is true for Q_2. Since errors restricted to Q_1 or Q_2 are correctable, it follows that each of these subsystems must be uncorrelated with A. This is proved in Ref. [15], though it would take us too far afield to introduce the set of ideas that underlie the proof. The interested reader is referred to Ref. [15] for futher details. To proceed further we will need a few properties of the von-Neumann entropy $S(\rho)$, which the reader is asked to prove in the following exercise.

Exercise 7.1 *Let the density operator ρ represent the state of a quantum system. The von-Neumann entropy $S(\rho)$ associated with this state (Section 1.3.5) is $S(\rho) = -\operatorname{tr} \rho \log \rho$.*

(a) Suppose two quantum systems E and F are uncorrelated so that their composite state is given by the density operator $\rho_{EF} = \rho_E \otimes \rho_F$. Show that

$$S(\rho_{EF}) = S(\rho_E) + S(\rho_F) \ .$$

(b) For arbitrary quantum systems E and F, show that

$$S(\rho_{EF}) \leq S(\rho_E) + S(\rho_F) \ .$$

This property is referred to as the subadditivity of the von-Neumann entropy. Here $\rho_E = \operatorname{tr}_F \rho_{EF}$ is the partial trace of ρ_{EF} over F, and similarly, ρ_F is the partial trace over E.

(c) Suppose four quantum systems A, Q_1, Q_2, and Q_3 are in a pure state. Show that

$$S(\rho_{AQ_1}) = S(\rho_{Q_2 Q_3})$$
$$S(\rho_{AQ_2}) = S(\rho_{Q_1 Q_3}) \ ,$$

where $\rho_{AQ_1} = \operatorname{tr}_{Q_2 Q_3}(\rho_{AQ_1 Q_2 Q_3})$ is the partial trace of $\rho_{AQ_1 Q_2 Q_3}$ over Q_2 and Q_3, and similar definitions apply for the other partial traces.

Since A is uncorrelated with Q_1 and Q_2, we have

$$S(\rho_{AQ_1}) = S(\rho_A) + S(\rho_{Q_1}) \ , \qquad (7.22)$$
$$S(\rho_{AQ_2}) = S(\rho_A) + S(\rho_{Q_2}) \ . \qquad (7.23)$$

Since the composite system AQ is in a pure state, it follows that

$$S(\rho_{AQ_1}) = S(\rho_{Q_2 Q_3}) \qquad (7.24)$$
$$S(\rho_{AQ_2}) = S(\rho_{Q_1 Q_3}) \ . \qquad (7.25)$$

Combining eqs. (7.22) and (7.24), and eqs. (7.23) and (7.25), gives, respectively,

$$S(\rho_A) + S(\rho_{Q_1}) = S(\rho_{Q_2 Q_3}) \qquad (7.26)$$
$$S(\rho_A) + S(\rho_{Q_2}) = S(\rho_{Q_1 Q_3}) \ . \qquad (7.27)$$

Finally, using subadditivity on the RHS of eqs. (7.26) and (7.27) gives

$$S(\rho_A) + S(\rho_{Q_1}) \leq S(\rho_{Q_2}) + S(\rho_{Q_3}) \qquad (7.28)$$
$$S(\rho_A) + S(\rho_{Q_2}) \leq S(\rho_{Q_1}) + S(\rho_{Q_3}) \ . \qquad (7.29)$$

Adding eqs. (7.28) and (7.29) gives

$$S(\rho_A) \leq S(\rho_{Q_3}) \ . \qquad (7.30)$$

Taking the partial trace of the density operator $|AQ\rangle\langle AQ|$ over Q_1, Q_2, and Q_3 gives ρ_A. Plugging ρ_A into the von-Neumann entropy and evaluating gives $S(\rho_A) = k$. Since the maximum value of $S(\rho_{Q_3})$ is the number of qubits in Q_3 (Exercise 1.6), we have that $S(\rho_{Q_3}) \leq n - 2(d-1)$. Inserting these results into eq. (7.30) gives

$$k \leq n - 2(d-1) \ ,$$

or

$$\frac{k}{n} \leq 1 - \frac{2}{n}(d-1) \ . \qquad (7.31)$$

Eq. (7.31) is the quantum Singleton bound [13]. Solved for d, it gives an upper bound on the code distance d. Note that the quantum Singleton bound applies to all QECCs since the above analysis is valid for both degenerate and non-degenerate codes. For codes that correct t errors, this bound can be rewritten as

$$\frac{k}{n} \leq 1 - \frac{4t}{n} \ . \qquad (7.32)$$

Combining eq. (7.31) with eq. (7.19) gives upper and lower bounds, respectively, on the code rate k/n:

$$1 - \frac{2t}{n}\log 3 - H(2t/n) \leq \frac{k}{n} \leq 1 - \frac{2}{n}(d-1) \ . \qquad (7.33)$$

As we have seen, the upper bound applies for all n, while the lower bound only applies for $n \gg 1$. With this restriction on the lower bound, both bounds in eq. (7.33) apply for both degenerate and non-degenerate QECCs.

7.4 Linear Programming Bounds for QECCs

This section presents a powerful approach for determining bounds on QECCs based on linear programming [11, 16–19]. This approach is inspired by earlier applications of linear programming to classical error correcting codes [20–22]. Our discussion in this section will be restricted to quantum stabilizer codes.

Ref. [17] shows how linear programming bounds can be established for general QECCs. We begin by introducing two weight enumerators for a quantum stabilizer code (Section 7.4.1); then a quantum analog of the classical MacWilliams identity is derived [23, 24] (Section 7.4.2). The shadow of the stabilizer is introduced in Section 7.4.3, and a MacWilliams-like identity is derived for the shadow enumerator. Finally, Section 7.4.4 introduces a powerful approach based on linear programming for determining bounds on the code distance d.

7.4.1 Weight Enumerators

Let $\mathcal{C}_q$ be an $[n, k, d]$ quantum stabilizer code with computational basis (CB) states $\{|j\rangle\}$, and let E_w be an element of the Pauli group $\mathcal{G}_n$ of weight w. The operator

$$P = \sum_j |j\rangle\langle j| \tag{7.34}$$

projects an arbitrary n-qubit state onto the code space spanned by the CB states $\{|j\rangle\}$. Note that j is shorthand for the bit-string $j = j_1 \cdots j_k$, where $j_i = 0, 1$.

We begin by introducing

$$A_w = \frac{1}{2^{2k}} \sum_{E_w} \mathrm{tr}\,(E_w P)\,\mathrm{tr}\,(E_w P) \tag{7.35}$$

$$= \frac{1}{2^{2k}} \sum_{E_w} \left|\sum_j \langle j|E_w|j\rangle\right|^2 \tag{7.36}$$

and

$$B_w = \frac{1}{2^k} \sum_{E_w} \mathrm{tr}\,(E_w P E_w P) \tag{7.37}$$

$$= \frac{1}{2^k} \sum_{E_w} \sum_{j,k} |\langle k|E_w|j\rangle|^2 \quad. \tag{7.38}$$

In eqs. (7.35)–(7.38) the sum over E_w is over all $E_w \in \mathcal{G}_n$ with weight w. The following theorem establishes the combinatorial significance of A_w and B_w.

THEOREM 7.1
Let $\mathcal{C}_q$ be an $[n,k,d]$ quantum stabilizer code with stabilizer $\mathcal{S}$ and centralizer $\mathcal{C}(\mathcal{S})$. The number of elements in $\mathcal{S}$ ($\mathcal{C}(\mathcal{S})$) with weight w is equal to A_w (B_w).

PROOF We begin with A_w. Three cases must be considered. Note that from eq. (7.34): $P = sP = Ps = sPs$ for all $s \in \mathcal{S}$.

(1) $\underline{E_w \in \mathcal{G}_n - \mathcal{C}(\mathcal{S})}$:

In this case, E_w anticommutes with at least one generator (say g_i) of $\mathcal{S}$. Thus

$$\begin{aligned} tr\,(E_w P) &= tr\,(E_w g_i P g_i) \\ &= tr\,(g_i E_w g_i P) \\ &= -tr\,(E_w P) \quad, \end{aligned}$$

where we used that $tr\,AB = tr\,BA$ and $g_i^2 = I$. It follows that $tr\,(E_w P) = 0$ and so from eq. (7.35), no $E_w \in \mathcal{G}_n - \mathcal{C}(\mathcal{S})$ contributes to A_w.

(2) $\underline{E_w \in \mathcal{C}(\mathcal{S}) - \mathcal{S}}$:

From Section 3.1.4 we can write E_w in terms of the encoded Pauli operators and the generators of $\mathcal{S}$:

$$E_w = i^\lambda X(a) Z(b) g(c) \quad, \tag{7.39}$$

where $a = a_1 \cdots a_k$, $b = b_1 \cdots b_k$, and $c = c_1 \cdots c_{n-k}$. Then

$$\left| \sum_j \langle j | E_w | j \rangle \right|^2 = \left| \sum_j (-1)^{j \cdot b} \langle j | X(a) | j \rangle \right|^2 \quad, \tag{7.40}$$

where $j = j_1 \cdots j_k$. Since $X(a)|j\rangle = |j \oplus a\rangle$ (where $\oplus$ signifies bit-wise addition modulo 2), it follows that $\langle j | X(a) | j \rangle = \delta_{j, j \oplus a}$. Then from eqs. (7.36), (7.39), and (7.40), it follows that no $E_w \in \mathcal{C}(\mathcal{S}) - \mathcal{S}$ with $a \neq 0$ contributes to A_w. For $E_w \in \mathcal{C}(\mathcal{S}) - \mathcal{S}$ with $a = 0$ and $b \neq 0$, eq. (7.40) becomes

$$\left| \sum_j \langle j | E_w | j \rangle \right|^2 = \left| \sum_j (-1)^{j \cdot b} \right|^2 \quad. \tag{7.41}$$

Eq. (7.41) vanishes since the contribution from $j = j_1 \cdots j_k$ is cancelled by the contribution from $j' = (j_1 \oplus 1) \cdots j_k$ for all j. Thus A_w receives no contribution from $E_w \in \mathcal{C}(\mathcal{S}) - \mathcal{S}$.

(3) $\underline{E_w \in \mathcal{S}}$:

This is the last possibility for E_w. Since E_w belongs to the stabilizer, it fixes the CB states so that $\langle j | E_w | j \rangle = 1$ for all $|j\rangle$. Thus eq. (7.36) becomes

$$\begin{aligned} A_w &= \frac{1}{2^{2k}} \sum_{E_w \in \mathcal{S}} \left| \sum_j 1 \right|^2 \\ &= \sum_{E_w \in \mathcal{S}} 1 \quad. \end{aligned} \tag{7.42}$$

Since the sum is over all $E_w \in \mathcal{S}$ of weight w, A_w is equal to the number of such elements in $\mathcal{S}$.

The proof for B_w only requires examination of two cases. Recall that $P = sP = Ps = sPs$ for all $s \in \mathcal{S}$.

(1) $E_w \in \mathcal{G}_n - \mathcal{C}(\mathcal{S})$:

Here again, E_w anticommutes with at least one generator g_i of $\mathcal{S}$ and so

$$tr\,(E_w P E_w P) = tr\,(E_w P g_i E_w P)$$
$$= -tr\,(E_w P E_w g_i P)$$
$$= -tr\,(E_w P E_w P) \ .$$

Thus $tr\,(E_w P E_w P) = 0$ in this case and so from eq. (7.37), B_w receives no contribution from $E_w \in \mathcal{G}_n - \mathcal{C}(\mathcal{S})$.

(2) $E_w \in \mathcal{C}(\mathcal{S})$:

Again we can write $E_w = i^\lambda X(a) Z(b) g(c)$ and so

$$\sum_{jk} |\langle k | E_w | j \rangle|^2 = \sum_{jk} |\delta_{k, j \oplus a}|^2 = 2^k \ .$$

Plugging this into eq. (7.38) gives

$$B_w = \frac{1}{2^k} \sum_{E_w \in \mathcal{C}(\mathcal{S})} 2^k = \sum_{E_w \in \mathcal{C}(\mathcal{S})} 1 \ , \qquad (7.43)$$

and so B_w is equal to the number of elements in $\mathcal{C}(\mathcal{S})$ of weight w. ∎

A number of results now follow that will be useful later. First, from eqs. (7.36) and (7.38), we have that $A_w \geq 0$ and $B_w \geq 0$. Next, from Theorem 7.1 it follows that the sum of all the A_w must give the number of elements in $\mathcal{S}$. Thus

$$\sum_{w=0}^{n} A_w = 2^{n-k} \ . \qquad (7.44)$$

Note also that $A_0 = B_0 = 1$ (see eqs. (7.36) and (7.38)) since the only E_w of weight 0 is the identity I. It follows from the Cauchy-Schwarz inequality and eqs. (7.36) and (7.38) that $A_w \leq B_w$ for $0 \leq w \leq n$. Finally, since $\mathcal{C}_q$ has distance d, we know from Section 3.1.2-(6) that $\mathcal{C}(\mathcal{S}) - \mathcal{S}$ has no elements of weight less than d. It follows that $A_w = B_w$ for $w = 0, \ldots, d-1$ since B_w only receives contributions from $\mathcal{S}$ for these weight values and so A_w and B_w count exactly the same operators. Note that for a non-degenerate QECC, $A_w = B_w = 0$ for $w = 0, \ldots, d-1$ since $\mathcal{S}$ has no elements of weight less than

Bounds on Quantum Error Correcting Codes

d for such codes (Theorem 3.3). These results will be needed later and so we collect them here for easy reference:

$$A_w \geq 0 \ ; \ B_w \geq 0 \qquad (7.45)$$

$$A_0 = 1 \ ; \ B_0 = 1 \qquad (7.46)$$

$$A_w = B_w \qquad (w = 0, \ldots, d-1) \qquad (7.47)$$

$$A_w \leq B_w \qquad (w = d, \ldots, n) \qquad (7.48)$$

$$\sum_{w=0}^{n} A_w = 2^{n-k} \ . \qquad (7.49)$$

We can now define the weight enumerators $A(z)$ and $B(z)$, which are polynomials with, respectively, A_w and B_w as coefficients:

$$A(z) = \sum_{w=0}^{n} A_w z^w \qquad (7.50)$$

$$B(z) = \sum_{w=0}^{n} B_w z^w \ . \qquad (7.51)$$

We show in the following subsection that $A(z)$ and $B(z)$ are related through the quantum MacWilliams identity.

7.4.2 Quantum MacWilliams Identity

Here we state and prove the quantum MacWilliams identity and derive from it a set of constraints that the $\{A_w\}$ must satisfy.

THEOREM 7.2
Let C_q be an $[n, k, d]$ quantum stabilizer code with stabilizer S and centralizer $C(S)$. The weight enumerator $B(z)$ can be determined from $A(z)$ by the following relation:

$$B(z) = \frac{1}{2^{n-k}} (1 + 3z)^n A\left(\frac{1-z}{1+3z}\right) \ . \qquad (7.52)$$

This relation is known as the quantum MacWilliams identity.

PROOF It proves useful to re-express the quantum MacWilliams identity in terms of the A_w and B_w. To that end we focus on the RHS of eq. (7.52) and use eq. (7.50) to write

$$\frac{1}{2^{n-k}} (1+3z)^n A\left(\frac{1-z}{1+3z}\right) = \frac{1}{2^{n-k}} \sum_{w'=0}^{n} A_{w'} (1+3z)^{n-w'} (1-z)^{w'} \ . \qquad (7.53)$$

The polynomial $(1+\gamma z)^{n-x}(1-z)^x$ is the well-known generating function for the Krawtchouk polynomials $P_w(x)$ [1]:

$$(1+\gamma z)^{n-x}(1-z)^x = \sum_{w=0}^{n} P_w(x) z^w, \qquad (7.54)$$

where

$$P_w(x) = \sum_{r=0}^{w} (-1)^r \gamma^{w-r} \binom{x}{r}\binom{n-x}{w-r} \qquad (w=0,\ldots,n). \qquad (7.55)$$

Plugging eq. (7.54) with $\gamma = 3$ into eq. (7.53) and switching the order of the sums gives

$$\frac{1}{2^{n-k}}(1+3z)^n A\left(\frac{1-z}{1+3z}\right) = \frac{1}{2^{n-k}} \sum_{w=0}^{n} \left[\sum_{w'=0}^{n} A_{w'} P_w(w')\right] z^w. \qquad (7.56)$$

Inserting eq. (7.56) into the RHS of eq. (7.52) and using eq. (7.51) on the LHS gives an alternative statement of the quantum MacWilliams identity:

$$B_w = \frac{1}{2^{n-k}} \sum_{w'=0}^{n} A_{w'} P_w(w') \qquad (w=0,\ldots,n). \qquad (7.57)$$

We will prove Theorem 7.2 by deriving eq. (7.57).

In Problem 7.3 it was shown that

$$\sum_{s\in\mathcal{S}} (-1)^{<s,E_w>} = \begin{cases} 2^{n-k} & \text{for } E_w \in \mathcal{C}(\mathcal{S}) \\ 0 & \text{for } E_w \in \mathcal{G}_n - \mathcal{C}(\mathcal{S}) \end{cases}.$$

From this it follows that

$$\frac{1}{2^{n-k}} \sum_{E_w \in \mathcal{G}_n} \sum_{s\in\mathcal{S}} (-1)^{<s,E_w>} = \sum_{E_w \in \mathcal{C}(\mathcal{S})} 1. \qquad (7.58)$$

Comparing the RHS of eq. (7.58) with eq. (7.43), we see that

$$B_w = \frac{1}{2^{n-k}} \sum_{E_w \in \mathcal{G}_n} \sum_{s\in\mathcal{S}} (-1)^{<s,E_w>}, \qquad (7.59)$$

where the first sum is over $E_w \in \mathcal{G}_n$ with weight w. Switching the order of the sums, and breaking up the sum over $s \in \mathcal{S}$ into a sum over all elements $s_{w'} \in \mathcal{S}$ of weight w', followed by a sum over all w', allows eq. (7.59) to be written as

$$B_w = \frac{1}{2^{n-k}} \sum_{w'=0}^{n} \sum_{s_{w'} \in \mathcal{S}} \sum_{E_w \in \mathcal{G}_n} (-1)^{<s_{w'},E_w>}. \qquad (7.60)$$

Bounds on Quantum Error Correcting Codes

By definition, since $s_{w'}$ (E_w) has weight w' (w), it acts non-trivially on w' (w) qubits. Let v be the number of qubits that are acted on non-trivially by both $s_{w'}$ and E_w. Said another way, there are v qubits that are the targets of v of the w' operators in $s_{w'}$ and v of the w operators in E_w. Now partition these v qubits into t qubits upon which $s_{w'}$ and E_w apply different Pauli operators, and $v - t$ qubits upon which they apply identical Pauli operators. Note that if t is even (odd), $s_{w'}$ and E_w will commute (anticommute). Thus

$$(-1)^{<s_{w'},E_w>} = (-1)^t \,. \tag{7.61}$$

The number of $E_w \in \mathcal{G}_n$ that agree with $s_{w'}$ on $v - t$ qubits and disagree on t qubits is

$$1^{v-t} 2^t 3^{w-v} \binom{v}{t}\binom{w'}{v}\binom{n-w'}{w-v}. \tag{7.62}$$

To see this, note that the first binomial coefficient counts the number of ways that t qubits can be chosen from the v qubits upon which both E_w and $s_{w'}$ act. The factor of 2^t counts the number of ways that Pauli operators can be assigned to E_w so that it will disagree with $s_{w'}$ on t qubits. The second binomial coefficient counts the number of ways that v qubits can be chosen from the w' qubits upon which $s_{w'}$ acts. Note that this factor is the same for all $s_{w'}$ of weight w'. The final binomial coefficient counts the number of ways that $w - v$ qubits can be chosen from the $n - w'$ qubits that are acted on by E_w, but not by $s_{w'}$. This factor is also the same for all $s_{w'}$ of weight w'. The factor 3^{w-v} appears because any of the three Pauli operators can be applied to each of these $w - v$ qubits. Finally, it proves convenient to include the trival factor 1^{v-t}. Plugging eqs. (7.61) and (7.62) into eq. (7.60) and rewriting the sum over E_w as sums over v and t gives

$$B_w = \frac{1}{2^{n-k}} \sum_{w'=0}^{n} \sum_{s_{w'} \in \mathcal{S}} \sum_{v=0}^{w} \left[\sum_{t=0}^{v} 1^{v-t}(-2)^t \binom{v}{t} \right] 3^{w-v} \binom{w'}{v}\binom{n-w'}{w-v}. \tag{7.63}$$

From the binomial theorem, the term in square brackets is simply $(1-2)^v$, and so

$$B_w = \frac{1}{2^{n-k}} \sum_{w'=0}^{n} \sum_{s_{w'} \in \mathcal{S}} \sum_{v=0}^{w} (-1)^v 3^{w-v} \binom{w'}{v}\binom{n-w'}{w-v}. \tag{7.64}$$

As noted earlier, the factors in the summand are independent of $s_{w'}$. Thus the sum over $s_{w'}$ can be moved all the way through to the right:

$$B_w = \frac{1}{2^{n-k}} \sum_{w'=0}^{n} \sum_{v=0}^{w} (-1)^v 3^{w-v} \binom{w'}{v}\binom{n-w'}{w-v} \sum_{s_{w'} \in \mathcal{S}} 1. \tag{7.65}$$

From eq. (7.42), with $E_w \to s_{w'}$, we see that the sum over $s_{w'}$ in eq. (7.65) is simply $A_{w'}$ and so

$$B_w = \frac{1}{2^{n-k}} \sum_{w'=0}^{n} \left[\sum_{v=0}^{w} (-1)^v 3^{w-v} \binom{w'}{v} \binom{n-w'}{w-v} \right] A_{w'} . \qquad (7.66)$$

From eq. (7.55), we recognize the term in square brackets as the Krawtchouk polynomial $P_w(w')$. Thus eq. (7.66) reduces to

$$B_w = \frac{1}{2^{n-k}} \sum_{w'=0}^{n} A_{w'} P_w(w') \qquad (w = 0, \ldots, n) , \qquad (7.67)$$

which completes the proof of the quantum MacWilliams identity. ∎

The quantum MacWilliams identity (in the form given in eq. (7.67)), in combination with eqs. (7.47) and (7.48), generate the following constraints on the $\{A_w\}$:

$$\frac{1}{2^{n-k}} \sum_{w'=0}^{n} A_{w'} P_w(w') = A_w \qquad (w = 0, \ldots, d-1) , \qquad (7.68)$$

$$\frac{1}{2^{n-k}} \sum_{w'=0}^{n} A_{w'} P_w(w') \geq A_w \qquad (w = d, \ldots, n) . \qquad (7.69)$$

We shall return to these constraints in Section 7.4.4.

7.4.3 Shadow Enumerator

In this subsection we introduce the shadow enumerator $S(z)$ and show that it satisfies a relation similar to the quantum MacWilliams identity. This new identity generates new constraints on the $\{A_w\}$ that are independent of those derived in Section 7.4.2. The new constraints further restrict the domain of the $\{A_w\}$, leading to tighter bounds on the code distance d than would be possible working solely with $A(z)$ and $B(z)$.

We begin by introducing the shadow of the stabilizer $Sh(\mathcal{S})$. Recall that $<s, E>$ was defined to be 0 (1) when s and E commute (anticommute).

DEFINITION 7.1 Let $\mathcal{C}_q$ be an $[n, k, d]$ quantum stabilizer code with stabilizer $\mathcal{S}$. The shadow of the stabilizer $Sh(\mathcal{S})$ is the set of operators $E \in \mathcal{G}_n$ which, for all $s \in \mathcal{S}$, satisfy

$$<s, E> = wt(s) \pmod{2} . \qquad (7.70)$$

Bounds on Quantum Error Correcting Codes

In other words, the elements of $Sh(S)$ commute (anticommute) with the even (odd) weight elements of S.

Let S_w be the number of elements in the shadow having weight w. The shadow enumerator $S(z)$ is defined to be the polynomial

$$S(z) = \sum_{w=0}^{n} S_w z^w \; . \tag{7.71}$$

The following theorem relates $S(z)$ to the weight enumerator $A(z)$.

THEOREM 7.3
Let $S(z)$ be the shadow enumerator for an $[n, k, d]$ quantum stabilizer code C_q and $A(z)$ the weight enumerator associated with the stabilizer S. $S(z)$ can be determined from $A(z)$ through the relation

$$S(z) = \frac{1}{2^{n-k}} (1 + 3z)^n A\left(\frac{z-1}{1+3z}\right) \; . \tag{7.72}$$

PROOF Two cases need to be considered: (1) all elements of S have even weight, and (2) the elements of S have even and odd weight.

(1) Suppose all elements of S have even weight. As noted above, all $E \in Sh(S)$ commute with the even weight elements of S, which in this case, is all of S. Thus $Sh(S) \subset C(S)$. On the other hand, since all $e \in C(S)$ commute with all $s \in S$, and $\text{wt}(s) = 0$ by assumption, eq. (7.70) is satisfied and so $C(S) \subset Sh(S)$. Thus $Sh(S) = C(S)$, and so S_w and B_w must be equal. It follows that $S(z) = B(z)$, and using the quantum MacWilliams identity, this becomes

$$S(z) = \frac{1}{2^{n-k}} (1 + 3z)^n A\left(\frac{1-z}{1+3z}\right) \; . \tag{7.73}$$

Since all elements of S have even weight, $A_w = 0$ for w odd and so $A(z) = A(-z)$. Using this in eq. (7.73) gives

$$S(z) = \frac{1}{2^{n-k}} (1 + 3z)^n A\left(\frac{z-1}{1+3z}\right) \; . \tag{7.74}$$

(2) Suppose S contains elements of even and odd weight. This means that some of the generators of S must have odd weight. Without loss of generality we can assume they are the first h generators $g_1, \ldots, g_h$. Now redefine the generators so that $g_i \to g_1 g_i$ for $i = 2, \ldots, h$ and $g_i \to g_i$ otherwise. After this redefinition, g_1 is the only generator with odd weight.

Let $S_e \subset S$ be the set of elements in S that have even weight. Clearly, the generators for S_e are $g_2, \ldots, g_{n-k}$ and so S_e contains $|S_e| = 2^{n-k-1}$ elements. Notice that S partitions into $S = g_1 S_e \cup S_e$. By definition, the centralizer

$C(\mathcal{S}_e)$ is the set of elements of $\mathcal{G}_n$ that commute with all elements of $\mathcal{S}_e$. Ignoring factors of i^λ, $C(\mathcal{S}_e)$ contains 2^{n+k+1} elements (Section 3.1.2-(5)). It is simple to show that $C(\mathcal{S})$ is a subgroup of $C(\mathcal{S}_e)$ and so its cosets can be used to partition $C(\mathcal{S}_e)$. By Lagrange's Theorem (Appendix A), the number of such cosets is $|C(\mathcal{S}_e)|/|C(\mathcal{S})| = 2$. We thus can write

$$C(\mathcal{S}_e) = EC(\mathcal{S}) \cup C(\mathcal{S}) \ . \tag{7.75}$$

Here $E \in \mathcal{G}_n - C(\mathcal{S})$, which belongs to $C(\mathcal{S}_e)$ but not to $C(\mathcal{S})$. Thus $<s, E> = 0$ for all $s \in \mathcal{S}_e$, and in particular, for $g_2, \ldots, g_{n-k}$. Since $E \notin C(\mathcal{S})$, it must anticommute with g_1 since we have just seen that E commutes with all the other generators of $\mathcal{S}$. Thus E has error syndrome $S(E) = 10\cdots0$ and all elements in $EC(\mathcal{S})$ have the same syndrome value (Theorem 3.2). Notice that if $s \in \mathcal{S}_e$ $(g_1\mathcal{S}_e)$, then $<s, E> = 0$ (1) and $\text{wt}(s) = 0$ (1) and so, from eq. (7.70), $E \in Sh(\mathcal{S})$. Now if $e \in EC(\mathcal{S})$, we can write $e = Ec$ where $c \in C(\mathcal{S})$. Thus if $s \in \mathcal{S}_e$ $(g_1\mathcal{S}_e)$, then $<s, e> = 0$ (1) and $\text{wt}(s) = 0$ (1). It follows from eq. (7.70) that $e \in Sh(\mathcal{S})$ and so $EC(\mathcal{S}) \subset Sh(\mathcal{S})$. On the other hand, if $e \in Sh(\mathcal{S})$, then $<g_1, e> = 1$ and $<g_i, e> = 0$ for $i = 2, \ldots, n-k$. Thus e and E have the same syndrome and so $e \in EC(\mathcal{S})$, and consequently, $Sh(\mathcal{S}) \subset EC(\mathcal{S})$. It follows from both inclusion results that $Sh(\mathcal{S}) = EC(\mathcal{S})$. From eq. (7.75) we see that $EC(\mathcal{S}) = C(\mathcal{S}_e) - C(\mathcal{S})$ and so

$$Sh(\mathcal{S}) = C(\mathcal{S}_e) - C(\mathcal{S}) \ . \tag{7.76}$$

Now let $B^e(z)$ and $A^e(z)$ be the weight enumerators associated with $C(\mathcal{S}_e)$ and $\mathcal{S}_e$, respectively. From eq. (7.76) it follows that $S_w = B_w^e - B_w$ and so

$$S(z) = B^e(z) - B(z)$$
$$= \frac{1}{2^{n-k-1}} (1+3z)^n A^e \left(\frac{1-z}{1+3z}\right) - \frac{1}{2^{n-k}} A\left(\frac{1-z}{1+3z}\right), \tag{7.77}$$

where we used the quantum MacWilliams identity and that $|\mathcal{S}_e| = 2^{n-k-1}$ in going from the first to the second line. Since all elements in $\mathcal{S}_e$ have even weight, (i) $A_w^e = 0$ for odd w, and (ii) $A_w^e = A_w$ for even w since (by definition) the even weight elements of $\mathcal{S}_e$ and $\mathcal{S}$ are the same. Thus

$$2A^e(z) = A(z) + A(-z) \ . \tag{7.78}$$

Inserting eq. (7.78) into eq. (7.77) gives

$$S(z) = \frac{1}{2^{n-k}} (1+3z)^n A\left(\frac{z-1}{1+3z}\right), \tag{7.79}$$

which completes the proof. ∎

As with the quantum MacWilliams identity, we can use eq. (7.79) to relate the $\{S_w\}$ to the $\{A_w\}$. The derivation of this relation follows the derivation of eq. (7.57) and we leave it to the reader to work out the details (Problem 7.4). The result is

$$S_w = \frac{1}{2^{n-k}} \sum_{w'=0}^{n} (-1)^{w'} A_{w'} P_w(w') \quad (w=0,\ldots,n) \ . \tag{7.80}$$

Here the $P_w(w')$ are the Krawtchouk polynomials encountered in Section 7.4.2. Since S_w counts the elements of weight w in the shadow $Sh(S)$, it follows that $S_w \geq 0$. This, combined with eq. (7.80), $A_0 = 1$, and that (see eq. (7.55))

$$P_w(0) = 3^w \binom{n}{w}, \tag{7.81}$$

yields a new set of constraints on the $\{A_w\}$ that are independent of those in eqs. (7.68) and (7.69). Specifically, they are

$$\sum_{w'=1}^{n} (-1)^{w'} A_{w'} P_w(w') \geq -3^w \binom{n}{w} \quad (w=0,\ldots,n) \ . \tag{7.82}$$

We now have all the results needed to obtain bounds on the code distance d through linear programming.

7.4.4 Bounds via Linear Programming

The standard problem in linear programming [3–5] is to find values for a set of non-negative variables $x_1, \ldots, x_n$ that (i) satisfy a set of linear constraints, and (ii) minimize (or maximize) a linear function of these variables known as the objective function. To formulate our linear programming problem, we begin by choosing $\{A_w : w = 1, \ldots, n\}$ to be the non-negative variables (recall $A_0 = 1$); and eqs. (7.68), (7.69), and (7.82) to be the constraints. To complete the problem specification we need an objective function. Recall from Theorem 3.3 that an [n,k,d] quantum stabilizer code is degenerate if the stabilizer S contains elements of weight less than d (other than the identity). With this theorem in mind, Ref. [11] defined the objective function to be

$$\sum_{w=1}^{d-1} A_w, \tag{7.83}$$

and required that it be minimized. Since the $\{A_w\}$ are non-negative, the smallest possible value for this objective function is zero, corresponding to a non-degenerate quantum stabilizer code. Requiring that the objective function be minimized thus requires the resulting quantum stabilizer code to be as minimally degenerate as possible. Thus we arrive at the following linear programming problem.

Problem: For given values of n, k, and d, find values for the non-negative variables $A_1, \ldots, A_n$ that (1) minimize the objective function $\sum_{w=1}^{d-1} A_w$, and (2) satisfy the following constraints:

$$A_0 = 1 \tag{7.84}$$

$$A_w \geq 0 \qquad (w = 1, \ldots, n) \tag{7.85}$$

$$\sum_{w=0}^{n} A_w = 2^{n-k} \tag{7.86}$$

$$\frac{1}{2^{n-k}} \sum_{w'=0}^{n} A_{w'} P_w(w') = A_w \qquad (w = 0, \ldots, d-1) \tag{7.87}$$

$$\frac{1}{2^{n-k}} \sum_{w'=0}^{n} A_{w'} P_w(w') \geq A_w \qquad (w = d, \ldots, n) \tag{7.88}$$

$$\sum_{w'=1}^{n} (-1)^{w'} A_{w'} P_w(w') \geq -3^w \binom{n}{w} \qquad (w = 0, \ldots, n). \tag{7.89}$$

Here the $\{P_w(w')\}$ are the Krawtchouk polynomials. ∎

Linear programming problems can be solved using the simplex method [3–5, 25]. When a solution to the above problem exists, the resulting values for the $\{A_w\}$ give the distribution of weights for elements of the stabilizer S and determine whether the code is degenerate or not. When no solution exists, then no quantum stabilizer code exists for the given values of n, k, and d. The analysis of quantum stabilizer codes via linear programming allows a systematic study of the code distance d to be carried out. For example, for given n and k, one can solve the above linear programming problem successively for $d = 1, 2, 3, \ldots$ to find the first value of the code distance d_* for which no solution exists. Then $d_* - 1$ is the largest possible value for the code distance for this choice of n and k. Such a study was done in Ref. [11] for $3 \leq n \leq 30$ and $0 \leq k \leq 23$. Having found the $\{A_w\}$, the quantum MacWilliams identity (eq. (7.57)) then determines the $\{B_w\}$. For further applications of this approach, see Refs. [11, 16, 17].

7.5 Entanglement Purification and QECCs

The context for the discussion in this section is the problem of quantum communication in which messages are encoded onto quantum states and transmitted over a noisy quantum channel from a sender (Alice) to a receiver (Bob). In Section 7.5.1 the idea of entanglement purification is introduced and two protocols for implementing such a purification are presented. Section 7.5.2 then demonstrates the close connection between entanglement purification protocols that are supported by one-way classical communication and QECCs used for reliable transmission of messages over noisy quantum channels. This section is based on work by Bennett and collaborators [6,7].

7.5.1 Purifying Entanglement

Consider n pairs of entangled qubits, with all pairs in the same mixed state (Appendix B). If M denotes the density operator for the mixed state, then the density operator for the composite-system ρ_o is

$$\rho_0 = \otimes_{i=1}^n M$$
$$\equiv (M)^n \ . \qquad (7.90)$$

Here $(\cdots)^n$ will be shorthand for an n-fold direct product. Alice and Bob each hold one qubit from each pair, and can be imagined to be in separate locations connected by a classical communication channel. Classical communication can be one-way (Alice $\to$ Bob) or two-way (Alice $\leftrightarrow$ Bob). Both Alice and Bob can apply local unitary operations and measurements to their respective qubits. A useful set of unitary operations and measurements will be specified below. Each can also introduce ancilla qubits and interact them with their own qubits. Non-local operations, however, are not allowed.

The goal of an entanglement purification protocol (EPP) is to take n non-maximally entangled qubit pairs and produce from them $m < n$ pairs, each in a state that closely approximates some desired maximally entangled state (Section 1.3.5) $|\psi_{max}\rangle$. As we shall see, this can be done by having Alice and Bob locally measure $n - m$ pairs, and then combine the measurement results with classical communication and local unitary operations to leave each of the remaining m pairs in the desired final state. If we denote the density operator for the final m pairs by ρ_f, the purification fidelity $\mathcal{F}$ is defined to be

$$\mathcal{F} = Tr[\rho_f (|\psi_{max}\rangle\langle\psi_{max}|)^m]$$
$$= (\langle\psi_{max}|)^m [\rho_f] (|\psi_{max}\rangle)^m \ . \qquad (7.91)$$

Formally then, the goal of an EPP is to have $\rho_f \to (|\psi_{max}\rangle\langle\psi_{max}|)^m$ so that $\mathcal{F} \to 1$ as closely as possible. The yield for an EPP is defined to be

$$D_P(M) = \lim_{n \to \infty} \frac{m}{n} \ . \qquad (7.92)$$

Note that since it is the pair entanglement that is being purified, the argument of $D_P(M)$ is the initial pair density operator M.

As a prelude to carrying out an EPP, one can imagine sending n maximally entangled pairs through a quantum environment χ, which then applies a quantum operation (Section 2.1) to the pairs, leaving them in a mixed state with density operator ρ_0. Such a situation would also occur if we wanted to carry out quantum teleportation [27]. In quantum communication theory, the quantum environment χ is called a quantum channel.

In the discussion to follow, we will often consider two-qubit density operators. It proves convenient to work with these operators in the representation spanned by the Bell basis states:

$$|\Phi^+\rangle = \tfrac{1}{\sqrt{2}}(|\uparrow\uparrow\rangle + |\downarrow\downarrow\rangle) \equiv |00\rangle$$
$$|\Psi^+\rangle = \tfrac{1}{\sqrt{2}}(|\uparrow\downarrow\rangle + |\downarrow\uparrow\rangle) \equiv |01\rangle$$
$$|\Phi^-\rangle = \tfrac{1}{\sqrt{2}}(|\uparrow\uparrow\rangle - |\downarrow\downarrow\rangle) \equiv |10\rangle$$
$$|\Psi^-\rangle = \tfrac{1}{\sqrt{2}}(|\uparrow\downarrow\rangle - |\downarrow\uparrow\rangle) \equiv |11\rangle \ . \qquad (7.93)$$

Here $|\uparrow\rangle$ ($|\downarrow\rangle$) is the eigenstate of σ_z with eigenvalue $+1$ (-1); and the notation on the far right will prove convenient later. Direct calculation shows that the entropy of entanglement (Section 1.3.5) for each Bell state is $\log_2 2 = 1$ so that the Bell basis is composed of maximally entangled states. An important entangled mixed state is the Werner state [26]. Its form in the Bell basis is

$$W_F = F|\Psi^-\rangle\langle\Psi^-| + \frac{1-F}{3}\left[|\Psi^+\rangle\langle\Psi^+| + |\Phi^+\rangle\langle\Phi^+| + |\Phi^-\rangle\langle\Phi^-|\right], \quad (7.94)$$

where $0 \leq F \leq 1$. If the desired maximally entangled state $|\psi_{max}\rangle$ for an EPP is chosen to be $|\Psi^-\rangle$, then from eqs. (7.91) and (7.94), the purification fidelity for the Werner state is $\mathcal{F} = F$. The Werner state will appear often in our discussions of EPPs [6].

Finally, we present some properties of the Bell states that are utilized by EPPs. For lack of space, we simply state the results and refer the reader to Ref. [6] for further details.

(1) Suppose that initially each of the n pairs used in an EPP is in an arbitrary entangled mixed state with density operator M. It is possible to use a twirl operation T to transform M into the Werner state W_F. A twirl operation is carried out by first having Alice and Bob choose, independently for each pair i ($i = 1, \ldots, n$), a random operation U_i from a particular 12-element subgroup of $SU(2)$. Then Alice and Bob each apply U_i to their respective qubit from pair i for all i. Ref. [6] shows that this process transforms $M \to W_F$.

(2) A number of local unitary operations produce useful permutations of the Bell basis states. These operations fall into two types. The first type consists of unilateral operations in which Alice or Bob (though not both) apply a local unitary operation. The EPPs discussed below will only need unilateral operations that apply 180° rotations about $\hat{\mathbf{x}}$, $\hat{\mathbf{y}}$, or $\hat{\mathbf{z}}$. The second type of operation is made up of bilateral operations in which Alice and Bob apply the same local unitary operation (viz. $U \otimes U$). Two classes of bilateral operations prove useful: (i) bilateral 90° rotations about $\hat{\mathbf{x}}$, $\hat{\mathbf{y}}$, or $\hat{\mathbf{z}}$; and

(ii) bilateral application of CNOT gates (BCNOT). [Note that Ref. [6] uses BXOR to denote the BCNOT operation.] To apply a BCNOT operation, Alice and Bob pick out two pairs from the original n pairs, labeling one the source pair and the other the target pair. Each then applies a local CNOT gate using their source qubit as control and their target qubit as target. The local CNOT gate action is to flip the target qubit ($|\uparrow\rangle \leftrightarrow |\downarrow\rangle$) only when the source qubit is in the spin-up state ($|\uparrow\rangle$). For example, if initially the source pair is in the Bell state $|\Psi^-\rangle$ and the target pair in the state $|\Phi^+\rangle$, it follows from eqs. (7.93) and the CNOT gate action that

$$\begin{aligned}U_{BCNOT}|\Psi^-\rangle|\Phi^+\rangle &= U_{BCNOT}\left[\frac{1}{\sqrt{2}}\{|\uparrow\downarrow\rangle - |\downarrow\uparrow\rangle\}\right]\left[\frac{1}{\sqrt{2}}\{|\uparrow\uparrow\rangle + |\downarrow\downarrow\rangle\}\right] \\ &= \frac{1}{2}U_{BCNOT}\{|\uparrow\downarrow\rangle[|\uparrow\uparrow\rangle + |\downarrow\downarrow\rangle] - |\downarrow\uparrow\rangle[|\uparrow\uparrow\rangle + |\downarrow\downarrow\rangle]\} \\ &= \frac{1}{2}\{|\uparrow\downarrow\rangle[|\downarrow\uparrow\rangle + |\uparrow\downarrow\rangle] - |\downarrow\uparrow\rangle[|\uparrow\downarrow\rangle + |\downarrow\uparrow\rangle]\} \\ &= |\Psi^-\rangle|\Psi^+\rangle \ . \end{aligned} \quad (7.95)$$

Ref. [6] provides look-up tables for the action of all unilateral and bilateral operations of interest. They can be found in Table I of that paper, which we have reproduced as Table 7.1 on page 254.

(3) Alice and Bob can use local measurements of σ_z^A and σ_z^B, respectively, plus classical communication to distinguish the $|\Phi^\pm\rangle$ Bell states from the $|\Psi^\pm\rangle$. If the measurement outcomes agree (disagree), the measurement has collapsed the pair state to a mixture of $|\Phi^\pm\rangle$ ($|\Psi^\pm\rangle$).

(4) It proves convenient to use $|\Phi^+\rangle$ as a standard state. For some purposes, the singlet state $|\Psi^-\rangle$ is more convenient; however, it can be obtained from $|\Phi^+\rangle$ by applying a unilateral σ_y operation (see Table 7.1). Note that using $|\Phi^+\rangle$ as the standard state requires that a modified twirl operation T' be introduced. This is because the twirl operation T leaves the singlet state $|\Psi^-\rangle$ invariant, while randomizing the other three Bell basis states. The modified twirl T' is implemented by first applying a unilateral σ_y operation that maps $|\Phi^+\rangle \to |\Psi^-\rangle$ (as well as permuting the other three Bell basis states (Table 7.1)). Next the original twirl operation T is applied, which leaves $|\Psi^-\rangle$ invariant and randomizes the remaining three Bell basis states. Finally, a unilateral σ_y operation is applied to map $|\Psi^-\rangle \to |\Phi^+\rangle$ (with a simultaneous permutation of the other three Bell basis states). The net result is that T' leaves $|\Phi^+\rangle$ invariant, while randomizing the other three Bell basis states.

Example: Recurrence Method (2-EPP)

The recurrence method is an EPP that requires two-way classical communication between Alice and Bob (2-EPP). If M is the density operator for a two-qubit entangled mixed state, the two-way distillable entanglement $D_2(M)$ is the maximum yield of pairs $D_P(M)$ that can be produced by any two-way entanglement purification protocol P:

$$D_2(M) \equiv \max\{ D_P(M) : P \text{ is a 2-EPP}\} \quad . \tag{7.96}$$

Our starting point is a collection of n pairs of qubits, with Alice and Bob each holding one qubit from each pair. All pairs are in the same entangled mixed state, and the associated density operator M is assumed to be diagonal in the Bell basis (eq. (7.93)):

$$M = \sum_{i,j=0}^{1} p_{ij} |ij\rangle\langle ij| \quad . \tag{7.97}$$

Here p_{ij} is the probability that a pair is in the Bell state $|ij\rangle$. By our earlier convention, the standard state is $|00\rangle = |\Phi^+\rangle$, and so the fidelity $F = p_{00}$. By prior agreement, Alice and Bob form $n/2$ pairs of pairs. In each two-pair set, one pair is labeled the source pair and the other the target pair. The recurrence method iterates the following three-step procedure. (1) Alice and Bob apply a BCNOT operation to each two-pair set, using the source qubits as the controls. (2) For each two-pair set, Alice and Bob then take the target pair and locally measure $\sigma_z^A \otimes \sigma_z^B$. (3) Finally Alice and Bob compare measurement results. It is in this last step that two-way classical communication is required. For each two-pair set, if the target qubits are found to have parallel spins, Alice and Bob label the associated source pair a "passed" pair. For anti-parallel spins, the (associated) source pair is labeled a "failed" pair. At this point all target pairs are discarded, and Alice and Bob separate the $n/2$ source pairs into a set of passed pairs and another of failed pairs.

Table II of Ref. [6] contains the information needed to analyze the performance of the recurrence method. For each possible two-pair state $|ij\rangle|kl\rangle$ it lists (i) the probability $p_{ij}p_{kl}$ that this state will be the input to the BCNOT operation, (ii) the corresponding BCNOT output state, and (iii) which final output states result in passed/failed source pairs. With this data in hand, much useful information can be calculated (Problem 7.5):

(1) the probability p_{pass} (p_{fail}) that the three-step procedure produces a passed (failed) source pair:

$$p_{pass} = \left(p_{00}^2 + p_{01}^2 + p_{10}^2 + p_{11}^2\right) + 2\left(p_{00}p_{10} + p_{01}p_{11}\right)$$
$$p_{fail} = 2p_{00}\left(p_{01} + p_{11}\right) + 2\left(p_{01}p_{10} + p_{10}p_{11}\right) \quad ; \tag{7.98}$$

(2) the probability p'_{ij} that $|ij\rangle$ is the final state for a passed source pair:

$$p'_{00} = \frac{p_{00}^2 + p_{10}^2}{p_{pass}} \quad ; \quad p'_{01} = \frac{p_{01}^2 + p_{11}^2}{p_{pass}}$$

$$p'_{10} = \frac{2\, p_{00}\, p_{10}}{p_{pass}} \quad ; \quad p'_{11} = \frac{2\, p_{01}\, p_{11}}{p_{pass}} \quad ; \tag{7.99}$$

(3) the probability p'_{ij} that $|ij\rangle$ is the final state for a failed source pair:

$$p'_{00} = \frac{p_{00}p_{01} + p_{10}p_{11}}{p_{fail}} \quad ; \quad p'_{01} = p'_{00}$$

$$p'_{10} = \frac{p_{00}p_{11} + p_{10}p_{01}}{p_{fail}} \quad ; \quad p'_{11} = p'_{10} \; . \qquad (7.100)$$

Now suppose that the initial state M is the modified Werner state W'_F:

$$W'_F = F|00\rangle\langle 00| + \frac{(1-F)}{3}[|01\rangle\langle 01| + |10\rangle\langle 10| + |11\rangle\langle 11|] \; . \qquad (7.101)$$

Comparing eqs. (7.97) and (7.101) gives

$$p_{00} = F \quad ; \quad p_{ij} = \frac{1-F}{3} \quad (ij \neq 00). \qquad (7.102)$$

From eqs. (7.98)–(7.100) and (7.102) it follows that

$$p_{pass} = \frac{8F^2 - 4F + 5}{9}$$

$$p_{fail} = \frac{4}{9}\left(1 + F - 2F^2\right) \; ; \qquad (7.103)$$

for passed pairs,

$$p'_{00} = \frac{F^2 + \left(\frac{1-F}{3}\right)^2}{p_{pass}} \quad ; \quad p'_{01} = \frac{2\left(\frac{1-F}{3}\right)^2}{p_{pass}}$$

$$p'_{10} = \frac{2F\left(\frac{1-F}{3}\right)}{p_{pass}} \quad ; \quad p'_{11} = p'_{01} \; ; \qquad (7.104)$$

and for failed pairs,

$$p'_{00} = p'_{01} = p'_{10} = p'_{11} = \frac{1}{4} \; . \qquad (7.105)$$

The final mixed state for the failed pairs can be shown to be non-entangled (Problem 7.6). They are thus of no use for entanglement purification and so are discarded. On the other hand, each of the passed pairs is in a Bell-diagonal entangled mixed state that can be transformed into the modified Werner state W'_F by a modified twirl operation T'. Since T' leaves the $|00\rangle$ state invariant while randomizing the other Bell basis states, the fidelity F' of the modified Werner state will be p'_{00} (eq. (7.104)):

$$F' = \frac{F^2 + \left(\frac{1-F}{3}\right)^2}{p_{pass}} \; . \qquad (7.106)$$

Figure 7.1 plots $\Delta F = F' - F$ versus the initial fidelity F. We see that for

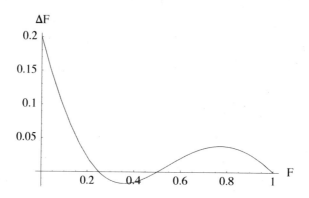

FIGURE 7.1
Plot of the change in fidelity $\Delta F = F' - F$ per iteration of the recurrence method versus the initial fidelity F.

$F > 1/2$, ΔF is positive and so the recurrence method increases the mixed state fidelity. Repeated iteration in this case will thus cause the fidelity to approach 1 as the number of iterations $r \to \infty$. [Note that although $\Delta F > 0$ for $0 < F < 0.25$, iterating the recurrence method in this case will only cause F' to oscillate about 0.25 as $r \to \infty$.] Unfortunately, the yield $D_P^r(W_F')$ after r iterations goes to zero as $r \to \infty$. To see this, let $D_P(W_F')$ be the yield after one iteration:

$$D_P(W_F') = \lim_{n \to \infty} \frac{m}{n}, \qquad (7.107)$$

where m is the number of purified pairs produced by a single iteration of the recurrence method. Recalling the relative frequency definition of probability, and that the initial number of source pairs is $n/2$, it follows that as $n \to \infty$,

$$\frac{m}{(n/2)} \to p_{pass}. \qquad (7.108)$$

Combining eqs. (7.107) and (7.108) gives

$$D_P(W_F') = \frac{p_{pass}}{2} \equiv \frac{1}{2}(1 - \epsilon)$$

for some $0 < \epsilon < 1$. The yield after r iterations $D_P^r(W_F')$ is then

$$\begin{aligned} D_P^r(W_F') &= [D_P(W_F')]^r \\ &= \frac{1}{2^r}(1 - \epsilon)^r \\ &= \frac{1}{2^r} \exp[r \ln(1 - \epsilon)]. \end{aligned}$$

Bounds on Quantum Error Correcting Codes

Since $\ln(1-\epsilon) < 0$, we see that $D_P^r(W_F')$ goes to zero exponentially with the number of iterations r. If, however, the recurrence method is combined with the hashing method described in the following example, an EPP is obtained that gives a positive yield of maximally entangled pairs.

Example: Hashing Method (1-EPP)

A one-way entanglement purification protocol (1-EPP) is one that uses one-way classical communication between Alice and Bob. If M is the density operator for a two-qubit entangled mixed state, the one-way distillable entanglement $D_1(M)$ is the maximum yield of pairs $D_P(M)$ that can be produced by any one-way entanglement purification protocol P:

$$D_1(M) \equiv \max\{D_P(M) : P \text{ is a 1-EPP}\} \ . \tag{7.109}$$

Our focus will be a system of n pairs of qubits, with Alice and Bob each holding one qubit from each pair. All pairs are in the same entangled mixed state with density operator M, which is assumed to be diagonal in the Bell basis (eq. (7.93)). The hashing method requires the entangled mixed state to be sufficiently purified that its von-Neumann entropy satisfies $S(M) < 1$. Said another way, the entangled mixed state has less than one bit of uncertainty. For example, the density operator

$$M = \frac{3}{4}|00\rangle\langle 00| + \frac{1}{4}|01\rangle\langle 01|$$

has $S(M) = 0.812$. We shall see that the hashing method has a yield of $1 - S(M)$ pairs and that the fidelity F of the final state approaches 1 as $n \to \infty$. The hashing protocol will be seen to be one-way as it only requires Alice to send Bob a classical message after having completed her part of the protocol. The information received allows Bob to complete his part of the protocol without having to send a message to Alice.

Some preliminary remarks are needed before we can present the hashing protocol and examine its performance. To produce our set of n pairs we can imagine that we have drawn n pairs from an ensemble of pairs whose state is described by the Bell-diagonal density operator M:

$$M = \sum_{i,j=0}^{1} p_{ij}|ij\rangle\langle ij| \ . \tag{7.110}$$

Here p_{ij} is the probability that a pair drawn from the ensemble is in the Bell basis state $|ij\rangle$. The probability P_x that the n drawn pairs are in the state

$$|\psi_x\rangle = |i_1 j_1\rangle \cdots |i_n j_n\rangle \tag{7.111}$$

is

$$P_x = p_{i_1 j_1} \cdots p_{i_n j_n} \ , \tag{7.112}$$

where x is the bit-string $x = i_1 j_1 \cdots i_n j_n$. Thus x labels the 2^{2n} possible states of the n drawn pairs. We define the parity π_x of x to be the modulo 2 sum of the bits $i_1, j_1; \ldots; i_n, j_n$:

$$\pi_x = \sum_{k=1}^{n} (i_k \oplus j_k) \; . \tag{7.113}$$

We will also need to consider the parity of a subset S of bits in x. We introduce the bit-string $s = s_1 \cdots s_{2n}$, which has s_k equal to 1 at the bit positions in x that are to contribute to the restricted parity, and zero at all other s_k. Then the restricted parity $\pi_{x|s}$ of x relative to s is the modulo 2 inner product $s \cdot x$:

$$\pi_{x|s} \equiv s \cdot x$$
$$= \sum_{k=1}^{n} (s_{2k-1} i_k \oplus s_{2k} j_k) \; . \tag{7.114}$$

The following three points are essential for the construction and analysis of the hashing method.

(1) The state of the kth pair drawn from the ensemble of pairs allows a classical random variable x_k to be defined. Let x_k take the value $i_k j_k$ when the state of the kth pair is $|i_k j_k\rangle$. Following Ref. [6] we shall refer to the first bit i_k in x_k as the phase bit and the second bit j_k as the amplitude bit. The probability distribution $p(x_k)$ associated with x_k is the set of $\{p_{i_k j_k}\}$ that appear in eq. (7.110). We can think of x_k as a symbol source whose alphabet is the set of bit-strings of length 2 and whose entropy is the von-Neumann entropy $S(M)$ of the pair ensemble. This construction thus yields n random variables $\{x_k : k = 1, \ldots, n\}$, which have identical independent (probability) distributions (i. i. d.). When necessary, we will write $p(\chi)$ for the common probability distribution and χ as a generic stand-in for any of the $\{x_k\}$. Now let $X = x_1 \cdots x_n$ be a random variable obtained by concatenating the x_k. Since the x_k are i. i. d., X has the asymptotic equipartition property (AEP) [28], which states that if $x = (i_1 j_1) \cdots (i_n j_n)$ is a possible assignment of values for the components $x_1, \ldots, x_n$ of X, then for any $\delta, \epsilon > 0$, there exists an integer $n_0(\delta, \epsilon)$ such that for $n > n_0(\delta, \epsilon)$,

$$\text{Prob}\left[\left|-\frac{1}{n}\log P_x - S(M)\right| < \delta\right] > 1 - \epsilon \; . \tag{7.115}$$

Because of the AEP, the probability distribution P_x receives effectively all of its support from a set $A_\delta^{(n)}$ of typical strings $x = x_1 \cdots x_n$.

DEFINITION 7.2 *Let $X = x_1 \cdots x_n$ be a random variable in which the $\{x_i\}$ are i. i. d. random variables, and let P_x be the probability distribution*

Bounds on Quantum Error Correcting Codes

for X. The typical set $A_\delta^{(n)}$ is the set of strings $x = x_1 \cdots x_n$ for which

$$2^{-n(S(M)+\delta)} \le P_x \le 2^{-n(S(M)-\delta)} \ . \tag{7.116}$$

It is possible to show [28] that (i) the size $\left|A_\delta^{(n)}\right|$ of the typical set $A_\delta^{(n)}$ satisfies the upper bound

$$\left|A_\delta^{(n)}\right| \le 2^{n(S(M)+\delta)} \ ; \tag{7.117}$$

and (ii) the probability $Prob\left[x \in A_\delta^{(n)}\right]$ that a randomly sampled string x belongs to $A_\delta^{(n)}$ satisfies

$$Prob\left[x \in A_\delta^{(n)}\right] > 1 - \epsilon \ . \tag{7.118}$$

By Definition 7.2, a string x is non-typical if

$$\left|-\frac{1}{n}\log P_x - S(M)\right| > \delta \ . \tag{7.119}$$

The probability that x is non-typical,

$$Prob\left[\left|-\frac{1}{n}\log P_x - S(M)\right| > \delta\right] \ ,$$

can be upper bounded using the Chebyshev inequality [29],

$$Prob\left[\left|-\frac{1}{n}\log P_x - S(M)\right| > \delta\right] \le \frac{var\left[\frac{1}{n}\log P_x\right]}{\delta^2} \ , \tag{7.120}$$

where the numerator on the RHS is the variance of $(1/n)\log P_x$. From the definition of variance, and recalling that $P_x = p(x_1)\cdots p(x_n)$, we have

$$var\left[\frac{1}{n}\log P_x\right] = \frac{1}{n^2} var\left[\sum_{k=1}^{n} \log p(x_k)\right]$$

$$= \frac{1}{n^2}\sum_{k=1}^{n} var\left[\log p(x_k)\right] \ . \tag{7.121}$$

Since the x_k have i. i. d. probabilities, we can write eq. (7.121) as

$$var\left[\frac{1}{n}\log P_x\right] = \frac{1}{n} var\left[\log p(\chi)\right] \ . \tag{7.122}$$

Using eq. (7.122) in eq. (7.120) gives

$$\text{Prob}\left[\left|-\frac{1}{n}\log P_x - S(M)\right| > \delta\right] \le \frac{\text{var}\left[\log p(\chi)\right]}{n\delta^2} \tag{7.123}$$

Thus the probability that a string x is non-typical is $\mathcal{O}\left(1/n\delta^2\right)$.

(2) As will be described in greater detail below, the hashing protocol uses local unitary operations and measurements to determine the restricted parity $\pi_{x^l|s_l}$, where x^l is a string that describes the current unknown state of the pairs and s_l is a string supplied by Alice. In the course of determining $\pi_{x^l|s_l}$, the string x^l is mapped to a string denoted $f_{s_l}(x^l)$ whose length is two shorter than that of x^l. Once the details of the hashing protocol have been presented it will be clear how to determine $f_{s_l}(x^l)$.

(3) Let x and y be two distinct bit-strings, and s a bit-string of non-zero weight. The probability that x and y have the same restricted parity relative to s is $1/2$. To see this, fix x and s. Let A_x be the set of bit-strings y for which $s \cdot y = s \cdot x$; and let D_x be the bit-strings y for which $s \cdot y = (s \cdot x) \oplus 1$. It is clear that y must belong to one of these two sets. Let i be an integer that specifies a component s_i of s such that $s_i = 1$. Define a bit-string e whose components satisfy $e_k = \delta_{k,i}$. It follows that $s \cdot e = 1$. Now suppose $y \in A_x$. Then the bit-string $y \oplus e \in D_x$ since

$$s \cdot (y \oplus e) = s \cdot y \oplus s \cdot e$$
$$= s \cdot x \oplus 1 \ .$$

Thus it is possible to map A_x into D_x and so the size of A_x is less than or equal to the size of D_x: $|A_x| \le |D_x|$. Next suppose $y \in D_x$. Then $y \oplus e \in A_x$ since $s \cdot (y \oplus e) = s \cdot x$. It follows that $|D_x| \le |A_x|$. Combining these two inequalities tells us that A_x and D_x have exactly the same number of bit-strings and so the probability that a string $y \in A_x$ is $1/2$. It follows from the definition of A_x that the probability that x and y have the same restricted parity relative to s is $1/2$.

We are now ready to describe the details of the hashing protocol. The protocol iterates a two-step procedure $n - m$ times to produce m pairs whose final state is $(|00\rangle)^m$ with fidelity $F \to 1$ as the number of initial pairs $n \to \infty$. We will discuss how to choose m after explaining how the two-step procedure works. As described above, Alice and Bob begin the protocol with n pairs, each in an unknown Bell basis state $|i_k j_k\rangle$. The 2^{2n} possible states of the n pairs $|\psi_{x^0}\rangle$ can be labeled by a bit-string $x^0 = x_1^0 \cdots x_n^0$ of length $2n$, where $x_k^0 = i_k^0 j_k^0$ is the 2-bit label for the kth pair state $|i_k^0 j_k^0\rangle$. The probability that the n pairs are in a state labeled by a particular bit-string x^0 is P_{x^0} (eq. (7.112)). As will be seen below, each iteration of the two-step procedure will require that one of the pairs be measured and discarded. Thus after l iterations, Alice and Bob will have $n - l$ pairs left in an unknown state labeled by a bit-string $x^l = x_1^l \cdots x_{n-l}^l$ of length $2(n - l)$. At the beginning of the $(l+1)^{\text{th}}$ iteration of the two-step procedure, Alice is required in step 1 to choose a non-zero weight bit-string s_l of length $2(n - l)$ and send it to Bob. This is the first of two places where the hashing method requires one-way classical communication. In step 2, Alice and Bob perform local unitary operations based on the bit-string s_l, and then locally measure a particular pair that is also determined from s_l. Alice then sends the result of her measurement to Bob, who is then able to determine the restricted parity $\pi_{x^l|s_l} = s_l \cdot x^l$. This is the second place where one-way classical communication is required. (Note that Alice can complete all of her tasks in the protocol, then send all the $\{s_l\}$ and measurement results to Bob in one communication.) As mentioned above, the measured pair is discarded and the remaining $n - l - 1$ pairs are in a state labeled by $x^{l+1} = f_{s_l}(x^l)$ of length $2(n - l - 1)$. The specific form of f_{s_l} depends on the local unitary operations Alice and Bob applied and on s_l. We will spell out what the unitary operations are after examining the hashing protocol's performance.

The effectiveness of the hashing protocol is based on the AEP of the random variable X and the consequent existence of the typical set $A_\delta^{(n)}$. From eq. (7.118), the probability that a bit-string x belongs to $A_\delta^{(n)}$ satisfies

$$Prob\left[x \in A_\delta^{(n)}\right] > 1 - \epsilon .$$

Thus with probability approaching 1, the bit-string associated with the initial unknown state of the n pairs belongs to $A_\delta^{(n)}$. The hashing protocol assumes this to be the case. Now suppose that Alice and Bob carry out r iterations of the two-step procedure just described. This generates a sequence of bit-strings $x^0, \ldots, x^r$ which label the state of the unmeasured pairs throughout the protocol; and a second sequence $s_0, \ldots, s_{r-1}$ used to specify the restricted parities that are to be measured $s_l \cdot x^l : l = 0, \ldots, r - 1$. The protocol succeeds if (i) x^0 is typical, and (ii) the sequence of measurements produce a unique final bit-string x^r. The hashing method error probability P_e is thus the sum of the probabilities that x^0 is non-typical and that x^r is not unique. We already know that the probability for x^0 to be non-typical is $\mathcal{O}(1/n\delta^2)$ from eq. (7.123). It remains to determine the probability for a non-unique x^r. To that end, let x^0 and y^0 be two distinct elements of $A_\delta^{(n)}$. Suppose the r measurements based on $s_0, \ldots, s_{r-1}$ map x^0 and y^0 into the sequences $(x^0, \ldots, x^r)$ and $(y^0, \ldots, y^r)$, respectively. By assumption, the measurement outcomes are the same ($s_l \cdot x^l = s_l \cdot y^l$ for $l = 0, \ldots, r - 1$) and yet $x^r \neq y^r$. At the start of each iteration, either x^l and y^l are equal or they are not. Thus the probability that $x^l \neq y^l$ is less than or equal to 1. We also know that the probability that $s_l \cdot x^l = s_l \cdot y^l$ and $x^l \neq y^l$ is $1/2$. Thus the probability that x^l and y^l are distinct and yet have the same restricted parity relative to s_l is less than or equal to $1/2$. The probability that this is true for all r measurements and yet $x^r \neq y^r$ is thus upper bounded by $1/2^r$. The argument so far is for specific, distinct $x^0, y^0 \in A_\delta^{(n)}$. Since the number of $y^0 \in A_\delta^{(n)}$ is upper bounded by $2^{n[S(M)+\delta]}$, the total probability that the r measurements do not uniquely determine x^r is less than or equal to $2^{n[S(M)+\delta]} \cdot 2^{-r}$. We have thus shown that

$$P_e \sim \mathcal{O}\left(2^{n[S(M)+\delta]-r}\right) + \mathcal{O}\left(\frac{1}{n\delta^2}\right) . \tag{7.124}$$

If we choose $\delta = n^{-1/4}$ and $r = n[S(M) + 2\delta]$, then

$$P_e \sim \mathcal{O}\left(2^{-n^{0.75}}\right) + \mathcal{O}\left(n^{-0.5}\right) , \tag{7.125}$$

which vanishes as $n \to \infty$. The number of unmeasured pairs m remaining upon completion of the protocol is then $m = n - r$. Thus the yield $D_H(M)$ is

$$D_H(M) = \lim_{n \to \infty} \frac{m}{n} = 1 - S(M) . \tag{7.126}$$

Since $P_e \to 0$ asymptotically, the final state of the m unmeasured pairs approaches $|\psi_{x^r}\rangle$ as $n \to \infty$. Knowing the bit-string x^r, Bob can then apply unilateral unitary operations (Table 7.1) to map $|\psi_{x^r}\rangle$ to $|\psi_{0\cdots 0}\rangle = (|00\rangle)^m$ with a fidelity that approaches 1 asymptotically.

Finally, we present the details of how Alice and Bob implement the two-step procedure that measures the restricted parity $s_l \cdot x^l$. Step 1 had Alice choosing a non-zero weight bit-string s_l and then sending it to Bob. Since we are dealing with sets of pairs, we will write s_l as a product of random variables $\{s_{k,l}\}$, where $s_{k,l} = (s_l^{2k-1} s_l^{2k})$, and s_l^{2k-1} and s_l^{2k} are binary random variables:

$$s_l = s_{1,l} \cdots s_{n-l,l}$$
$$= \left(s_l^1 s_l^2\right) \cdots \left(s_l^{2(n-l)-1} s_l^{2(n-l)}\right) . \tag{7.127}$$

Alice and Bob examine the $s_{k,l}$ to determine the smallest value of k for which $s_{k,l} \neq 00$. Let k_* denote this k value. By convention, Alice and Bob set pair k_* to be the destination pair in whose amplitude bit they will accumulate the restricted parity $s_l \cdot x^l$. Local measurement plus one-way classical communication will then allow Bob to determine the value of $s_l \cdot x^l$. Adapting eq. (7.114) to $n - l$ pairs and writing $x^l = x_1^l \cdots x_{n-l}^l$, with $x_k^l = (i_k^l j_k^l)$ gives

$$s_l \cdot x^l = \sum_{k=1}^{n-l} \left(s_l^{2k-1} i_k^l \oplus s_l^{2k} j_k^l\right) . \tag{7.128}$$

To systematize the determination of $s_l \cdot x^l$, the contribution of the kth pair to eq. (7.128) will be transferred to the pair k amplitude bit. (The procedure for transferring it from there to the destination pair's amplitude bit will be explained shortly.) The operation needed to carry out the transfer depends on the value of $s_{k,l}$:

1. $s_{k,l} = 00$: Pair k does not contribute to eq. (7.128) in this case. Thus it will not be involved in the operations that determine $s_l \cdot x^l$.

2. $s_{k,l} = 01$: Pair k contributes j_k^l to eq. (7.128). Since j_k^l is already stored in the pair k amplitude bit, no transfer operation is needed in this case.

3. $s_{k,l} = 10$: The kth pair contributes i_k^l to eq. (7.128). A transfer operation is needed in this case since i_k^l is stored in the pair k phase bit. Alice and Bob must apply a SWAP operation to pair k to transfer i_k^l to the amplitude bit.

4. $s_{k,l} = 11$: The pair k contribution to eq. (7.128) is $i_k^l \oplus j_k^l$. To store this in the pair k amplitude bit, Alice and Bob must apply a CNOT to pair k using the phase bit as the control.

Exercise 7.2 shows how Alice and Bob can use local unitary operations to apply SWAP and CNOT operations to a pair. With these operations the set of intra-pair transfer operations can be carried out. Soon we will need to know the action of a BCNOT operation on a product of Bell basis states and so this is also worked out in Exercise 7.2.

Exercise 7.2

(a) Use Table 7.1 on page 254 to show that bilateral rotation B_y applies a SWAP operation to the Bell basis states: $|ij\rangle \to |ji\rangle$.

(b) Show that the composite operation $B_x \sigma_x = \sigma_x B_x$ applies a CNOT operation to a pair in which the phase bit acts as the control: $|ij\rangle \to |i(j \oplus i)\rangle$.

(c) Consider two pairs s and t. Show that a BCNOT operation in which s acts as the source pair and t as the target pair produces the following map:

$$|i_s j_s\rangle |k_t l_t\rangle \to |(i_s \oplus k_t) j_s\rangle |k_t (l_t \oplus j_s)\rangle . \qquad (7.129)$$

Thus a BCNOT operation applies (i) a CNOT to the amplitude bits with the source-pair amplitude bit acting as the control, and (ii) a CNOT to the phase bits with the target-pair phase bit acting as the control.

With the transfer operations completed, the amplitude bit of each pair that has $s_{k,l} \neq 00$ contains the contribution of that pair to $s_l \cdot x^l$. To accumulate all these contributions in the amplitude bit of the destination pair, Alice and Bob apply a separate BCNOT operation from each of the non-destination pairs with $s_{k,l} \neq 00$ to the destination pair. For each BCNOT, the non-destination pair acts as the source-pair and the destination pair as the target-pair. From eq. (7.129), a BCNOT operation is seen to apply a CNOT operation from the source-pair amplitude bit to the target-pair amplitude bit. Thus, once all BCNOT operations are completed, $s_l \cdot x^l$ has been accumulated in the destination pair's amplitude bit as desired. Finally, Alice and Bob locally measure $\sigma_z^A \otimes \sigma_z^B$ on the destination pair. If the spins are parallel, then the

TABLE 7.1
Unilateral and bilateral operations used by Alice and Bob to map Bell states to Bell states. For brevity, the Bell basis states $|ij\rangle$ are denoted ij with $i, j = 0, 1$. Each entry of the BCNOT table has two lines, the first showing the output state for the source-pair, the second showing the output state of the target-pair. Reprinted Table 1 with permission from C. H. Bennett et. al., *Phys. Rev. A* **54**, 3824 (1996). ©American Physical Society, 1996.

		\multicolumn{4}{c}{Source Input}			
	Operation	11	10	00	01		
	I	11	10	00	01		
Unilateral π rotations:	σ_x	10	11	01	00		
	σ_y	00	01	11	10		
	σ_z	01	00	10	11		
		\multicolumn{4}{c}{Source Input}			
	Operation	11	10	00	01		
	I	11	10	00	01		
Bilateral $\pi/2$ rotations:	B_x	11	10	01	00		
	B_y	11	01	00	10		
	B_z	11	00	10	01		
			\multicolumn{4}{c}{Source Input}		
	Target Input	11	10	00	01	Output Pair
		01	00	10	11	(source)
	11	10	11	11	10	(target)
		01	00	10	11	(source)
BCNOT:	10	11	10	10	11	(target)
		11	10	00	01	(source)
	00	01	00	00	01	(target)
		11	10	00	01	(source)
	01	00	01	01	00	(target)

pair state has collapsed to a mixture of $|\Phi^{\pm}\rangle$ states. From eq. (7.93), this means the destination pair amplitude bit is 0 and so $s_l \cdot x^l = 0$. Similarly, antiparallel spins indicate that $s_l \cdot x^l = 1$. It is important to notice that eq. (7.129) also indicates that there is a back-action from the target-pair phase bit to the source-pair phase bit. This has two consequences. First, the above procedure for accumulating $s_l \cdot x^l$ is not fault-tolerant. A bit-flip error produced by a single BCNOT operation on the phase bit of the target-pair

Bounds on Quantum Error Correcting Codes

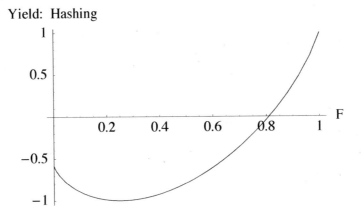

FIGURE 7.2
Plot of the pair yield $D_H(W_F)$ for the hashing method when the pair state is the Werner state W_F. The yield is positive for $F > 0.8107$.

will spread via the remaining BCNOT operations to all subsequent source-pairs. Secondly, when determining the map $x^{l+1} = f_{s_l}(x^l)$, one must track the changes to each x_k^l produced by (i) the B_y and $B_x\sigma_x$ operations, and (ii) the back-action of the BCNOT operations.

It is a simple matter to work out the hashing method yield $D_H(M) = 1 - S(M)$ when the entangled mixed state is the Werner state $M = W_F$. The von-Neumann entropy $S(W_F)$ is easily found to be

$$S(W_F) = -F \log F - (1-F) \log \left[\frac{1-F}{3}\right] . \qquad (7.130)$$

The yield is then

$$D_H(W_F) = 1 + F \log F + (1-F) \log \left[\frac{1-F}{3}\right] , \qquad (7.131)$$

which is plotted in Figure 7.2. We see that $D_H(W_F) > 0$ if the fidelity F is greater than (approximately) 0.8107. Note that the recurrence method could be used to improve a Werner state with $F > 0.5$ until its fidelity $F > 0.8107$ and then the hashing method could be used to produce a positive yield of (asymptotically) maximally entangled pairs.

7.5.2 QECCs and 1-EPPs

In this subsection we show that a close connection exists between QECCs and 1-EPPs. It is assumed that Alice and Bob can create maximally entangled pairs, and can carry out quantum teleportation [27]. We first show that, given an $[m, n, d]$ QECC and one-way classical communication, it is possible to implement a 1-EPP whose yield is equal to the rate m/n of the QECC. We then show that a 1-EPP with yield $D_P(M)$ can be used to implement a QECC that uses n qubits to reliably transmit $m = nD_P(M)$ qubits through a noisy quantum teleportation channel. The rate m/n of the resulting QECC is thus equal to the yield $D_P(M)$ of the 1-EPP.

(1) By assumption Alice and Bob are able to encode and decode an $[m, n, d]$ QECC. Initially Alice and Bob share n impure pairs. All pairs are in the same *non-maximally* entangled state described by the density operator M. These pairs will be used as a noisy quantum teleportation channel. Alice begins by creating m maximally entangled pairs. She separates off m qubits, one from each pair, and encodes them with $n - m$ ancilla qubits into a fiducial n-qubit encoded state $|\overline{f}_0\rangle$. She then teleports the state $|\overline{f}_0\rangle$ to Bob via the noisy teleportation channel. Due to the impurity of the channel, Bob receives an erroneous version of $|\overline{f}_0\rangle$. Assuming the errors are correctable, Bob can recover $|\overline{f}_0\rangle$ and decode it. Once this is done, Alice and Bob share m *maximally* entangled pairs. This procedure has allowed Alice and Bob to use n shared impure pairs, and from them, produce and share m maximally entangled pairs. It thus acts like a 1-EPP whose yield $D_P(M)$ is equal to the code rate m/n.

(2) Here Alice and Bob are assumed to be able to carry out a 1-EPP whose yield is $D_P(M)$. The process begins with Alice creating n maximally entangled pairs and sending half of each pair to Bob. By the time Bob receives his n qubits, the pair states are mixed and described by the density operator M. Next Alice carries out her half of the 1-EPP, sending Bob any classical information required by the protocol. Bob receives the information, then carries out his half of the 1-EPP. This leaves Alice and Bob sharing $m = nD_P(M)$ maximally entangled pairs. Alice takes her half of the m pairs and uses them to teleport an m-qubit state $|\psi_m\rangle$ to Bob. She does her part of the teleportation protocol and sends Bob her measurement results via the one-way classical communication channel. Bob receives her message and carries out his part of the teleportation protocol. This leaves his half of the m pairs in the state $|\psi_m\rangle$. This procedure has allowed the n qubits Alice sent to Bob to (eventually) provide the means to reliably send an m-qubit state. It thus acts like a QECC whose rate m/n is equal to the yield $D_P(M)$ of the 1-EPP.

Ref. [6] examines the connection between QECCs and 1-EPPs in much greater detail. The reader is referred to that discussion for (inter alia) the derivation of inportant inequalities connecting the quantum channel capacity $Q(\chi)$ and the one-way distillable entanglement $D_1(M)$.

> *Now this is not the end.*
> *It is not even the beginning of the end.*
> *But it is, perhaps, the end of the beginning.*
>
> Winston Churchill

Problems

7.1 *Fill in the missing steps in the derivation of the asymptotic form of the quantum Hamming bound (eq. (7.11)).*

7.2 *Fill in the missing steps in the derivation of the asymptotic form of the quantum Gilbert-Varshamov bound (eq. (7.19)).*

7.3 *Let C_q be an $[n,k,d]$ quantum stabilizer code with stabilizer S and centralizer $C(S)$, and let E be an element of the Pauli group $\mathcal{G}_n$. Suppose $s \in S$. Define $<s,E>$ to be 0 (1) if s and E commute (anticommute). From Section 3.1.2 we know that E belongs to either $C(S)$ or to $\mathcal{G}_n - C(S)$.*

(a) Suppose $E \in C(S)$. Show that

$$\sum_{s \in S} (-1)^{<s,E>} = 2^{n-k} \ .$$

(b) Suppose $E \in \mathcal{G}_n - C(S)$ so that it anticommutes with a subset of the generators of S. Without loss of generality we can suppose the anticommuting generators are $g_1, \ldots, g_h$. Redefine the generators so that $g_i \to g_1 g_i$ for $i = 2, \ldots, h$ and $g_i \to g_i$ otherwise. Note that for this new set of generators, only g_1 anticommutes with E. Show that E anticommutes with half the elements of S and commutes with the other half. Show that, consequently, for $E \in \mathcal{G}_n - C(S)$,

$$\sum_{s \in S} (-1)^{<s,E>} = 0 \ .$$

7.4 *Adapt the derivation of eq. (7.57) to derive eq. (7.80).*

7.5 *Here the basic formulas for the Recurrence Method for 2-EPP are derived.*

(a) Using Table II of Ref. [6], derive the formulas for p_{pass} and p_{fail} appearing in eq. (7.98).

(b) In eq. (7.99) the probability p'_{ij} is the conditional probability that a source pair is in the state $|ij\rangle$ upon completion of the BCNOT operation, given that it is a passed pair. Conditional and joint probabilities, respectively, $p(a|b)$ and $p(a,b)$ satisfy the well-known relation $p(a,b) = p(a|b)p(b)$. Use this result from probability theory to derive eq. (7.99).

(c) Repeat part (b) for failed pairs to derive eq. (7.100).

(d) Use eqs. (7.98)-(7.102) to derive eqs. (7.103)-(7.105).

7.6 Two definitions must be introduced before we can begin. (1) Let ξ be an ensemble of bipartite pure states $|\psi_i\rangle$ which occur with probabilities p_i. The entanglement of formation $E(\xi)$ for the ensemble ξ is defined to be the ensemble average of the entropy of entanglement $E(|\psi_i\rangle)$: $\sum_i p_i E(|\psi_i\rangle)$. (The entropy of entanglement was defined in Section 1.3.5.) (2) Let M be the density operator for an entangled mixed state. The entanglement of formation $E(M)$ for this mixed state is the minimum value of $E(\xi)$ over all ensembles ξ that have M as their density operator.

After one iteration of the recurrence method for 2-EPP, the failed pairs are in a mixed state with density operator

$$M_f = \frac{1}{4} \sum_{i,j=0}^{1} |ij\rangle\langle ij| ,$$

where the $|ij\rangle$ are the Bell basis states. Use eqs. (7.93) to show that M_f can be rewritten as

$$M_f = \frac{1}{4}[|\uparrow\uparrow\rangle\langle\uparrow\uparrow| + |\uparrow\downarrow\rangle\langle\uparrow\downarrow| + |\downarrow\uparrow\rangle\langle\downarrow\uparrow| + |\downarrow\downarrow\rangle\langle\downarrow\downarrow|] . \quad (7.132)$$

Use eq. (7.132) to show that the entanglement of formation $E(M_f) = 0$, and consequently, that the failed pairs are in a non-entangled mixed state.

References

[1] MacWilliams, F. J., and Sloane, N. J. A., *The Theory of Error Correcting Codes*, North-Holland, New York, 1977, chap. 5.

[2] Conway, J. H., and Sloane, N. J. A., A new upper bound on the minimal distance of self-dual codes, *IEEE Trans. Inform. Theory* **36**, 1319, 1990.

[3] Dantzig, G. B., *Linear Programming and Extensions*, Princeton University Press, Princeton, NJ, 1963.

[4] Gass, S. I., *Linear Programming*, 5th ed., McGraw-Hill, New York, 1985.

[5] Simmonard, M., *Linear Programming*, Prentice-Hall, Englewood Cliffs, NJ, 1966.

[6] Bennett, C. H., DiVincenzo, D. P., Smolin, J. A., and Wooters, W. K., Mixed-state entanglement and quantum error correction, *Phys. Rev. A* **54**, 3824, 1996.

[7] Bennett, C. H. et al., Purification of noisy entanglement and faithful teleportation via noisy channels, *Phys. Rev. Lett.* **76**, 722, 1996.

[8] Bennett, C. H. and Shor, P. W., Quantum information theory, *IEEE Trans. Inform. Theory* **44**, 2724, 1998.

[9] Bennett, C. H. and DiVincenzo, D. P., Quantum information and computation, *Nature* **404**, 247, 2000.

[10] Ekert, A. and Macchiavello, C., Quantum error correction for communication, *Phys. Rev. Lett.* **77**, 2585, 1996.

[11] Calderbank, A. R. et al., Quantum error correction via codes over GF(4), *IEEE Trans. Info. Theor.* **44**, 1369, 1998.

[12] Peterson, W. W. and Weldon, E. J., *Error Correcting Codes*, MIT Press, Cambridge, MA, 1972.

[13] Knill, E. and Laflamme, R., Theory of quantum error-correcting codes, *Phys. Rev. A* **55**, 900, 1997.

[14] Preskill, J., See http://www.theory.caltech.edu/people/preskill/ph229/.

[15] Nielsen, M. A. and Chuang, I. L., *Quantum Computation and Quantum Information*, Cambridge University Press, London, 2000.

[16] Shor, P. W. and Laflamme, R., Quantum analog of the MacWilliams identities for classical coding theory, *Phys. Rev. Lett.* **78**, 1600, 1997.

[17] Rains, E., Quantum shadow enumerators, downloadable at http://arXiv.org/abs/quant-ph/9611001, 1996.

[18] Rains, E., Quantum weight enumerators, downloadable at http://arXiv.org/abs/quant-ph/9612015, 1996.

[19] Rains, E., Polynomial invariants of quantum codes, downloadable at http://arXiv.org/abs/quant-ph/9704042, 1997.

[20] Delsarte, P., Bounds for unrestricted codes, by linear programming, *Philips Res. Reports* **27**, 272, 1972.

[21] Delsarte, P., Four fundamental parameters of a code and their combinatorial significance, *Info. and Control* **23**, 407, 1973.

[22] Delsarte, P., An algebraic approach to the association schemes of coding theory, *Philips Res. Reports Suppl.* No. 10, 1973.

[23] MacWilliams, F. J., Combinatorial problems of elementary group theory, Ph. D. thesis, Harvard University, Cambridge, MA, 1962.

[24] MacWilliams, F. J., A theorem on the distribution of weights in a systematic code, *Bell Syst. Tech. J.* **42**, 79, 1963.

[25] Press, W. H. et al., *Numerical Recipes: The Art of Scientific Computing*, Cambridge University Press, London, 1992.

[26] Werner, R. F., Quantum states with Einstein-Podolsky-Rosen correlations admitting a hidden-variable model, *Phys. Rev. A* **40**, 4277, 1989.

[27] Bennett, C. H. et al., Teleporting an unknown quantum state via dual classical and Einstein-Podolsky-Rosen channels, *Phys. Rev. Lett.* **70**, 1895, 1993.

[28] Cover, T. M. and Thomas, J. A., *Elements of Information Theory*, Wiley&Sons, New York, 1991.

[29] Davenport, W. B., *Probability and Random Processes*, McGraw-Hill, New York, 1970.

A

Group Theory

This appendix provides a brief review of group theory. Section A.1 defines a group, then presents some useful developments of this key idea. Section A.2 considers the action of a group on a set, including the important case where the group acts on itself. Finally, Section A.3 discusses the mapping of one group to another. This leads to the idea of a homomorphism and to a statement of the Fundamental Homomorphism Theorem. The material in this appendix is standard and can be found in any good text on group theory. Refs. [1] and [2] are aimed at a more mathematically oriented reader, while Refs. [3–7] are well-known presentations aimed at physicists and chemists. Note that the theorems in this appendix will be stated without proof. In all cases, proofs can be found in Refs. [1] and/or [2].

A.1 Fundamental Notions

This section presents the definition of a group, and introduces a number of properties that make it possible to classify groups into broad categories (Section A.1.1). In Section A.1.2 we define a subgroup and develop theoretical ideas that open the door to a study of group structure, and under appropriate conditions, to construction of a new group called the quotient group.

A.1.1 Groups

A group is one of the central objects in abstract algebra. The importance of this idea for physics, chemistry, and classical coding theory is well-known.

DEFINITION A.1 *A group is a set of elements G together with an operation $*$ that satisfies the following requirements:*

G1. Closure: *To each pair of elements a and b in G, there is a unique element $a*b$ called the product of a and b that also belongs to G.*

G2. Identity: *There is an element e in G called the identity that satisfies*

$$a*e = a = e*a ,$$

for all a in G.

G3. Inverse: *For every a in G there is an element a^{-1} in G such that*

$$a * a^{-1} = e = a^{-1} * a \ .$$

G4. Associativity: *For any elements a, b, and c in G,*

$$(a * b) * c = a * (b * c) \ .$$

To make the language a little less clumsy, throughout this appendix we will refer to the $*$ operation as multiplication. Also, to simplify notation, we will usually suppress the $*$ symbol when writing a product. Any set, whether a group or not, which has the closure property for multiplication is said to be closed under multiplication. Note that it follows from Definition A.1 that the identity e is unique, and for each $a \in G$, the inverse element a^{-1} is unique [1].

A group G is abelian if for all elements a and b in G: $ab = ba$. Otherwise G is said to be a non-abelian group. The number of elements in a group G is called the order of G and denoted $|G|$. If $|G|$ is finite (infinite), G is said to be a finite (infinite) group. Suppose a is an element of a group G. If there is an integer $m > 1$ such that $a^m = e$, then the order of a is defined to be the smallest positive integer n for which $a^n = e$. If there is no integer $m > 1$ for which $a^m = e$, then a is defined to have order infinity.

A.1.2 Subgroups

We begin with the definition of a subgroup.

DEFINITION A.2 *Let G be a group and H a nonempty subset that is closed under the $*$ operation in G. If H is also a group under $*$, H is called a subgroup of G.*

The following three theorems establish useful properties of subgroups.

THEOREM A.1
Let G be a group and H be a nonempty subset of G. H is a subgroup of G if and only if $ab^{-1} \in H$ for all $a, b \in H$.

Theorem A.1 provides a simple test to determine whether a subset H of G is a subgroup of G.

THEOREM A.2
Let G be a group, and $\{H_i \,|\, i \in I\}$ a nonempty collection of subgroups with index set I. The intersection $\bigcap_{i \in I} H_i$ is a subgroup of G.

Group Theory

The following definition will allow us to introduce the generators of a subgroup.

DEFINITION A.3 *Let G be a group and X a subset of G. Let $\{H_i \mid i \in I\}$ be the collection of all subgroups of G that contain X. Then the intersection $\bigcap_{i \in I} H_i$ is called the subgroup of G generated by X and is denoted by $<X>$.*

The elements of X are the generators of the subgroup $<X>$. If X has a finite number of elements, then $<X>$ is said to be finitely generated. If X has only one element a, then $<a>$ is the cyclic subgroup generated by a.

THEOREM A.3
Let G be a group and X a nonempty subset of G with elements $\{x_i \mid i \in I\}$. The subgroup $<X>$ generated by X consists of all finite products of the x_i.

A quantum stabilizer code with stabilizer group $\mathcal{S}$ and generators $g_1, \ldots, g_{n-k}$ provides an example of Theorem A.3. As seen in Section 1.4.1, $\mathcal{S}$ is abelian, the generators have order 2, and for all $s \in \mathcal{S}$,

$$s = g_1^{p_1} \cdots g_{n-k}^{p_{n-k}} \quad , \tag{A.1}$$

with the $p_i = 0, 1$. Since $\mathcal{S}$ is abelian, the order of the products in eq. (A.1) is unimportant.

If H is a subgroup of G, Lagrange's theorem relates the order of G to the order of H. We state the theorem for finite groups, which is the case of interest in this book.

THEOREM A.4 (Lagrange's Theorem)
Let G be a finite group and H an arbitrary subgroup. The order of H divides the order of G.

Next we introduce the important idea of a coset.

DEFINITION A.4 *Let G be a group and H a subgroup of G. For any element $a \in G$, the set $aH = \{ah \mid h \in H\}$ is called a left coset of H in G. Similarly, the set $Ha = \{ha \mid h \in H\}$ is called a right coset of H in G.*

Note that whether an argument is made solely in terms of left or right cosets is usually not important. We will typically use left cosets and will refer to them simply as cosets.

Recall that a partition of a set A is a family $\{A_i \mid i \in I\}$ of nonempty subsets of A that are mutually disjoint and whose union is all of A. The following theorem states an important property of cosets.

THEOREM A.5

Let G be a group and H a subgroup of G. The collection of left cosets of H $\{aH \mid a \in G\}$ forms a partition of G

The number of cosets of H in G is called the index of H in G and is denoted $[G : H]$. It follows from Lagrange's theorem that

$$|G| = [G : H]\,|H| \ .$$

If a and x are elements of a group G, then xax^{-1} is called a conjugate of a and the operation that maps $a \to xax^{-1}$ is called conjugation by x. If H is any subset of G, we say that H is closed with respect to conjugates if every conjugate of every element of H is also in H. This leads to the definition of a normal subgroup.

DEFINITION A.5 *Let G be a group and H a subgroup of G. H is a normal subgroup of G if it is closed with respect to conjugates: $xax^{-1} \in H$ for any $a \in H$ and $x \in G$.*

An equivalent way of defining a normal subgroup H is to say that H is fixed under conjugation by the elements of G: $xHx^{-1} = H$ for all $x \in G$.

If G is a group and H is a normal subgroup of G, it is possible to form a new group out of the cosets of H. This requires the definition of coset multiplication. To that end, if aH and bH are cosets of H, we define their product to be

$$aH * bH = abH \ ,$$

where ab is the product of a and b in G. The requirement that H be a normal subgroup insures that coset multiplication is well-defined. This leads us to the following theorem.

THEOREM A.6

Let G be a group and H a normal subgroup of G. If G/H denotes the set of cosets of H in G, then G/H with coset multiplication is a group.

G/H is known as the quotient group of G by H. From Lagrange's Theorem the order of G/H is the index of H in G: $|G/H| = [G : H]$. We emphasize that the elements of G/H are cosets of H.

A.2 Group Action

In physical applications one is often interested in a set S that is acted on by a set of transformations that form a group. We first examine the action of a group G on an arbitrary set S (Section A.2.1), and then consider the case where G acts on itself so that $S = G$ (Section A.2.2). Both cases are encountered in our study of quantum stabilizer codes.

A.2.1 On a Set

A group G is said to act on a set S if:

1. each element $g \in G$ implements a map $s \to g(s)$, where $s, g(s) \in S$;

2. the identity e in G produces the identity map
$$e(s) = s \ ;$$

3. the map produced by $g_1 g_2$ is the composition of the maps produced by g_1 and g_2:
$$g_1 g_2(s) = g_1(g_2(s)) \ .$$

Let s be an element of S. The orbit of s (denoted $orb(s)$) is the set of elements of S that are images of s under the action of G:
$$orb(s) = \{g(s) \,|\, g \in G\} \ .$$

The stabilizer of s (denoted $\mathcal{S}_s$) is the set of elements of G that fix s:
$$\mathcal{S}_s = \{g \in G \,|\, g(s) = s\} \ .$$

It is possible to show that (i) each orbit defines an equivalence class on S and so the collection of orbits partitions S, and (ii) the stabilizer $\mathcal{S}_s$ is a subgroup of G.

In our discussion of quantum stabilizer codes, the group G can be identified with the Pauli group $\mathcal{G}_n$ and the set S with the subspace spanned by the basis codewords. Each basis codeword $|i\rangle$ then defines its own stabilizer $\mathcal{S}_i$, and the code stabilizer S is the intersection of all the $\mathcal{S}_i$.

A.2.2 On Itself

It is not uncommon to encounter situations where the set S is also the group G. In quantum mechanical applications, the action is usually conjugation: $g \to xgx^{-1}$ for $g, x \in G$. In this case the orbit of $g \in G$ is called the conjugacy

class of g, and the stabilizer S_g is called the centralizer of g in G (denoted $C_G(g)$):
$$C_G(g) = \{x \in G \,|\, xgx^{-1} = g\} \ . \tag{A.2}$$
It follows from eq. (A.2) that $C_G(g)$ contains all elements in G that commute with g. In our discussions of quantum stabilizer codes we encountered the centralizer $C(S)$ that contains all elements of the Pauli group $\mathcal{G}_n$ that commute with all elements of the code stabilizer S. If g_i is a generator of S, then $C_{\mathcal{G}_n}(g_i)$ is the set of elements in $\mathcal{G}_n$ that commute with g_i. It follows that the centralizer $C(S)$ is the intersection of all the $C_{\mathcal{G}_n}(g_i)$.

Finally, let H be a subgroup of G. The normalizer of H in G (denoted $N_G(H)$) is the set of elements of G that fix H under conjugation:
$$N_G(H) = \{x \in G \,|\, xHx^{-1} = H\} \ .$$
Thus $x \in N_G(H)$ if, for all $h \in H$, $xhx^{-1} \in H$. We have encountered normalizers a number of times in our study of quantum stabilizer codes and fault-tolerant quantum computing. For example, the normalizer $N_\mathcal{G}(S)$ was defined in Exercise 3.2 as the set of errors in the Pauli group $\mathcal{G}_n$ that fix the code stabilizer S. Another example is the Clifford group $N_U(\mathcal{G}_n)$ (Section 5.3), which is the normalizer of the Pauli group $\mathcal{G}_n$ in the unitary group $U(n)$.

A.3 Mapping Groups

If G and H are groups, it proves fruitful to consider the different mappings of $G \to H$. We begin in Section A.3.1 with a discussion of homomorphisms, along with a number of refinements of this idea. This is followed in Section A.3.2 with a statement of the Fundamental Homomorphism Theorem, which establishes an isomorphism between the image of G under a homomorphism f and the quotient group $G/Ker(f)$ of G by the kernel of f.

A.3.1 Homomorphisms

Before discussing the mapping of one group to another we remind the reader of some terminology associated with functions. Suppose A and B are sets and f is a function from A to B ($f : A \to B$). If each element of B is the image of no more than one element of A, then f is said to be injective or one-to-one. If each element of B is the image of at least one element of A, then f is said to be surjective or onto. Finally, if f is both injective and surjective (one-to-one and onto), then f is said to be bijective (one-to-one correspondence).

Functions that are especially interesting when discussing mappings from one group to another are those that preserve the input group's multiplication table. This leads us to the definition of a homomorphism.

Group Theory

DEFINITION A.6 *Let G and H be groups, and f a function that maps $G \to H$. f is a homomorphism from G to H if, for any two elements $a, b \in G$,*

$$f(ab) = f(a)f(b) \ . \tag{A.3}$$

If f maps G onto H, H is said to be a homomorphic image of G.

Eq. (A.3) says that if f is a homomorphism, the image of a product ab is the product of the images of a and b.

A few refinements of the definition of a homomorphism prove useful. If f is a homomorphism and f is bijective, then f is said to be an isomorphism from G to H (denoted $G \cong H$). If f is a homomorphism (isomorphism) that maps $G \to G$, then f is called an endomorphism (automorphism) of G.

Let f be a homomorphism from $G \to H$. The kernel of f (denoted $Ker(f)$) is the set of elements belonging to G that are mapped to the identity element e_H of H:

$$Ker(f) = \{x \in G \, | \, f(x) = e_H\} \ .$$

The image of f (denoted $Im(f)$) is the set of elements in H that are the image of some $a \in G$:

$$Im(f) = \{b \in H \, | \, b = f(a) \ \text{for some} \ a \in G\} \ .$$

THEOREM A.7
Let G and H be groups and f a homomorphism that maps $G \to H$. Then (a) the kernel of f is a normal subgroup of G, and (b) the image of f is a subgroup of H.

By Theorem A.7-(b) and Definition A.6, $Im(f)$ is a homomorphic image of G.

A.3.2 Fundamental Homomorphism Theorem

As noted earlier, the Fundamental Homomorphism Theorem establishes an isomorphism between a homomorphic image of a group G and one of its quotient groups. The essential insight leading to the theorem is that, given a homomorphism f, equality of images $f(a) = f(b)$ is possible if and only if a and b belong to the same coset of $Ker(f)$. The proof of this is straightforward and can be found in Ref. [2]. It follows that f induces a bijection between the cosets of $Ker(f)$ and the elements of $Im(f)$. The Fundamental Homomorphism Theorem is then proved by showing that this bijection is also an isomorphism. The proof of this can be found in Refs. [1, 2].

THEOREM A.8
Let G and H be groups. If $f : G \to H$ is a homomorphism, then f induces

an isomorphism between $Im(f)$ and the quotient group $G/Ker(f)$:

$$Im(f) \cong G/Ker(f) \ .$$

It thus follows that one can study the homomorphic images of a group G by studying its quotient groups, and vice versa.

References

[1] Hungerford, T. W., *Algebra*, Springer, New York, 1984.

[2] Pinter, C. C., *A Book of Abstract Algebra*, McGraw-Hill, New York, 1982.

[3] Cornwell, J. F., *Group Theory in Physics*, Vols. I–III, Academic Press, New York, 1984.

[4] Hammermesh, M., *Group Theory*, Addison-Wesley, New York, 1964.

[5] Wigner, E. P., *Group Theory*, Academic Press, New York, 1959.

[6] Weyl, H., *Theory of Groups and Quantum Mechanics*, Dover, New York, 1950.

[7] Tinkham, M., *Group Theory and Quantum Mechanics*, McGraw-Hill, New York, 1964.

B

Quantum Mechanics

In this appendix we state the axioms upon which quantum mechanics is founded. It will be sufficient for purposes of this book to restrict our discussion to the non-relativistic theory. In quantum mechanics, the primitive undefined concepts are physical system, observable, and state. A physical system will be considered to be any sufficiently isolated thing, say an electron, a molecule, or a photon. An observable will be identified with a measurable property of a physical system, say energy or z-component of spin. The state of a physical system proves to be a trickier concept in quantum mechanics than in classical mechanics. Subtleties arise when considering the state of a composite physical system. In particular, states exist for a bipartite physical system in which neither of the subsystems is in a definite state. Such states are known as entangled states (Section 1.3.5) and are inherently quantum mechanical. Even in cases where a physical system can be described as being in a state, two classes of state are possible: pure and mixed. These two types of states will defined later in this appendix.

As we shall see, the predictions of quantum mechanics are intrinsically statistical in character. As such, when discussing the measurement of an observable $\mathcal{O}$, an ensemble of N identically prepared identical physical systems must be introduced. Each of the N physical systems is said to be an element of the ensemble. Quantum mechanics predicts the probability p_i that an outcome $\mathcal{O}_i$ will be found when the observable $\mathcal{O}$ is measured. It thus predicts, for N sufficiently large, the number of elements n_i in the ensemble expected to have outcome $\mathcal{O}_i$: $n_i \to p_i N$. As might be anticipated from these remarks, ensembles will appear throughout the discussions in this appendix.

To complete this introduction, we outline the structure of this appendix. Sections B.1–B.5 present the axioms for non-relativistic quantum mechanics for the case of a physical system in a pure state. Section B.6 begins by introducing the density operator, mixed states, and the reduced density operator. With these ideas, the quantum axioms are then re-expressed so that they apply to physical systems in mixed states. Many books describe the material presented in this appendix. Refs. [1–5] represent an incomplete sampling.

B.1 States

The first quantum axiom associates a Hilbert space $\mathcal{H}$ with the description of an ensemble E of identical physical systems.

Axiom 1 *To every ensemble E of identical physical systems there corresponds a Hilbert space $\mathcal{H}$. An element $|\psi\rangle$ of $\mathcal{H}$ is called a state vector, and for purposes of predicting experimental results, situations occur in which the ensemble can be described by a definite state vector.*

A Hilbert space $\mathcal{H}$ is a complete complex vector space with strictly positive inner product (see Refs. [1,2,6]). A Hilbert space may possess additional properties. It may, for example, be separable or non-separable. In quantum computing the physical systems of interest are associated with finite-dimensional Hilbert spaces. For such spaces, separability and completeness follow from the requirement that $\mathcal{H}$ be a complex vector space with strictly positive inner product. For our purposes then, it is sufficient to think of a Hilbert space as being a vector space over the complex numbers with appropriate inner product.

An ensemble E is said to be in a pure state when it can be described by a state vector. Section B.6 will (*inter alia*) discuss pure states in more detail, and will also introduce the idea of a mixed state.

Suppose the state vectors $|\psi_1\rangle$ and $|\psi_2\rangle$ describe possible pure states of an ensemble. The superposition principle states that their sum $|\psi\rangle = |\psi_1\rangle + |\psi_2\rangle$ also describes a possible pure state. The state vector $|\psi\rangle$ is said to be a linear superposition of $|\psi_1\rangle$ and $|\psi_2\rangle$. The superposition principle is extremely important in quantum mechanics, although it is not universally valid. It is violated whenever a superselection rule is present [7]. In this case, the Hilbert space splits up into subspaces; and even though two state vectors may each individually describe a possible pure state, their superposition, when they belong to different subspaces, will not be a possible pure state. Superselection rules arise in the context of quantum field theories. Since we only consider physical systems with finite-dimensional Hilbert spaces, superselection rules will not be relevant for our purposes, and can be safely ignored.

Finally, it will be seen in Section B.5 that two state vectors $|\psi\rangle$ and $e^{i\chi}|\psi\rangle$ that are identical up to a phase factor are physically equivalent. Said another way, the phase χ has no physical significance. Note however that the *relative* phase ϕ between two terms in a superposition $|\psi_1\rangle + e^{i\phi}|\psi_2\rangle$ *is* physically significant and gives rise to observable quantum interference effects.

B.2 Composite Systems

It is clear from looking at the world around us that physical systems often come together to form larger composite physical systems. Axiom 2 identifies the Hilbert space of the composite system with the tensor product of the component system Hilbert spaces.

Axiom 2 *Let S and S' be two non-identical non-interacting physical systems that are, respectively, elements of ensembles described by state vectors $|\psi\rangle$ and $|\psi'\rangle$. Furthermore, let $\mathcal{H}$ and $\mathcal{H}'$ be the respective Hilbert spaces for the two ensembles. The ensemble of composite systems (formed from pairing up the S and S') is described by the state vector $|\psi\rangle \otimes |\psi'\rangle$, which belongs to the tensor product space $\mathcal{H} \otimes \mathcal{H}'$.*

The restriction to non-identical physical systems is for convenience. With identical physical systems, the composite state vector must have a definite permutation symmetry that depends on whether the component systems are fermions or bosons. This complication is avoided by restricting ourselves to non-identical physical systems. The question of observables for the composite system is taken up in the following section.

B.3 Observables

The following two axioms, respectively, associate to each observable $\mathcal{O}$ an Hermitian operator $\hat{O}$, and identify the possible outcomes of measuring $\mathcal{O}$ with the eigenvalues of $\hat{O}$.

Axiom 3 *Let S be a physical system with Hilbert space $\mathcal{H}$. To every observable $\mathcal{O}$ of S there corresponds an Hermitian operator $\hat{O}$ that acts on the state vectors in $\mathcal{H}$.*

Note that the converse of Axiom 3 (to every Hermitian operator acting on $\mathcal{H}$ there corresponds an observable of S) is not true. Superselection rules have been shown to lead to Hermitian operators that are not observable [7].

Two observables whose corresponding Hermitian operators commute are said to be compatible observables.

Axiom 4 *The measurement of an observable $\mathcal{O}$ of a physical system S will always yield an eigenvalue of the associated Hermitian operator $\hat{O}$ as its outcome.*

Note that when a Hilbert space is finite-dimensional, observables will always have a finite number of eigenvalues, and so measurements will only have a finite number of possible outcomes. This is the usual case encountered in quantum computing.

Suppose C is a composite system with component systems A and B, and let $\mathcal{H}_A$ and $\mathcal{H}_B$ be their respective Hilbert spaces. We know from Axiom 2 that the Hilbert space for C is $\mathcal{H}_A \otimes \mathcal{H}_B$, and from Axiom 3 that to each observable $\mathcal{O}_C$ of C there corresponds an Hermitian operator $\hat{O}_C$ that acts on $\mathcal{H}_A \otimes \mathcal{H}_B$. Often the operators $\hat{O}_C$ that are encountered in practice are linear combinations of tensor products of Hermitian operators $\hat{O}_A$ and $\hat{O}_B$. Here the tensor product operator $\hat{O}_A \otimes \hat{O}_B$ is defined to have the following action on the state vector $|\psi_A\rangle \otimes |\psi_B\rangle \in \mathcal{H}_A \otimes \mathcal{H}_B$:

$$\left(\hat{O}_A \otimes \hat{O}_B\right)|\psi_A\rangle \otimes |\psi_B\rangle = \left(\hat{O}_A|\psi_A\rangle\right) \otimes \left(\hat{O}_B|\psi_B\rangle\right) \quad (B.1)$$

The Hermiticity of $\hat{O}_A$ and $\hat{O}_B$ insures that $\hat{O}_A \otimes \hat{O}_B$ is Hermitian. The action of $\hat{O}_A \otimes \hat{O}_B$ on an arbitrary state vector follows from requiring that it act linearly on $\mathcal{H}_A \otimes \mathcal{H}_B$. To complete the definition, one requires the action of a linear combination of operators $\hat{A}_i \otimes \hat{B}_i$ to be

$$\left(\sum_i c_i \hat{A}_i \otimes \hat{B}_i\right)|\psi_A\rangle \otimes |\psi_B\rangle = \sum_i c_i \left(\hat{A}_i|\psi_A\rangle\right) \otimes \left(\hat{B}_i|\psi_B\rangle\right) \quad (B.2)$$

B.4 Dynamics

The following axiom specifies the dynamical rule governing the time evolution of state vectors. This corresponds to the Schrodinger representation of the dynamics. For other representations, see for example Ref. [1].

Axiom 5 *Let E be an ensemble of physical systems that are only subjected to external forces. Furthermore, let E be described by the state vector $|\psi\rangle$. The time evolution of $|\psi\rangle$ is determined by the Schrodinger equation*

$$i\hbar \frac{\partial}{\partial t}|\psi(t)\rangle = H(t)|\psi(t)\rangle , \quad (B.3)$$

where $H(t)$ is the system Hamiltonian and $\hbar = 1.055 \times 10^{-34}$ J-s is Planck's constant.

In this axiom, a force is considered to be an "external force" if the back-action of the system on the physical agency that produces the force is negligible.

The Schrodinger dynamics can be shown to implement a unitary transformation on the initial state vector $|\psi(t_0)\rangle$:

$$|\psi(t)\rangle = U(t, t_0)|\psi(t_0)\rangle ,$$

where $U^\dagger(t, t_0) = U^{-1}(t, t_0)$. For a time-independent Hamiltonian H, eq. (B.3) can be easily integrated to give

$$U(t, t_0) = \exp\left[-\frac{i}{\hbar}H(t - t_0)\right] \, , \tag{B.4}$$

while for a time-dependent Hamiltonian $H(t)$, the integration leads to a time-ordered exponential:

$$U(t, t_0) = T\left\{\exp\left[-\frac{i}{\hbar}\int_{t_0}^{t} dt'\right]\right\}$$
$$\equiv \lim_{N \to \infty} \prod_{n=0}^{N-1} \exp\left[-\frac{i}{\hbar}H(t_0 + n\Delta_N)\Delta_N\right] \, , \tag{B.5}$$

where $\Delta_N = (t - t_0)/N$. A derivation of these formulas can be found in many textbooks on quantum mechanics; see for example Ref. [8].

B.5 Measurement and State Preparation

The remaining axioms have to do with measurements. Before stating Axiom 6 we note that the Hilbert space inner product allows the length of a state vector to be defined. The inner product maps every pair of state vectors $|a\rangle$ and $|b\rangle$ to a complex number denoted $\langle a|b\rangle$ in the Dirac notation [9]. (The inner product has other properties that need not be enumerated here.) The length or norm l_a of $|a\rangle$ is then defined to be $l_a = \sqrt{|\langle a|a\rangle|}$. When $l_a = 1$, $|a\rangle$ is said to be normalized.

Axiom 6 *Let S be a physical system and $\mathcal{O}$ an observable with eigenvalues and normalized eigenvectors $\{\mathcal{O}_i\,;|\mathcal{O}_i;r\rangle\}$, where r is a degeneracy index. Let E be an ensemble of physical systems S that, immediately before making an $\mathcal{O}$ measurement, is described by the normalized state vector $|\psi\rangle$. Immediately after the measurement, the probability p_i that eigenvalue $\mathcal{O}_i$ is observed is*

$$p_i = \sum_r |\langle \mathcal{O}_i; r|\psi\rangle|^2 \, . \tag{B.6}$$

This axiom states that quantum mechanics only predicts probabilities for the outcomes of measurements on ensembles. Except when $|\psi\rangle$ is an eigenvector of $\mathcal{O}$, the outcome of a measurement on an individual physical system is not predictable. It follows from eq. (B.6) that the states $|\psi\rangle$ and $e^{i\chi}|\psi\rangle$ are physically equivalent since they give rise to the same probabilities p_i. Axiom 6 is stated in a form appropriate for a discrete set of eigenvalues $\{\mathcal{O}_i\}$. This is the case of usual interest in quantum computing. Treatment of observables with a continuous set of eigenvalues is a standard topic in texts on quantum mechanics; see for example Refs. [1,8,9]. The presence of a degeneracy index r in the eigenvectors of $\mathcal{O}$ often signals the existence of at least one compatible observable $\mathcal{O}'$. Usually r is related to the eigenvalues of the compatible observables. We will return to this point below. Finally, note that Axiom 6 makes no statement about how the measurement apparatus functions; in particular, whether it also obeys the quantum axioms. Studies aimed at understanding the measurement process when the apparatus is treated as a quantum system lead directly to the difficult and fascinating quantum measurement problem. This problem was first considered by von Neumann [2] and continues to be a subject of vigorous debate. A discussion of this problem lies outside the scope of this book. Useful introductions to this subject can be found in Refs. [3, 4, 10, 11].

Axiom 7 introduces a measurement of the first kind (see Refs. [1, 2]).

Axiom 7 *For any observable $\mathcal{O}$ it is possible in principle to construct a measurement apparatus such that if $\mathcal{O}$ is measured twice on a given system S, and the time separating the two measurements is negligibly small, then both measurements give the same outcome.*

As will be seen shortly, measurements of the first kind are simply the projective measurements introduced in Section 1.3.3.

Axiom 8 *Let $\mathcal{O}$ and $\mathcal{O}'$ be compatible observables. Suppose that on a given physical system S, three successive measurements of the first kind are made: first $\mathcal{O}$, then $\mathcal{O}'$, then $\mathcal{O}$ again. If the time separating two successive measurements is negligibly small, then the outcome of the last measurement is the same as that of the first.*

It follows from Axiom 8 that the order in which a pair of compatible observables is measured is unimportant so long as the time between measurements is negligibly small. This axiom gives meaning to the phrase "simultaneous measurement" of compatible observables. The extension of Axiom 8 to larger sets of compatible observables is straightforward.

A set S of pairwise compatible observables $\{\mathcal{O}_1, \mathcal{O}_2, \ldots\}$ is said to be complete if any other observable $\mathcal{O}'$ compatible with each $\mathcal{O}_i \in S$ is a function of these observables: $\mathcal{O}' = F(\mathcal{O}_1, \mathcal{O}_2, \ldots)$. A measurement is said to be complete if it is a simultaneous measurement of a complete set of compatible observables.

Axiom 9 *Let $\{\mathcal{O}_1, \mathcal{O}_2, \ldots\}$ be a complete set of compatible observables for a physical system S. Suppose that by some means (see below) the results of simultaneous measurements of $\mathcal{O}_1, \mathcal{O}_2, \ldots$ on an ensemble E of physical systems S is known in advance, and the results are the same for all $S \in E$. Then E can be described by a state vector $|\psi\rangle$.*

Let $\lambda_1, \lambda_2, \ldots$ denote the results of the simultaneous measurement described in Axiom 9. By Axiom 4, λ_i is an eigenvalue of $\mathcal{O}_i$. Let $|\lambda_1 \lambda_2 \cdots\rangle$ denote the simultaneous eigenvector of $\mathcal{O}_1, \mathcal{O}_2, \ldots$ that corresponds to the eigenvalues $\lambda_1, \lambda_2, \ldots$. It follows from the assumptions in Axiom 9 and the Cauchy-Schwarz inequality that $|\psi\rangle = |\lambda_1 \lambda_2 \cdots\rangle$ to within a constant numerical factor that can be removed by normalization. The following axiom is a special case of Axiom 9 that indicates how a complete measurement of the first kind can be used for state preparation. It is the axiom that introduces the collapse/reduction of the wavefunction.

Axiom 10 *Let E be an ensemble of physical systems and let* $\mathcal{S} = \{\mathcal{O}_1, \mathcal{O}_2, \ldots\}$ *be a complete set of compatible observables. Suppose a simultaneous measurement of the first kind is made of* $\mathcal{S}$. *The measurement partitions E into subensembles* $E = E_1 \cup E_2 \cup \cdots$, *with each subensemble* E_k *made up of systems for which the measurement produce identical results* $\{\lambda_1^k, \lambda_2^k, \ldots\}$. *For each k, subensemble* E_k *is described by the state vector* $|\lambda_1^k \lambda_2^k, \ldots\rangle$.

Thus a simultaneous measurement of the first kind of $\mathcal{S}$ can be used to prepare a set of subensembles, each described by a simultaneous eigenvector of the observables in $\mathcal{S}$. If we take one of these subensembles and use it as the ensemble E in Axiom 9, we will know in advance the results of the simultaneous measurement described in that axiom, and that all systems in E will yield the same measurement results.

For the case of a simultaneous measurement of the first kind of an *incomplete* set of compatible observables $\mathcal{I} = \{\mathcal{O}_1, \ldots, \mathcal{O}_s\}$, the measurement is again postulated to produce subensembles E_k made up of systems for which the measurement produced identical results $\{\lambda_1^k, \ldots, \lambda_s^k\}$. However, this time the state vector $|\psi_k\rangle$ that describes E_k is only constrained to lie in the subspace $\mathcal{M}$ spanned by the simultaneous eigenvectors $\{|\lambda_1^k \cdots \lambda_s^k; r\rangle\}$ of $\mathcal{I}$, where $r = 1, \ldots, \dim \mathcal{M}$ is a degeneracy index. *We will refer to the remarks in this paragraph as Axiom 10'.*

Axioms 4, 6, and 10' justify the discussion of measurements given in Section 1.3.3. Eqs. (1.35) and (1.36) follow from Axioms 6 and 10'. We see that the measurements described in Section 1.3.3 are actually measurements of the first kind. As shown there, such measurements can be described using projection operators. Thus measurements of the first kind are simply projective measurements. A more general discussion of measurements is possible [1]. It is known, however, that these more general measurements are equivalent to projective measurements supplemented with unitary operations. It is thus sufficient to restricted our discussion to projective measurements. A concise, readable account of general measurements can be found in Ref. [12].

B.6 Mixed States

In this section we consider a class of quantum states known as mixed states which cannot be represented by a state vector. We begin by defining the density operator and enumerating some of its properties. After re-expressing our earlier definition of a pure state in terms of the density operator, we give the definition of a mixed state. The reduced density operator for a subsystem of a composite system is then introduced, and we close by modifying the quantum axioms so that they apply to physical systems in mixed states.

DEFINITION B.1 *Let E be an ensemble containing N copies of a physical system S. Suppose that E breaks up into s subensembles $E = E_1 \cup \cdots \cup E_s$, with each subensemble E_k described by a normalized state vector $|\psi_k\rangle$ and containing N_k copies of S ($N = N_1 + \cdots + N_s$). The density operator ρ associated with E is*

$$\rho = \sum_{k=1}^{s} \frac{N_k}{N} |\psi_k\rangle\langle\psi_k| \ . \tag{B.7}$$

Note that the states $\{|\psi_k\rangle\}$ need not be linearly independent.

The density operator ρ has a number of properties that follow from its definition. We will simply state them and refer the reader to Refs. [1, 3] for proofs.

1. ρ is Hermitian: $\rho^\dagger = \rho$.

2. ρ is positive definite: $\langle\psi|\rho|\psi\rangle \geq 0$ for all $|\psi\rangle$.

3. The *trace* of an operator O is the sum of its diagonal matrix elements: $Tr\, O = \sum_k O_{kk} = \sum_k \langle k|O|k\rangle$. For all density operators ρ: $Tr\,\rho = 1$.

4. The eigenvalues p_i of ρ satisfy $0 \leq p_i \leq 1$.

5. If ρ is a *projection operator* (Hermitian and $\rho^2 = \rho$), it projects onto a one-dimensional subspace.

6. $Tr\,\rho^2 \leq 1$, with equality holding when ρ is a projection operator.

7. Let ρ be written as in eq. (B.7). Then ρ is a projection operator if and only if all the $|\psi_k\rangle$ are identical up to a phase factor so that eq. (B.7) reduces to one term.

Recall that an ensemble E is in a pure state when it can be described by a state vector $|\psi\rangle$. In this case there is only one subensemble in Definition B.1 and so $s = 1$, $N_1 = N$, and from eq. (B.7),

$$\rho = |\psi\rangle\langle\psi| \ . \tag{B.8}$$

Quantum Mechanics

It follows from eq. (B.8) that $\rho^2 = \rho$. In conjunction with Property 1 above, we see that ρ is a projection operator when E is in a pure state. In fact, more can be said. From Property 7 above it follows that ρ is a projection operator *if and only if* E is in a pure state. Now let $\mathcal{O}$ be an observable with eigenvalues $\{\mathcal{O}_i\}$ and normalized eigenvectors $\{|\mathcal{O}_i; r\rangle\}$, where r is a degeneracy index. From Axiom 6, the probability p_i that a measurement of $\mathcal{O}$ yields $\mathcal{O}_i$ when E is described by the state vector $|\psi\rangle$ is

$$p_i = \sum_r |\langle \mathcal{O}_i; r|\psi\rangle|^2$$
$$= \sum_r \langle \mathcal{O}_i; r|\psi\rangle\langle\psi|\mathcal{O}_i; r\rangle \quad . \tag{B.9}$$

Here we have used that $\langle \mathcal{O}_i; r|\psi\rangle^* = \langle\psi|\mathcal{O}_i; r\rangle$, which is a property of the inner product [9]. The expectation value of $\mathcal{O}$ is then

$$\overline{\mathcal{O}} = \sum_i p_i \mathcal{O}_i$$
$$= \sum_i \sum_r \langle\mathcal{O}_i; r|\psi\rangle\langle\psi|\mathcal{O}_i; r\rangle \mathcal{O}_i$$
$$= \sum_{i,r} \langle\mathcal{O}_i; r|\rho\,\mathcal{O}|\mathcal{O}_i; r\rangle$$
$$= Tr\,\rho\mathcal{O} \quad . \tag{B.10}$$

(1) In going from the second to the third line we have used that $\rho = |\psi\rangle\langle\psi|$ and that $|\mathcal{O}_i; r\rangle$ is an eigenvector of $\mathcal{O}$ with eigenvalue $\mathcal{O}_i$; and (2) in going from the third line to the fourth, that the eigenvectors $\{|\mathcal{O}_i; r\rangle\}$ form a basis for the Hilbert space $\mathcal{H}$, and so the RHS of line 3 is actually the trace of $\rho\mathcal{O}$. Now let $\{|n\rangle\}$ be another basis for $\mathcal{H}$. Using the completeness property $\sum_n |n\rangle\langle n| = I$ of the basis $\{|n\rangle\}$ in eq. (B.9) gives

$$p_i = \sum_r \langle\mathcal{O}_i; r|\left[\sum_n |n\rangle\langle n|\right]|\psi\rangle\langle\psi|\mathcal{O}_i; r\rangle$$
$$= \sum_n \langle n|\psi\rangle\langle\psi|\left[\sum_r |\mathcal{O}_i; r\rangle\langle\mathcal{O}_i; r|\right]|n\rangle$$
$$= Tr\,\rho\,P_i \quad . \tag{B.11}$$

Here we have used that $\rho = |\psi\rangle\langle\psi|$, and introduced

$$P_i = \sum_r |\mathcal{O}_i; r\rangle\langle\mathcal{O}_i; r| \quad , \tag{B.12}$$

which projects onto the subspace spanned by the eigenvectors of $\mathcal{O}$ with eigenvalue $\mathcal{O}_i$. We see that for pure states, the density operator ρ contains the same information as the state vector $|\psi\rangle$. We go on now to examine the case of mixed states that can only be described by density operators.

In a mixed state, the ensemble E in Definition B.1 contains *at least* two subensembles whose associated state vectors are linearly independent. By definition then, mixed states *cannot* be described by a single state vector. We now show that eq. (B.10) continues to be true for mixed states. Let the observable $\mathcal{O}$ be the same as in that discussion. The ensemble average of $\mathcal{O}$ is then

$$\overline{\mathcal{O}} = \sum_{k=1}^{s} \frac{N_k}{N} \overline{\mathcal{O}}_k \ , \tag{B.13}$$

where $\overline{\mathcal{O}}_k = Tr\,|\psi_k\rangle\langle\psi_k|\mathcal{O}$ is the expectation value of $\mathcal{O}$ over the subensemble E_k described by the state vector $|\psi_k\rangle$. Thus,

$$\begin{aligned}\overline{\mathcal{O}} &= \sum_{k=1}^{s} \frac{N_k}{N} Tr\,|\psi_k\rangle\langle\psi_k|\mathcal{O} \\ &= Tr\left[\sum_{k=1}^{s} \frac{N_k}{N} |\psi_k\rangle\langle\psi_k|\mathcal{O}\right] \\ &= Tr\,\rho\,\mathcal{O} \ .\end{aligned} \tag{B.14}$$

A similar calculation shows that eq. (B.11),

$$p_i = Tr\,\rho\,P_i \ , \tag{B.15}$$

is also valid for an ensemble in a mixed state.

Recall from Theorem 2.1 that two ensembles E_ϕ and E_ψ will have the same density operator ρ if and only if any state vector representing a subensemble in E_ϕ (E_ψ) can be written as a linear combination of the state vectors representing the subensembles in E_ψ (E_ϕ). Thus, although an ensemble specifies a particular density operator, a given density operator does not single out a particular ensemble.

Since the state vectors $\{|\psi_k\rangle\}$ describing the subensembles $\{E_k\}$ in the definition of ρ obey the Schrodinger equation, it is straightforward to show that the dynamical rule for $\rho(t)$ is

$$i\hbar\frac{d\rho}{dt} = [H(t), \rho] \ . \tag{B.16}$$

Here $H(t)$ is the system Hamiltonian and $[H(t), \rho] = H(t)\rho - \rho H(t)$ is the commutator of $H(t)$ and ρ. As with the Schrodinger equation, eq. (B.16) produces a unitary transformation of ρ:

$$\rho(t) = U(t, t_0)\,\rho(t_0)\,U^\dagger(t, t_0) \ , \tag{B.17}$$

where $U(t, t_0)$ is given by eq. (B.4) (eq. (B.5)) when the Hamiltonian is time-independent (time-dependent). Note that eq. (B.17) shows that $\rho^2(t) =$

$U(t,t_0)\rho^2(t_0)U(t,t_0)$, and so if $\rho(t_0)$ is a projection operator (so that $\rho^2(t_0) = \rho(t_0)$), then so is $\rho(t)$. Thus if E is initially in a pure state, the Schrodinger dynamics insures that E remains in a pure state (viz. the Schrodinger dynamics cannot take a pure state to a mixed state).

Our definition of the density operator placed no restrictions on the physical system S. Thus Definition B.1 applies to simple as well as composite systems. Suppose S is a bipartite composite system with component systems A and B. Let $\{|\alpha_i\rangle\}$ ($\{|\beta_j\rangle\}$) be a basis for the Hilbert space $\mathcal{H}_A$ ($\mathcal{H}_B$), and let E be an ensemble of physical systems S described by the density operator ρ. The reduced density operator ρ_A for subsystem A is defined to be the partial trace of ρ over B:

$$\rho_A \equiv Tr_B(\rho) \equiv \sum_j \langle \beta_j | \rho | \beta_j \rangle \ . \tag{B.18}$$

Similarly, the reduced density operator ρ_B is defined to be

$$\rho_B \equiv Tr_A(\rho) \equiv \sum_i \langle \alpha_i | \rho | \alpha_i \rangle \ . \tag{B.19}$$

As shown in Ref. [12], the reduced density operator ρ_A (ρ_B) is the correct tool to use when analyzing physical properties that belong solely to A (B).

To apply to mixed states, five of the quantum Axioms must be modified. To make it simpler to spot the changes, only the modified axioms will be stated here.

Axiom 1 *To every ensemble E of identical physical systems S there corresponds a Hilbert space $\mathcal{H}$. An element $|\psi\rangle$ of $\mathcal{H}$ is called a state vector, and for purposes of predicting experimental results, situations occur in which the ensemble can be described by a definite state vector. If E breaks up into s subensembles $E = E_1 \cup \cdots \cup E_s$, with each subensemble E_k described by a normalized state vector $|\psi_k\rangle$ and containing N_k copies of S ($N = N_1 + \cdots + N_s$), then E can be described by the density operator ρ:*

$$\rho = \sum_{k=1}^{s} \frac{N_k}{N} |\psi_k\rangle\langle\psi_k| \ .$$

Axiom 2 *Let S and S' be two non-identical non-interacting physical systems that are, respectively, elements of ensembles described by density operators ρ and ρ'. Furthermore, let $\mathcal{H}$ and $\mathcal{H}'$ be the respective Hilbert spaces for the two ensembles. The ensemble of composite systems (formed from pairing up the S and S') is described by the density operator $\rho \otimes \rho'$ that acts on the tensor product Hilbert space $\mathcal{H} \otimes \mathcal{H}'$.*

Axiom 5 *Let E be an ensemble of physical systems that are only subjected to external forces. Furthermore, let E be described by the density operator ρ. The time evolution of ρ is determined by the equation*

$$i\hbar \frac{d\rho}{dt} = [H(t), \rho] \ ,$$

where $H(t)$ is the system Hamiltonian; $[H(t), \rho] = H(t)\rho - \rho H(t)$ is the commutator of $H(t)$ and ρ; and $\hbar = 1.055 \times 10^{-34} J\text{-}s$ is Planck's constant.

Axiom 6 *Let S be a physical system and $\mathcal{O}$ an observable with eigenvalues and normalized eigenvectors $\{\mathcal{O}_i; |\mathcal{O}_i; r\rangle\}$, where r is a degeneracy index. Let E be an ensemble of physical systems S which, immediately before an $\mathcal{O}$ measurement, is described by the density operator ρ. Immediately after the measurement, the probability p_i that eigenvalue $\mathcal{O}_i$ is observed is*

$$p_i = Tr \, \rho \, P_i \; ,$$

where

$$P_i = \sum_r |\mathcal{O}_i; r\rangle\langle \mathcal{O}_i; r|$$

is the projection operator for the subspace spanned by the eigenvectors of $\mathcal{O}$ with eigenvalue $\mathcal{O}_i$.

Axiom 10 *Let E be an ensemble of physical systems and let $\mathcal{S} = \{\mathcal{O}_1, \mathcal{O}_2, \ldots\}$ be a complete set of compatible observables. Suppose a simultaneous measurement of the first kind is made of $\mathcal{S}$. The measurement partitions E into subensembles $E = E_1 \cup E_2 \cup \cdots$, with each subensemble E_k made up of systems for which the measurement produced identical results $\{\lambda_1^k, \lambda_2^k, \ldots\}$. For each k, subensemble E_k is described by the density operator*

$$\rho_k = |\lambda_1^k \lambda_2^k \cdots \rangle\langle \lambda_1^k \lambda_2^k \cdots | \; .$$

References

[1] Peres, A., *Quantum Theory: Concepts and Methods*, Kluwer, Boston, 1995.

[2] von Neumann, J., *Mathematical Foundations of Quantum Mechanics*, Princeton University Press, Princeton, NJ, 1955.

[3] d'Espagnat, B., *Conceptual Foundations of Quantum Mechanics*, 2nd ed., Addison-Wesley, New York, 1989.

[4] Jammer, M., *Philosophy of Quantum Mechanics*, Wiley, New York, 1974.

[5] Weyl, H., *Theory of Groups and Quantum Mechanics*, Dover, New York, 1950.

[6] Jauch, J. M., *Foundations of Quantum Mechanics*, Addison-Wesley, Reading, MA, 1968.

[7] Wick, G., Wightman, A., and Wigner, E., The intrinsic parity of elementary particles, *Phys. Rev.* **88**, 101, 1952.

[8] Shankar, R., *Principles of Quantum Mechanics*, 2nd ed., Plenum, New York, 1994.

[9] Dirac, P. A. M., *The Principles of Quantum Mechanics*, 4th ed., Oxford University Press, Oxford, 1958.

[10] Wheeler, J. A. and Zurek, W. H., *Quantum Theory and Measurement*, Princeton University Press, Princeton, NJ, 1983.

[11] Omnès, R., *The Interpretation of Quantum Mechanics*, Princeton University Press, Princeton, NJ, 1994.

[12] Nielsen, M. A. and Chuang, I. L., *Quantum Computation and Quantum Information*, Cambridge University Press, London, 2000.

C

Quantum Circuits

In this appendix we provide a brief review of quantum circuit theory. In Section C.1 we introduce the idea of a quantum circuit, and then present the different quantum gates that appear in this book. The Gottesman-Knill theorem is stated and proved in Section C.2, and finally we present some well-known universal sets of quantum gates in Section C.3. For readers interested in learning more about quantum circuits, we recommend the excellent introduction that appears in Ref. [1].

C.1 Basic Circuit Elements

Deutsch introduced quantum circuits as a generalization of the classical logic circuits used in present-day digital computers [2]. A classical logic circuit is an acyclic circuit of Boolean gates that acts on a set of n bits. By acyclic is meant that no cycles/loops appear in the circuit diagram [3]. In a quantum circuit the classical bits are replaced by qubits and the Boolean gates by quantum gates. An n-qubit quantum gate implements a fixed unitary operation on a set of n qubits. A quantum circuit is then a sequence of synchronized applications of quantum gates to a register of N qubits. Quantum circuits can also include measurements. For example, the quantum circuits used to do syndrome extraction in Chapters 5 and 6 apply measurements to the ancilla qubits as the final step in determining the error syndrome. By appropriately choosing the quantum gates and measurements, and the sequence in which they are applied, quantum circuits are able to do different quantum computations.

One-Qubit Gates

The following one-qubit gates are presented: (i) the identity gate I, (ii) the Hadamard gate H, (iii) the NOT gate U_{NOT}, (iv) the phase gate P, and (v) the $\pi/8$ gate T. We specify the action of each gate on the computational basis states $\{|j\rangle : j = 0, 1\}$, and also give its circuit symbol. One-qubit measurements appear in many quantum circuits in this book. Such measurements project a one-qubit state vector $|\psi\rangle$ to an output state vector $|M\rangle$ that is an eigenvector of the observable being measured. The action and circuit symbol for a one-qubit measurement is also given below.

(i) IDENTITY GATE
Comment : Describes action of ideal wire in a quantum circuit.
Action : $I|j\rangle = |j\rangle$
Circuit symbol : ────────────

(ii) HADAMARD GATE
Comment : Converts eigenstates of σ_z (σ_x) to eigenstates of σ_x (σ_z).
Action : $H|j\rangle = \frac{1}{\sqrt{2}}\left[|0\rangle + (-1)^j|1\rangle\right]$
Circuit symbol : ──[H]──

(iii) NOT GATE
Comment : Quantum analog of classical NOT gate.
Action : $U_{NOT}|i\rangle = |i \oplus 1\rangle$ where $\oplus$ is addition modulo 2.
Circuit symbol : ──⊕──

(iv) PHASE GATE
Comment : Multiplies computational basis state $|j\rangle$ by i^j.
Action : $P|j\rangle = i^j|j\rangle$
Circuit symbol : ──[P]──

(v) $\pi/8$ GATE
Comment : Multiplies computational basis state $|j\rangle$ by $\exp(i\pi j/4)$.
Action : $T|j\rangle = \exp(i\pi j/4)|j\rangle$
Circuit symbol : ──[T]──

(vi) MEASUREMENT OF ONE-QUBIT OBSERVABLE $\hat{M}$
Comment : Eigenvalue M observed; final state vector $|M\rangle$.
Action : $P_M|\psi\rangle = |M\rangle$
Circuit symbol : ──────⊣ M

Two-Qubit Gates

The two-qubit gates presented are (i) the CNOT gate, (ii) the controlled-phase gate, and (iii) the SWAP gate. The format is the same as for the one-qubit gates.

(i) CNOT GATE
Comment : First (second) qubit is control (target) qubit.
Action : $U_{CNOT}|ij\rangle = |i(i \oplus j)\rangle$ where $\oplus$ is addition modulo 2.

Quantum Circuits

Circuit symbol : ─●─
                    ─⊕─

(ii) CONTROLLED-PHASE GATE
Comment : Applies σ_z^2 when computational basis state is $|11\rangle$.
Action : $U_{CP}|ij\rangle = (-1)^{ij}|ij\rangle$.

Circuit symbol : ─●─
                    ─σ_z─

(iii) SWAP GATE
Comment : Swaps qubits.
Action : $U_{SWAP}^{12}|ij\rangle = |ji\rangle$.

Circuit symbol : ─⊕─
                    ─⊕─

Three-Qubit Gate

The three-qubit Toffoli gate is presented. The format is the same as for the one- and two-qubit gates.

TOFFOLI GATE
Comment : First two qubits are controls; last qubit is target.
Action : $U_T|ijk\rangle = |ij\,(ij \oplus k)\rangle$ where $\oplus$ is addition modulo 2.

Circuit symbol : ─●─
                    ─●─
                    ─⊕─

C.2 Gottesman-Knill Theorem

Chapter 5 introduced the Clifford group $N_U(\mathcal{G}_n)$. This group is generated by the Hadamard, phase, and CNOT gates. Gottesman [4] cites Knill as the source of the following theorem.

THEOREM C.1 *(Gottesman-Knill Theorem)*
A quantum computation that solely uses gates from the Clifford group $N_U(\mathcal{G}_n)$ and measurements of observables that belong to the Pauli group $\mathcal{G}_n$ can be simulated efficiently on a classical computer.

PROOF Suppose the quantum computation uses N_g gates from $N_U(\mathcal{G}_n)$ and N_m measurements of observables in $\mathcal{G}_n$. As shown in Chapter 5, an n-qubit state vector $|\psi\rangle$ can be specified by the stabilizer group $\mathcal{S}$ that fixes it, and $\mathcal{S}$ in turn is specified by its generators $\{G_i : i = 1,\ldots,n\}$. From Section 3.3, each generator G_i determines a $2n$-component binary vector $v(G_i)$. Suppose we initially prepare the n qubits in the state $|0\cdots 0\rangle$ by measuring $\sigma_z^1,\ldots,\sigma_z^n$. This can be done since the σ_z^i are observables belonging to $\mathcal{G}_n$. The initial stabilizer then has generators $\{G_i = \sigma_z^i\}$. Both measurements and gates in $N_U(\mathcal{G}_n)$ then map $\mathcal{S} \to \mathcal{S}' \subset \mathcal{G}_n$. This follows since (i) measurement of $\mathcal{O}$ maps the initial generators to $\{\mathcal{O}, G_2', \ldots, G_n'\}$, where the G_i' are given in eq. (5.67), and so the new generators all belong to $\mathcal{G}_n$; and (ii) the Clifford group is the normalizer of $\mathcal{G}_n$ in $U(n)$, and so each of its elements U maps $G_i \to UG_iU^\dagger \in \mathcal{G}_n$. Thus each operation in the quantum computation maps the $v(G_i)$ to new $v(G_i')$, and we can follow the evolution of the state $|\psi\rangle$ by tracking the evolution of its stabilizer, and hence of the $v(G_i)$. A classical computer can do this efficiently. To see this, note that Section 5.3.1 worked out the action of the generators of $N_U(\mathcal{G}_n)$ on the generators of $\mathcal{G}_1$ and $\mathcal{G}_2$, and hence on the generators of $\mathcal{G}_n$. This allows a look-up table to be set up from which the $v(G_i')$ can be computed in $\mathcal{O}(n^2)$ steps (n look-ups per G_i; $2n$ additions to construct $v(G_i)$; then repeat the $3n$ steps for all n generators). Tracking the action of a measurement of $\mathcal{O}$ also requires $\mathcal{O}(n^2)$ steps ($2n$ substitutions to do $v(G_1) \to v(\mathcal{O})$; for each $G_k' = G_1 G_k$, $2n$ additions to evaluate $v(G_k') = v(G_1) \oplus v(G_k)$, then do this no more than $n-1$ times since there are no more than $n-1$ generators of the form $G_1 G_k$). For N_g gates and N_m measurements, the total number of steps is $\mathcal{O}(N_g n^2) + \mathcal{O}(N_m n^2) = \mathcal{O}(n^2)$. Thus a classical computer can efficiently carry out the quantum computation described in the theorem statement. ∎

C.3 Universal Sets of Quantum Gates

A set of quantum gates $\mathcal{S}$ is said to be universal if it contains a finite number of elements and an arbitrary unitary operation (viz. quantum computation) can be done with arbitrarily small error probability using a quantum circuit that only uses gates from $\mathcal{S}$. The Gottesman-Knill theorem shows that the generators of the Clifford group $\mathcal{S}_C = \{H, P, CNOT\}$ cannot form a universal set of quantum gates. The reason is that their action can be simulated efficiently using a classical computer, whereas a quantum computation *cannot* be simulated efficiently on such a computer [5]. Thus the set $\mathcal{S}_C$ cannot implement a quantum computation! It is known, however, that $\mathcal{S}_C$ can be extended to a universal set of quantum gates by adding an appropriate fourth quantum gate (see below). In this section, a few well-known universal sets

Quantum Circuits

of quantum gates are presented. It goes beyond the scope of this book to demonstrate that these sets are universal. References to the literature are given so that the interested reader can track down the proofs if so desired.

(1) Adding the Toffoli gate U_T to the generators of the Clifford group produces the universal set $\mathcal{U}_1 = \{H, P, CNOT, U_T\}$ [6]. This is the universal set of quantum gates used in Chapters 5 and 6 to discuss fault-tolerant quantum computing and the accuracy threshold theorem.

(2) The set $\mathcal{U}_2 = \{H, P, \pi/8, CNOT\}$ is universal [7]. This set is presumably easier to implement experimentally than $\mathcal{U}_1$ since the $\pi/8$ gate is a one-qubit gate while the Toffoli gate is a three-qubit gate.

Universal sets of quantum gates are known that contain only one quantum gate.

(3) It has been shown that almost any two-qubit gate is universal [8,9], and Barenco has shown that the following two-qubit gate is universal [10]. The matrix elements are relative to the computational basis $\{|ij\rangle : i, j = 0, 1\}$.

$$A(\phi, \alpha, \theta) = \begin{pmatrix} 1 & 0 & 0 & 0 \\ 0 & 1 & 0 & 0 \\ 0 & 0 & e^{i\alpha} \cos\theta & -ie^{i(\alpha-\phi)} \sin\theta \\ 0 & 0 & -ie^{i(\alpha+\phi)} \sin\theta & e^{i\alpha} \cos\theta \end{pmatrix}$$

(4) The three-qubit Deutsch gate $D(\alpha)$ is another universal quantum gate [2]. By proving its universality, Deutsch demonstrated for the first time that universal sets of gates exist for quantum as well as classical computers. The Deutsch gate depends on a parameter α that can be any irrational number. In the computational basis, it has the form

$$D(\alpha) = \begin{pmatrix} 1 & 0 & 0 & 0 & 0 & 0 & 0 & 0 \\ 0 & 1 & 0 & 0 & 0 & 0 & 0 & 0 \\ 0 & 0 & 1 & 0 & 0 & 0 & 0 & 0 \\ 0 & 0 & 0 & 1 & 0 & 0 & 0 & 0 \\ 0 & 0 & 0 & 0 & 1 & 0 & 0 & 0 \\ 0 & 0 & 0 & 0 & 0 & 1 & 0 & 0 \\ 0 & 0 & 0 & 0 & 0 & 0 & i\cos(\pi\alpha/2) & \sin(\pi\alpha/2) \\ 0 & 0 & 0 & 0 & 0 & 0 & \sin(\pi\alpha/2) & i\cos(\pi\alpha/2) \end{pmatrix}.$$

References

[1] Nielsen, M. A. and Chuang, I. L., *Quantum Computation and Quantum Information*, Cambridge University Press, London, 2000.

[2] Deutsch, D., Quantum computational networks, *Proc. R. Soc. Lond. A* **425**, 73, 1989.

[3] Papadimitriou, C. H., *Computational Complexity*, Addison-Wesley-Longman, Reading, MA, 1994.

[4] Gottesman, D., Stabilizer codes and quantum error correction, Ph. D. thesis, California Institute of Technology, Pasadena, CA, 1997.

[5] Feynman, R. P., Simulating physics with computers, *Int. J. Theo. Phys.* **21**, 467, 1982.

[6] Shor, P. W., Fault-tolerant quantum computation, in *Proceedings, 37th Annual Symposium on Fundamentals of Computer Science*, IEEE Press, Los Alamitos, CA, 1996.

[7] Boykin, P. O. et al., On universal and fault-tolerant quantum computing, in *Proceedings, 40th Annual Symposium on Foundations of Computer Science*, IEEE Press, Los Alamitos, CA, 1999.

[8] Lloyd, S., Almost any quantum logic gate is universal, *Phys. Rev. Lett.* **75**, 346, 1996.

[9] Deutsch, D., Barenco, A., and Ekert, A., Universality in quantum computation, *Proc. R. Soc. Lond. A* **449**, 669, 1995.

[10] Barenco, A., A universal two-bit gate for quantum computation, *Proc. R. Soc. Lond. A* **449**, 679, 1995.

Index

accuracy threshold, 1, 197, 199
accuracy threshold theorem, 195–218
    statement, 199
amplitude damping channel, 71, 78
    generalized, 79
ancilla, 27
asymptotic equipartition property (AEP), 248
    typical set, 249
automorphism, 155, 267

Bell basis states, 242
binary symmetric channel, 7
bit, 2
bit-flip channel, 103
bit-wise complement, 143
Bloch vector, 30
bounds
    classical
        Gilbert-Varshamov, 15
        Hamming, 14
        Singleton, 14
    quantum
        Gilbert-Varshamov, 225–227
        Hamming, 223–225
        Linear Programming, 229–240
        Singleton, 227–229

cat-state, 149
Churchill, Winston, 256
classical error correcting codes
    coset leader, 11
    doubly-even, 159
    dual, 6
    extending, 49
    Hamming, 49
    linear, 4–16
        definition, 5
        dimension, 5
        length, 5
        minimum distance, 8
    non-linear, 4
    perfect, 14
    quasi-perfect, 14
    rate, 14
    simplex, 49
    standard array, 10
Clifford group, 153
compatible observables, 271
    complete set, 274
computational basis
    n-qubit, 18
    single-qubit, 17

decoding
    maximum likelihood, 7
    nearest neighbor, 8
decoherence, 31
density operator, 276
    reduced, 26, 29, 58, 61, 62
    definition, 279
depolarizing channel, 70

ensemble, 269
entanglement, 24
    definition, 25
    distillable
        one-way: $D_1(M)$, 247
        two-way: $D_2(M)$, 244
    entropy, 27
    formation
        ensemble, 258

mixed state, 258
maximally entangled state, 27
entanglement purification, 241–255
  1-EPP and QECC connection, 255
  bilateral operation, 242
  entanglement purification protocol (EPP), 241
    1-EPP, 247
    2-EPP, 243
  hashing method, 247–255
  recurrence method, 243–247
  twirl, 242
  unilateral operation, 242
  yield, 241
entropy
  binary, 225
  von-Neumann, *see* von-Neumann entropy
environment, 28
error operators, 58
  weight, 73, 97
error syndrome
  classical, 10
  quantum, 28, 36, 86
    syndrome extraction, 28, 144–148
    syndrome verification, 151–153
errors
  classical, 7
  quantum, 85–90
    correctable, 73–75, 86
    detectable, 75, 86
    non-detectable, 75, 87

fault-tolerant design, 139
fault-tolerant operation, 140
  error correction, 141–153
  SWAP gate, 140
  Toffoli gate, 173–180
  transversal operation, 141
fidelity
  entanglement purification, 241

quantum error correcting code, 43

generator matrix, 5
Gottesman-Knill theorem, 285
group, 261
  abelian/non-abelian, 262
  automorphism, 267
  coset, 263
  endomorphism, 267
  finite/infinite, 262
  fundamental homomorphism theorem, 267
  generators, 263
  homomorphism, 267
    image, 267
    kernel, 267
  isomorphism, 267
  quotient group, 264
  subgroup, 262
    index, 264
    normal, 264
group action
  centralizer, 266
  normalizer, 266
  orbit, 265
  stabilizer, 265

Hamming distance, 8
Hamming weight, 7
Hilbert space, 270

implicit measurement
  principle of, 61–62
intersection
  binary vectors, 48

Krawtchouk polynomials, 234

Lagrange's theorem, 263
leakage error, 72
linear superposition, 17, 270

measurement
  complete, 274
  computational basis, 22

Index 291

    first kind, 274
    projective, 21
mixed state, 278–280

no-cloning theorem, 20
noise, 29
non-selective dynamics, 63

order
    of a group, 262
    of an element, 262

parity check, 4
parity check matrix, 5
Pauli group, 73, 84
phase decoherence, *see* decoherence
phase-flip channel, 31
polarization, *see* Bloch vector
projection operator, 276
pure state, 270, 276–278

quantum channel, 242
quantum circuit, 283
quantum error correcting codes, 42, 73–75
    additive, 84
    basis codewords, 73, 83
    Calderbank-Shor-Steane, 50, 75–77, 158–162
    code space, 73, 83
    codewords, 73, 83
    concatenated codes, 100–110, 196–197
        effective error probability, 200
        primitive error probability, 200
    degenerate, 74, 89
    distance, 74
    error probability, 43
    non-degenerate, 74
    rate, 227
quantum gate, 283
    $\pi/8$, 284
    Barenco, 287
    CNOT, 156, 284
    controlled-phase, 157, 285
    Deutsch, 287
    Hadamard, 155, 284
    identity, 284
    NOT, 284
    phase, 156, 284
    SWAP, 140, 285
    Toffoli, 174, 285
quantum MacWilliams identity, 233–236
    alternative statement, 234
    statement, 233
quantum operations, 29, 57–73
    non-trace-preserving, 29, 58, 63
    operation elements, 58
    operator-sum representation, 58
    trace-preserving, 29, 58, 63
quantum parallelism, 19
quantum register, 17
quantum stabilizer codes, 32, 83
    [4,2,2] code, 95, 186–188
    [5,1,3] code, 93, 180–186
    [8,3,3] code, 95
    centralizer, 86
    cyclic codes, 94
    degenerate errors, 89
    distance, 88, 97
    normalizer
        $N_U(\mathcal{G}_n)$, *see* Clifford group
        $N_U(\mathcal{S})$, 153
        $N_{\mathcal{G}}(\mathcal{S})$, 90, 153, 266
    Shor [9,1,3] code, 79, 102
    stabilizer group, 32, 84
    standard form, 115–121
quantum teleportation, 191
qubit, 17

Schrodinger equation, 272
selective dynamics, 63
shadow, 236
shadow enumerator, 237
Shor state, 144
    verification, 148–151
state vector, 270
    normalized, 273

superposition principle, 270
superselection rule, 270
symplectic inner product, 97

Toffoli gate, *see* fault-tolerant operation, Toffoli gate
trace, 276

universal set of quantum gates, 286–287

von-Neumann entropy, 27
    subadditivity, 228

weight, *see* error operators, weight
weight enumerators, 230–233
    $A(z)$, 233
    A_w, 230
    $B(z)$, 233
    B_w, 230
Werner state, 242